AF834525

Advances in

CANCER RESEARCH

Volume 112

Intratumor Diversity and Clonal Evolution in Cancer

Advances in CANCER RESEARCH

Volume 112

Series Editors

Kenneth D. Tew

Cell and Molecular Pharmacology, John C. West Chair of Cancer Research, Department of Cell and Molecular Pharmacology & Experimental Therapeutics at MUSCCharleston, South Carolina, USA

Paul B. Fisher

Department of Human & Molecular Genetics, VCU Institute of Molecular Medicine Richmond, Virginia, USA

Edited by

David Gisselsson

Department of Clinical Genetics Lund University, University Hospital Lund, Sweden

ELSEVIER

AMSTERDAM • BOSTON • HEIDELBERG • LONDON
NEW YORK • OXFORD • PARIS • SAN DIEGO
SAN FRANCISCO • SINGAPORE • SYDNEY • TOKYO
Academic Press is an imprint of Elsevier

Academic Press is an imprint of Elsevier
525 B Street, Suite 1900, San Diego, CA 92101-4495, USA
225 Wyman Street, Waltham, MA 02451, USA
32 Jamestown Road, London, NW1 7BY, UK
Linacre House, Jordan Hill, Oxford OX2 8DP, UK
Radarweg 29, PO Box 211, 1000 AE Amsterdam, The Netherlands

First edition 2011

Copyright © 2011 Elsevier Inc. All rights reserved.

No part of this publication may be reproduced, stored in a retrieval system or transmitted in any form or by any means electronic, mechanical, photocopying, recording or otherwise without the prior written permission of the Publisher.

Permissions may be sought directly from Elsevier's Science & Technology Rights Department in Oxford, UK: phone (+44) (0) 1865 843830; fax (+44) (0) 1865 853333; email: permissions@elsevier.com. Alternatively you can submit your request online by visiting the Elsevier website at http://elsevier.com/locate/permissions, and selecting *Obtaining permission to use Elsevier material.*

Notice
No responsibility is assumed by the publisher for any injury and/or damage to persons or property as a matter of products liability, negligence or otherwise, or from any use or operation of any methods, products, instructions or ideas contained in the material herein. Because of rapid advances in the medical sciences, in particular, independent verification of diagnoses and drug dosages should be made.

ISBN: 978-0-12-387688-1
ISSN: 0065-230X

For information on all Academic Press publications visit our website at www.elsevierdirect.com

Printed and bound by CPI Group (UK) Ltd, Croydon, CR0 4YY

Transferred to Digital Print 2011

Working together to grow libraries in developing countries

www.elsevier.com | www.bookaid.org | www.sabre.org

ELSEVIER BOOK AID International Sabre Foundation

Contents

Nuclear Remodeling as a Mechanism for Genomic Instability in Cancer

Macoura Gadji, Rhea Vallente, Ludger Klewes, Christiaan Righolt, Landon Wark, Narisorn Kongruttanachok, Hans Knecht and Sabine Mai

Cytogenetic Analysis of Tumor Clonality

Manuel R. Teixeira and Sverre Heim

Genotyping Techniques to Address Diversity in Tumors

David Lindgren, Mattias Höglund and Johan Vallon-Christersson,

Prostate Cancer as a Model System for Genetic Diversity in Tumors

Jeremy A. Squire, Paul C. Park, Maisa Yoshimoto, Jennifer Alami, Julia L. Williams, Andrew Evans and Anthony M. Joshua

Cancer: Beyond Speciation

Mark D. Vincent

Contributors

Numbers in parentheses indicate the pages on which the authors' contributions begin.

Batoul Y. Abdallah, Center for Molecular Medicine and Genetics, Wayne State University School of Medicine, MI, USA (217)

Jennifer Alami, Department of Pathology and Molecular Medicine, Queen's University, Kingston, ON, Canada (183)

Steven W. Bremer, Center for Molecular Medicine and Genetics, Wayne State University School of Medicine, MI, USA (217)

Daniela Cimini, Department of Biological Sciences, Virginia Tech, Blacksburg, VA, USA (43)

Daniel Domínguez, Cell Biology Unit, Department of Cell Biology, Physiology and Immunology, Bioscience School, Universitat Autònoma de Barcelona, Bellaterra, Spain (11)

Andrew Evans, Department of Pathology, University Health Network, Toronto, Canada (183)

Cristina Frías, Cell Biology Unit, Department of Cell Biology, Physiology and Immunology, Bioscience School, Universitat Autònoma de Barcelona, Bellaterra, Spain (11)

Macoura Gadji, The University of Manitoba, Manitoba Institute of Cell Biology, Cancer Care Manitoba, The Genomic Centre for Cancer Research and Diagnosis, Winnipeg, Manitoba, Canada; The Cheikh Anta Diop University of Dakar, Laboratory of Haematology and Immunology, National Centre of Blood Transfusion of Dakar, Senegal (77)

Anna Genescá, Cell Biology Unit, Department of Cell Biology, Physiology and Immunology, Bioscience School, Universitat Autònoma de Barcelona, Bellaterra, Spain (11)

David Gisselsson, Departments of Clinical Genetics and Pathology, Lund, University, Lund, Sweden (1)

Sverre Heim, Centre for Cancer Biomedicine, University of Oslo, Oslo, Norway; Section for Cancer Cytogenetics, Institute for Medical Informatics, The Norwegian Radium Hospital, Oslo University Hospital, Oslo, Norway; Medical Faculty, University of Oslo, Oslo, Norway (127)

Henry H.Q. Heng, Center for Molecular Medicine and Genetics, Wayne State University School of Medicine, MI, USA; Karmanos Cancer Institute, Wayne State University School of Medicine, MI, USA; Department of Pathology, Wayne State University School of Medicine, MI, USA (217)

Mattias Höglund, Department of Oncology, Clinical Sciences, Lund University, Lund, Sweden (151)

Anthony M. Joshua, Department of Medical Biophysics, University of Toronto, Toronto, Canada; Division of Applied Molecular Oncology, Ontario Cancer Institute, Princess Margaret Hospital, Toronto, Canada; Department of Medical Oncology, Princess Margaret Hospital, Toronto, Canada (183)

Ludger Klewes, The University of Manitoba, Manitoba Institute of Cell Biology, Cancer Care Manitoba, The Genomic Centre for Cancer Research and Diagnosis, Winnipeg, Manitoba, Canada (77)

Hans Knecht, Division of Haematology/Oncology, CHUS University Hospital, Sherbrooke (Québec), Canada (77)

Narisorn Kongruttanachok, The University of Manitoba, Manitoba Institute of Cell Biology, Cancer Care Manitoba, The Genomic Centre for Cancer Research and Diagnosis, Winnipeg, Manitoba, Canada (77)

David Lindgren, Center for Molecular Pathology, Department of Laboratory Medicine, Lund University, SUS Malmö, Malmö, Sweden (151)

Guo Liu, Center for Molecular Medicine and Genetics, Wayne State University School of Medicine, MI, USA (217)

Sabine Mai, The University of Manitoba, Manitoba Institute of Cell Biology, Cancer Care Manitoba, The Genomic Centre for Cancer Research and Diagnosis, Winnipeg, Manitoba, Canada (77)

Joshua M. Nicholson, Department of Biological Sciences, Virginia Tech, Blacksburg, VA, USA (43)

Judit Pampalona, Cell Biology Unit, Department of Cell Biology, Physiology and Immunology, Bioscience School, Universitat Autònoma de Barcelona, Bellaterra, Spain (11)

Paul C. Park, Department of Pathology and Molecular Medicine, Queen's University, Kingston, ON, Canada (183)

Alexander Pietras, Center for Molecular Pathology, CREATE Health, Department of Laboratory Medicine, Lund University, Skåne University Hospital, Malmö, Sweden (255)

Jeremy A. Squire, Department of Pathology and Molecular Medicine, Queen's University, Kingston, ON, Canada (183)

Christiaan Righolt, The University of Manitoba, Manitoba Institute of Cell Biology, Cancer Care Manitoba, The Genomic Centre for Cancer Research and Diagnosis, Winnipeg, Manitoba, Canada; Department of Imaging Science and Technology, Delft University of Technology, Delft, The Netherlands (77)

Joshua B. Stevens, Center for Molecular Medicine and Genetics, Wayne State University School of Medicine, MI, USA (217)

Manuel R. Teixeira, Department of Genetics, Portuguese Oncology Institute, Porto, Portugal; Biomedical Sciences Institute (ICBAS), University of Porto, Porto, Portugal; Centre for Cancer Biomedicine, University of Oslo, Oslo, Norway (127)

Laura Tusell, Cell Biology Unit, Department of Cell Biology, Physiology and Immunology, Bioscience School, Universitat Autònoma de Barcelona, Bellaterra, Spain (11)

Rhea Vallente, The University of Manitoba, Manitoba Institute of Cell Biology, Cancer Care Manitoba, The Genomic Centre for Cancer Research and Diagnosis, Winnipeg, Manitoba, Canada (77)

Johan Vallon-Christersson, Department of Oncology, Clinical Sciences, Lund University, Lund, Sweden; CREATE Health Strategic Center for Translational Cancer Research, Lund University, Lund, Sweden (151)

Mark D. Vincent, Department of Medical Oncology, London Regional Cancer Centre, London Health Sciences Centre, Ontario, Canada (283)

Landon Wark, The University of Manitoba, Manitoba Institute of Cell Biology, Cancer Care Manitoba, The Genomic Centre for Cancer Research and Diagnosis, Winnipeg, Manitoba, Canada (77)

Julia L. Williams, Department of Pathology and Molecular Medicine, Queen's University, Kingston, ON, Canada (183)

Christine J. Ye, Department of Internal Medicine, Wayne State University School of Medicine, MI, USA (217)

Maisa Yoshimoto, Department of Pathology and Molecular Medicine, Queen's University, Kingston, ON, Canada (183)

Intratumor Diversity and Clonal Evolution in Cancer—A Skeptical Standpoint

David Gisselsson

Departments of Clinical Genetics and Pathology, Lund University, Lund, Sweden

Clonal evolution in cancer is intimately linked to the concept of intratumor cellular diversity, as the latter is a prerequisite for Darwinian selection at the micro-level. It has been frequently suggested in the literature that clonal evolution can be promoted by an elevated rate of mutation in tumor cells, so-called genomic instability, the mechanisms of which are now becoming increasingly well characterized. However, several issues need clarification before the presumably complex relationship between mutation rate, intratumor diversity, and clonal evolution can be understood sufficiently well to translate into models that predict the course of tumor disease. In particular, it has to be clarified which of the proposed mechanisms for genomic instability that are able to generate daughter cells with sufficient viability to form novel clones, how clones with different genomic changes differ phenotypically from each other, and what the selective forces are that guide competition among diverse clones in different microenvironments. Furthermore, standardized measurements of mutation rates at the chromosome level, as well as genotypic and phenotypic diversity, are essential to compare data from different studies. Finally, the relationship between clonal variation brought about by genomic instability, on the one hand, and cellular differentiation hierarchies, on the other hand, should be explored to put genomic instability in the context of the tumor stem cell hypothesis. © 2011 Elsevier Inc.

Advances in CANCER RESEARCH, Volume 112
Copyright 2011, Elsevier Inc. All right reserved.
0065-230X/10 $35.00
DOI: 10.1016/B978-0-12-387688-1.00001-6

I. INTRODUCTION

Intratumor diversity is currently discussed primarily in the contexts of genomic instability, clonal selection leading to chemotherapy resistance, and tumor cell differentiation hierarchies, including so-called tumor-initiating cells or cancer stem cells. The issue of intercellular heterogeneity in tumors is undoubtedly complex, not least because many of the terms used in this field lack consensus definitions. As a consequence, experimental work is challenged by the fact that few standardized approaches are available to measure core parameters such as the rate of chromosomal instability or the degree of intercellular heterogeneity. Yet another problem is that much of the published work has been performed on established cancer cell lines, representing systems optimized for *in vitro* growth in a stable microenvironment, in stark contrast to the challenging and shifting milieu in which human tumors develop. Few methods are currently available to thoroughly dissect mutation rates and intercellular variation *in vivo*. A third point that needs clarification is the presumed connection between genomic instability and clonal evolution in cancer. This is based on the premise that Darwinian evolution requires heritable and variable traits, and that cell populations with sufficient heritable variation for an evolutionary process to take place are more likely to arise with an increased mutation rate. Even though natural selection is a well-consolidated mechanism behind species evolution, a number of items still require clarification before it can be linked to genomic instability and widely applied to the development of neoplastic cells (Fig. 1). Some of the more evident ones are as follows.

A. Is Intratumor Diversity a Feature of All Neoplasms?

It is generally assumed that clonal selection is a key process in carcinogenesis, but the experimental basis for this remains to be extended. There are of course already several levels at which experimental studies have shown tumor cells to exhibit a higher degree of diversity than their non-neoplastic counterparts: (1) morphology, (2) gene expression including protein markers, and (3) genomic alterations. However, the extent to which cells exhibiting different morphologies and gene expression signatures in fact constitute functionally different cells with stable, distinct phenotypes over time is less clear. It is therefore possible that some of these ostensibly distinct cell populations may continually change characteristics and cross into each other's phenotypic spectra. That such a change of basic characteristics is not only limited to more stem cell-like tumor cells differentiating into more mature cells over time is indicated by recent

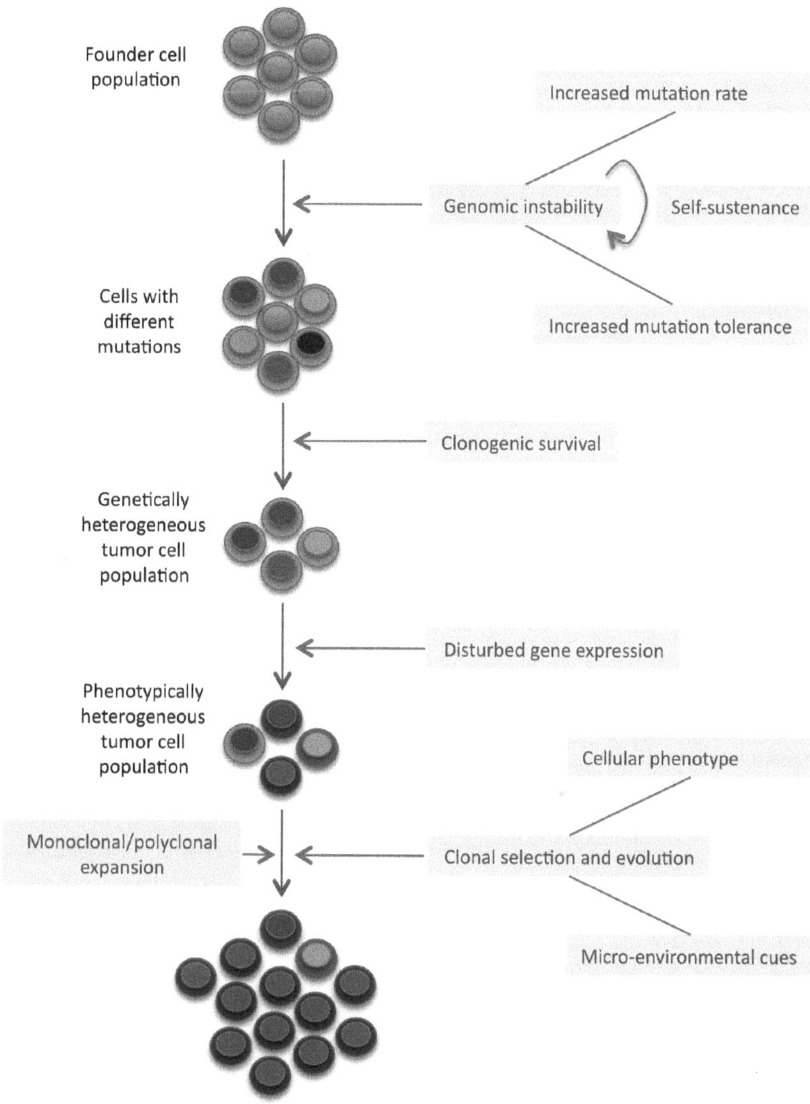

Fig. 1 A simplified model connecting genomic instability and clonal evolution in cancer. There are several crucial points in need of clarification in this process, including: What is the primary generator of genomic diversity—an increased mutation rate or an increased cellular tolerance for mutations? Is genomic instability self-sustaining, for example through aneuploidy giving rise to dysregulation of mitotic checkpoints, resulting in further aneuploidy? Which of the described mechanisms for genomic instability are compatible with clonogenic survival? Which of the genetic alterations brought about by genomic instability do in fact result in an altered cellular phenotype through an altered gene expression? What microenvironmental selective forces guide the selection and clonal expansion of phenotypically different cells? And, finally, is the process a result of clonal expansion from one or several founder cells? See text for details. (For color version of this figure, the reader is referred to the web version of this book.)

studies showing that nonstem cell-like tumor cells can spontaneously change to a more stem-like state (Chaffer *et al.*, 2011). Such bidirectional interconversions would make it difficult to confidently delineate stable and distinct tumor cell populations within a tumor and stress the importance of including a temporal aspect in studies of phenotypic diversity. In contrast, genetic variability constitutes a robust parameter to pinpoint the relationship between different cell populations in a tumor. However, not all tumors have been shown to contain cell populations that are genetically distinct. While many of the common cancers such as adenocarcinomas of the breast, colon, and lung frequently show cell populations with different genomic changes within the same tumor, this is far from always the case in leukemias, lymphomas, and embryonic childhood tumors (Bielas *et al.*, 2006; Mitelman *et al.*, 2011). Whether all of these latter tumor groups will in fact be shown to contain genetically diverse cell populations when subjected to novel high-resolution genetic analyses, such as next-generation sequencing, remains to be seen.

B. To What Extent Does Genomic Instability Lead to Genomic Intercellular Diversity?

A first prerequisite to be met for genomic instability to have a role in promoting variation and ultimately clonal evolution is the survival of cells having undergone mutational events, such as missegregation of one or several chromosomes. The few studies addressing the survival of cells with aneuploidy as a result of chromosomal missegregation have so far shown that the capacity for clonogenic survival varies greatly among cells having undergone different types of chromosome missegregations. For example, there is substantial evidence for multipolar mitosis giving rise to three or more aneuploid daughters not resulting in viable clones (Ganem, Godinho, and Pellman, 2009; Stewénius *et al.*, 2005). On the other hand, multipolar metaphase cells that reorient to a bipolar configuration or at least undergo bipolar cytokinesis have the capacity to generate aneuploid clonogenic daughters at least *in vitro* (Ganem *et al.*, 2009; Gisselsson *et al.*, 2010). This amply illustrates that not all types of so-called chromosomal instability can give rise to genetic diversity, thus disqualifying some of them as key mechanisms behind clonal evolution. This must be kept in mind when surveying the vast number of reports of molecular mechanisms giving rise to chromosomal segregation errors, few of which experimentally address the actual fate of cells having attained the respective types of genomic instability. It cannot be excluded that the tendency to missegregate chromosomes at a high rate, in some cancers, is simply a side effect of deregulation of cell cycle checkpoint systems (either epigenetically or by sporadic mutation), and that this checkpoint deregulation confers a growth advantage distinct from that of chromosomal variation. In that case, it is possible that cells having undergone gross genomic alterations rarely reproduce in comparison to cells managing to segregate their chromosomes correctly despite the checkpoint deficiency. If so, the latter cells will be the ones that predominantly contribute to tumor growth, while the former would continuously reach evolutionary dead ends. Hence, there could be little genomic variation in a tumor cell population despite an elevated rate of chromosome missegregation. Finally, one needs to consider whether the increased number of mutations observed in tumor cells could in some cases merely reflect an increased cellular tolerance to certain mutations, rather than a consequence of an increased mutation rate.

C. To What Extent Does Genetic Variation Correspond to Phenotypic Variation?

For Darwinian selection to function, heritable genetic variation must also result in different phenotypic traits, associated with variable degrees of fitness. The extent to which different genomic changes affect cellular fitness has been very little explored. Recent studies on murine cells have indicated that alterations in fitness resulting from whole chromosome aberrations are unpredictable and vary from chromosome to chromosome (Williams *et al.*, 2008). A similar complex picture has started to appear regarding the relationship between differentiation, on the one hand, and genomic alterations, on the other. At least in some tumors, for example, childhood tumors where differentiation hierarchies can be unequivocally assessed, a complex scenario is emerging, where some large-scale genomic alterations (such as trisomies) appear to exist without correlation to differentiation hierarchies, while smaller genomic imbalances (such as deletions of key embryonic genes) may exist only in cell populations at certain maturation levels (Gisselsson *et al.*, 2010; Holmquist Mengelbier *et al.*, 2010). Furthermore, in breast neoplasms it has been shown that stem-cell-like and more differentiated tumor populations can be highly genetically distinct (Park *et al.*, 2010), questioning a straightforward differentiation hierarchy in which maturing cells are invariably the offspring of stem cell-like cells. Evidently, a thorough examination of the relationship between genetic diversity and differentiation hierarchies, spanning multiple tumor types, would be of value to further assess the mechanisms responsible for phenotypic diversity among cells in a certain tumor.

D. Is genomic Instability a Self-Sustaining Process?

Many authors equal aneuploidy (the state of a chromosome number that is not a multiple of the haploid set, in humans $n = 23$) to genomic, or at least chromosomal instability (an increased rate of change of the chromosome complement). Some studies indicate that once an aneuploid state has been established, this confers a high probability of developing chromosomal instability because of the alterations in gene dosage brought about by abnormal chromosome copy numbers (Duesberg *et al.*, 1998, 2000). However, few studies have sought to reproduce these first experiments, and at least one study argues strongly against the idea of such autocatalytic aneuploidy, based on the finding that extra copies generated by chromosome transfer do not induce chromosomal instability (Lengauer *et al.*, 1997). The issue remains contested. A major objection that can be raised against a positive feedback loop for aneuploidy is that patients with constitutional aneuploidy, such as Down syndrome, do not typically show overt evidence of chromosomal instability such as mosaicism, when subjected to routine cytogenetic analysis. However, fairly recent findings indicate that some low-grade elevation of the chromosome mutation frequency may in fact exist in trisomic individuals (Lightfoot and Hoog, 2004; Thomas *et al.*, 2008) although the parameters studied mainly correlate to structural and not numerical chromosomal instability. Another objection is that several types of tumors, in particular those harboring gene fusions driving pathogenesis, are largely monoclonal cytogenetically also when they carry numerical changes in the stem line (Mitelman *et al.*, 2011). However, one may argue that routine cytogenetics is not sufficiently sensitive to detect small subclones harboring the additional numerical changes that would be expected if aneuploidy indeed led to additional chromosomal alterations. In fact, increasing the number of analyzed tumor cells have in some neoplasms been shown to result in detection of an unexpectedly high number of clones carrying different configurations of chromosome changes (Gorunova *et al.*, 1995, 1998). Furthermore, detailed *in situ* analyses of tumor tissue have revealed that chromosome aberrations assigned to the tumor stem line of ostensibly monoclonal tumors by cytogenetic and or genomic array analyses are in fact restricted to certain tumor subcompartments, indicating that there is a degree of genomic intercellular variation that failed detection by standard techniques (Gisselsson *et al.*, 2010; Holmquist Mengelbier *et al.*, 2010). Notwithstanding the fact that the degree of chromosomal variation may be underestimated both in tumor tissue and in patients with aneuploidy syndromes, it remains to be clarified to what extent aneuploidy in itself leads to chromosomal instability. Does this differ between different patterns of aneuploidy? Is there a threshold of aneuploidy that has to be overcome for the triggering of genomic instability? Presently, one cannot exclude that some aneuploid cancers obtain their abnormal chromosome number through a single event or a transient phase of instability, after which the mutation rate returns to that of nonneoplastic cells.

E. Do Diverse Tumor Cell Populations Always Stem from a Single Cell?

The notion that tumors originate from a single cell that acquires sequential somatic mutations is now well established in the cancer research community, and remains largely unchallenged. However, there is accumulating cytogenetic as well as molecular genetic (based on X chromsome inactivation) evidence that this scenario is not always applicable (Davidsson *et al.*, 2005; Heim *et al.*, 1997; Johansson *et al.*, 1999; Parsons, 2008; Teixeira *et al.*, 2001), as unrelated neoplastic clones have been found in several tumor types. This casts a new light on the presumed connection between genomic instability, intercellular diversity, and clonal selection. Genetic diversity in tumor tissue that occurs primarily because of parallel evolution of multiple clones is to some extent different from the scenario where the diversity stems from genomic instability in a single proliferating clone. On the other hand, some degree of instability may still be acquired for the development of somatic mutations within each unrelated clone. Evidently, the issue of mono- versus polyclonality needs to be thoroughly addressed before detailed models can be made of the role played by genomic instability in the generation of intercellular diversity and clonal evolution in tumors.

The current special volume of Advances in Cancer Research, for which this chapter serves as an introduction, is an attempt to deliver updated information on some of the key questions mentioned above. The chapters are arranged to cover the field from genotype to phenotype, including mechanisms behind chromosomal missegregation, genomic intratumor diversity, diversity in relation to differentiation hierarchies, and evolutionary models of carcinogenesis. It will be apparent to the reader that the field of genomic instability and clonal evolution is still a juvenile one, with mind-challenging hypotheses flourishing amidst the patches of raw experimental data.

ACKNOWLEDGMENT

Grant sponsors: Swedish Childhood Cancer Foundation, Swedish Cancer Society, Swedish Research Council, Crafoord Foundation, Gunnar Nilsson Cancer Foundation, the Royal Physiographic Society, and the Medical Faculty at Lund University.

REFERENCES

Bielas, J. H., Loeb, K. R., Rubin, B. P., True, L. D., and Loeb, L. A. (2006). Human cancers express a mutator phenotype. *Proc. Natl Acad. Sci. U.S.A.* **103**, 18238–18242.

Chaffer, C. L., Brueckmann, I., Scheel, C., Kaestli, A. J., Wiggins, P. A., Rodrigues, L. O., Brooks, M., Reinhardt, F., Su, Y., Polyak, K., Arendt, L. M., Kuperwasser, C., Bierie, B., and Weinberg, R. A. (2011). Normal and neoplastic nonstem cells can spontaneously convert to a stem-like state. *Proc. Natl Acad. Sci. U.S.A.* **108**, 7950–7955.

Davidsson, J., Paulsson, K., and Johansson, B. (2005). Multicolor fluorescence *in situ* hybridization characterization of cytogenetically polyclonal hematologic malignancies. *Cancer Genet. Cytogenet.* **163**, 180–183.

Duesberg, P., Li, R., Rasnick, D., Rausch, C., Willer, A., Kraemer, A., Yerganian, G., and Hehlmann, R. (2000). Aneuploidy precedes and segregates with chemical carcinogenesis. *Cancer Genet. Cytogenet.* **119**, 83–93.

Duesberg, P., Rausch, C., Rasnick, D., and Hehlmann, R. (1998). Genetic instability of cancer cells is proportional to their degree of aneuploidy. *Proc. Natl Acad. Sci. U.S.A.* **95**, 13692–13697.

Ganem, N. J., Godinho, S. A., and Pellman, D. (2009). A mechanism linking extra centrosomes to chromosomal instability. *Nature* **460**, 278–282.

Gisselsson, D., Jin, Y., Lindgren, D., Persson, J., Gisselsson, L., Hanks, S., Sehic, D., Mengelbier, L. H., Ora, I., Rahman, N., Mertens, F., Mitelman, F., and Mandahl, N. (2010). Generation of trisomies in cancer cells by multipolar mitosis and incomplete cytokinesis. *Proc. Natl Acad. Sci. U.S.A.* **107**, 20489–20493.

Gorunova, L., Höglund, M., Andrén-Sandberg, A., Dawiskiba, S., Jin, Y., Mitelman, F., and Johansson, B. (1998). Cytogenetic analysis of pancreatic carcinomas: intratumor heterogeneity and nonrandom pattern of chromosome aberrations. *Genes Chromosomes Cancer* **23**, 81–99.

Gorunova, L., Johansson, B., Dawiskiba, S., Andrén-Sandberg, A., Jin, Y., Mandahl, N., Heim, S., and Mitelman, F. (1995). Massive cytogenetic heterogeneity in a pancreatic carcinoma: fifty-four karyotypically unrelated clones. *Genes Chromosomes Cancer* **14**, 259–266.

Heim, S., Teixeira, M. R., Dietrich, C. U., and Pandis, N. (1997). Cytogenetic polyclonality in tumors of the breast. *Cancer Genet. Cytogenet.* **95**, 16–19.

Holmquist Mengelbier, L., Lindgren, D., Øra, I., Isaksson, M., Frigyesi, I., Frigyesi, A., Bras, J., Sandstedt, B., and Gisselsson, D. (2010). Deletions of 16q in Wilms tumors localize to blastemal-anaplastic cells and are associated with reduced expression of the IRXB renal tubulogenesis gene cluster. *Am. J. Pathol.* **177**, 2609–2621.

Johansson, B., Billstrom, R., Broberg, K., Fioretos, T., Nilsson, P. G., Ahlgren, T., Malm, C., Samuelsson, B. O., and Mitelman, F. (1999). Cytogenetic polyclonality in hematologic malignancies. *Genes Chromosomes Cancer* **24**, 222–229.

Lengauer, C., Kinzler, K. W., and Vogelstein, B. (1997). Genetic instability in colorectal cancers. *Nature* **386**, 623–627.

Lightfoot, D. A., and Hoog, C. (2004). Low level chromosome instability in embryonic cells of primary aneuploid mice. *Cytogenet. Genome Res.* **107**, 95–98.

Mitelman, F., Johansson, B., and Mertens, F., 2011. *Mitelman Database of Chromosome Aberration in Cancer.* http://cgap.nci.nih.gov/Chromosomes/Mitelman.

Park, S. Y., Gonen, M., Kim, H. J., Michor, F., and Polyak, K. (2010). Cellular and genetic diversity in the progression of *in situ* human breast carcinomas to an invasive phenotype. *J. Clin. Invest.* **120**, 636–644.

Parsons, B. L. (2008). Many different tumor types have polyclonal tumor origin: evidence and implications. *Mutat. Res.* **659**, 232–247.

Stewénius, Y., Gorunova, L., Jonson, T., Larsson, N., Höglund, M., Mandahl, N., Mertens, F., Mitelman, F., and Gisselsson, D. (2005). Structural and numerical chromosome changes in colon cancer develop through telomere-mediated anaphase bridges, not through mitotic multipolarity. *Proc. Natl Acad. Sci. U.S.A.* **102**, 5541–5546.

Teixeira, M. R., Tsarouha, H., Kraggerud, S. M., Pandis, N., Dimitriadis, E., Andersen, J. A., Lothe, R. A., and Heim, S. (2001). Evaluation of breast cancer polyclonality by combined chromosome banding and comparative genomic hybridization analysis. *Neoplasia* **3**, 204–214.

Thomas, P., Harvey, S., Gruner, T., and Fenech, M. (2008). The buccal cytome and micronucleus frequency is substantially altered in Down's syndrome and normal ageing compared to young healthy controls. *Mutat. Res.* **638**, 37–47.

Williams, B. R., Prabhu, V. R., Hunter, K. E., Glazier, C. M., Whittaker, C. A., Housman, D. E., and Amon, A. (2008). Aneuploidy affects proliferation and spontaneous immortalization in mammalian cells. *Science* **322**, 703–709.

Role of Telomere Dysfunction in Genetic Intratumor Diversity

Anna Genescà, Judit Pampalona, Cristina Frías, Daniel Domínguez, Laura Tusell

Cell Biology Unit, Department of Cell Biology, Physiology and Immunology, Bioscience School, Universitat Autònoma de Barcelona, Bellaterra, Spain

Most solid tumors are unable to maintain the stability of their genomes at the chromosome level. Indeed, cancer cells display highly rearranged karyotypes containing translocations, amplifications, deletions, and gains and losses of whole chromosomes, which reshuffle steadily. This chromosomal instability most likely occurs early in the development of cancer, and may represent an important step in promoting the multiple genetic changes required for the initiation and/or progression of the disease. Different mechanisms may underlie chromosome instability in cancer cells, but a prominent role for telomeres, the tip of linear chromosomes, has

Advances in CANCER RESEARCH, Volume 112
Copyright 2011, Elsevier Inc. All right reserved.
0065-230X/10 $35.00
DOI: 10.1016/B978-0-12-387688-1.00002-8

been determined. Telomeres are ribonucleoprotein structures that prevent natural chromosome ends being recognized as DNA double-strand breaks, by adopting a loop structure. Loss of telomere function appears from either alteration on telomere-binding proteins or from the progressive telomere shortening that normally occurs under physiological conditions in the majority of cells in tissues. Importantly, unmasked telomeres may either trigger the senescent phenotype that has been linked to the aging process or may initiate the chromosome instability needed for cancer development, depending on the integrity of the DNA damage checkpoint responses. Telomere dysfunction contributes to chromosome instability through end-to-end chromosome fusions entering breakage-fusion-bridge (BFB) cycles. Resolution of chromatin bridge intermediates is likely to contribute greatly to the generation of segmental chromosome amplification events, unbalanced chromosome rearrangements, and whole chromosome aneuploidy. Noteworthy is the fact that telomere length heterogeneity among individuals may directly influence the scrambling of the genome at tumor initiation. However, reiterated BFB cycles would randomly reorganize the cell karyotype, thus increasing the genetic diversity that characterizes tumor cells. Even though a direct link is still lacking, multiple evidence lead one to believe that telomere dysfunction directly contributes to cancer development in humans. The expansion of highly unstable cells due to telomere dysfunction enhances the genetic diversity needed to fuel specific mutations that may promote cell immortalization and the acquisition of a tumor phenotype. © 2011 Elsevier Inc.

I. INTRODUCTION

Several lines of evidence indicate that tumorigenesis in humans is a multistep process in which a succession of genetic changes leads to the progressive conversion of normal human cells into cancer cells. Within a developing tumor, mutations accumulate over time giving rise to variant cell populations that present the six biological endpoints that collectively dictate malignant growth: sustaining proliferative signaling, evading growth suppressors, resisting cell death, limitless replicative potential, inducing angiogenesis, and tissue invasion and metastasis (Hanahan and Weinberg, 2000, 2011).

Another feature that characterizes a high proportion of cancers is the presence of a highly complex karyotype. Studies over a century have shown that cancer cells experience gains and losses of whole chromosomes as well as changes in chromosome structure, including reciprocal and nonreciprocal translocations, inversions, deletions, duplications, and amplifications (Bayani *et al.*, 2007; Camps *et al.*, 2005; Roschke *et al.*, 2002). Hematologic malignancies and pediatric neoplasms are mainly characterized by relatively simple primary aberrations, which are usually distinctive, recurrent, and associated with a particular cancer type. In contrast, most human cancers in the adult population have an epithelial origin and display a myriad of complex chromosomal aberrations that are not always shared by cells of the same tumor or linked to a particular tumor type, which suggest constant genetic reshuffling. Thus, cancer cell heterogeneity in a tumor is the rule rather than an exception. This chromosomal diversity among the tumor cell population is thought to be acquired through chromosomal instability (CIN), which can be defined as the continuous formation of new structural and numerical chromosome aberrations, and is one of the most frequent forms of instability in human cancers (Lengauer *et al.*, 1998).

Genomic instability is a prominent characteristic of most cancer types but at what stage of cancer development it arises and what its molecular basis is are questions to which we are only beginning to get answers (Negrini *et al.*, 2010). By using mouse modeling approaches it is becoming clear that CIN is not simply a passenger phenotype but probably plays a causative role in the onset of a substantial proportion of malignancies (Schvartzman *et al.*, 2010). In humans, a large amount of data collected from tumor biopsies also suggest that CIN has a founder effect in tumorigenesis since it is present in all stages of cancer; from precancerous lesions, even before *TP53* mutations are acquired (Bartkova *et al.*, 2005; Gorgoulis *et al.*, 2005), to advanced cancers (Lengauer *et al.*, 1997; Nowell, 1976). In addition, the level of CIN and chromosomal aberrations correlates with tumor grade and prognosis (Carter *et al.*, 2006), that is, chromosomal aberrations tend to be more numerous in malignant tumors than in benign ones, and both aneuploidy and CIN are associated with poorer prognoses as well as aggressive histopathologic features (Blegen *et al.*, 2003; Habermann *et al.*, 2007; Heilig *et al.*, 2010; Nishizaki *et al.*, 2002; Reshmi and Gollin, 2005; Ried *et al.*, 1999; Risques *et al.*, 2003; Selvarajah *et al.*, 2008;). Collectively these data suggest that genomic instability enables evolving populations of premalignant cells to become malignant by accelerating the accumulation of genetic changes that are responsible for cancer cell evolution (Hanahan and Weinberg, 2000).

The presence of genomic instability reflects defects in cellular processes that maintain the integrity of the genome. To date, the specific molecular mechanisms that underlie CIN in cancer cells remain obscure. A network of genome surveillance mechanisms and DNA repair pathways ensures the integrity of the genome and cell viability. Moreover, equal chromosome segregation during cell division is controlled by mitotic checkpoints, which monitor the complement of chromosomes being passed to daughter cells. Accordingly, disruption of any of the multiple pathways that regulate mitotic fidelity would lead to an increased instability of the genome (Roschke and Kirsch, 2010). However, data analysis of different high-throughput sequencing studies to determine cancer-relevant genes suggested that mutations in DNA repair and mitotic checkpoint genes probably do not account for the presence of genomic instability in many sporadic cancers (Negrini *et al.*, 2010). In contrast, they propose oncogene-induced DNA replication stress or telomere erosion as responsible mechanisms for the presence of genomic instability in these tumors (Negrini *et al.*, 2010).

In this respect, growing evidence correlates the telomere shortening that occurs normally during the aging process with the onset and progression of neoplasia. The ongoing chromosome instability that occurs when telomere function is lost could increase the mutation rate and thus alter the copy numbers of oncogenes and tumor-suppressor genes, which in turn could promote the tumorigenic process (Maser and DePinho, 2002). Underscoring the importance of telomeres in cancer comes from a recent study where leukocyte telomere length was measured over a follow-up period of 10 years in 787 individuals (Willeit *et al.*, 2010). Remarkably, individuals who had shorter telomere length were more likely to develop cancer, even after adjusting for more conventional cancer risk factors, including age. Those with the shortest telomere length had more than a triplicated risk of developing cancer, and those in the middle group had twice the risk compared to those with the longest telomeres. And not only that, those in the short telomere group also had a higher risk of dying from their malignancy than those with longer telomeres, probably because reduced telomere length was also associated with more aggressive cancers, such as tumors of stomach, lung, and ovary (Willeit *et al.*, 2010). All of this indicates that telomere dysfunction is one of the sparks that lit the flame before the fire.

II. CHROMOSOME END PROTECTION: THE SHELTERIN COMPLEX

Telomeres cap the ends of linear chromosomes in a variety of different species to distinguish a chromosome's natural end from a DNA break. In eukaryotes, telomeric DNA consists of a double-stranded region rich in GC, which terminates at the 3′ end in a single-stranded overhang (Blackburn, 1984; Henderson and Blackburn, 1989) (Fig. 1). The telomeric sequence is highly conserved in evolution, suggesting that it is an efficient system of genome protection. All vertebrates share the repetitive sequence TTAGGG, but its length varies between different species; human telomeres range between 5 and 15 kb while laboratory mouse telomeres are remarkably longer at 40–60 kb (Lejnine *et al.*, 1995; Meyne *et al.*, 1989; Zijlmans *et al.*, 1997).

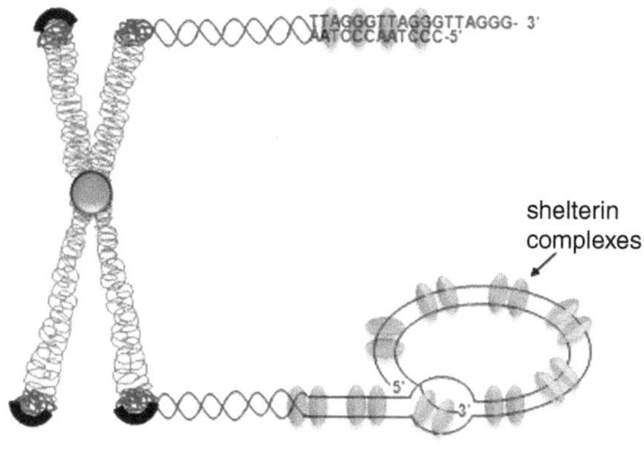

Fig. 1 **The structure of mammalian telomeres.** Telomeres consist of many kb of TTAGGG/AATCCC repeats ending in a G-rich overhang (top right). The telomeric DNA is decorated by shelterin proteins that modulate telomere functions and homeostasis. The protective function of the telomere is achieved through the formation of a lariat structure (bottom right). The 3′ overhang invades the double telomeric track with the aid of the shelterin proteins, thus avoiding being recognized as a double-strand break (DSB). (For color version of this figure, the reader is referred to the web version of this book.)

Telomere DNA sequences, including the overhang, serve as a platform for the assembly of specific protein complexes, which modulate the chromosome extremity properties (Liu *et al.*, 2004). Some of these proteins are maintained at the telomeres during all cell cycle phases and do not function anywhere else in the nucleus. This is the case for the shelterin complex, consisting of the following proteins: TRF1 (TTAGGG Repeat Factor 1), TRF2 (TTAGGG Repeat Factor 2), POT1 (Protection of Telomeres 1), TIN2 (TRF1-Interacting Nuclear protein 2), TPP1 (POT1 and TIN2 organizing protein), and RAP1 (Repressor-Activator Protein 1). This complex is crucial to form and maintain sealed chromosome ends, suppress DNA repair processes, and regulate telomeric DNA length (reviewed in Palm and de Lange, 2008). In particular, the protective function of telomeres is accomplished through remodeling the terminal linear DNA into a lariat conformation. Incubation of a linear telomeric model, consisting of double-stranded TTAGGG repeats ending in a 3′ overhang, with the shelterin protein TRF2 presumably produced the invasion of the single-stranded overhang inside the double helix of the telomeric track located further back (Griffith *et al.*, 1999). Indeed, at least *in vitro*, TRF2 induces positive supercoils that can promote unwinding and invasion (Amiard *et al.*, 2007). But most importantly, this closed chromatin loop was also observed *in vivo* after psoralen cross-linking purified telomeric DNA (Griffith *et al.*, 1999) and in telomere-enriched chromatin (Nikitina and Woodcock, 2004). As a whole, these studies elucidated for the first time the end structure of the chromosome terminus and provided a new vision of how the protective function of the telomere is achieved. Modeling of the linear telomeric track into a loop physically hides the chromosome ends from inappropriate DNA damage responses (DDRs).

In addition to structural functions, shelterin proteins can also regulate the telomere length by controlling the accessibility of telomeres for telomerase—the enzyme that synthesizes telomeres *de novo*. Telomerase does not act on every telomere in each cell cycle but instead exhibits an increasing preference for telomeres as their length declines (Teixeira *et al.*, 2004). Thus, it was proposed that availability for telomerase elongation would probably depend on the amount of shelterin on the telomere track (Lei *et al.*, 2004; Loayza *et al.*, 2004). In this regard, a model where the telomere alternates between an open and closed state has been proposed. In this model, the switch for the conformational change depends on the likelihood that POT1 components can position themselves on the 3′ terminus, through TPP1 interactions (de Lange, 2005).

Although telomeres accommodate a core of six proteins that coat and shape chromosome ends ensuring their integrity, telomeric proteome studies have listed a plethora of other proteins associated with mammalian telomeres (Dejardin and Kingston, 2009) that might also influence chromosome end protection and dynamics. Importantly, some of them function in DNA damage signaling and repair such as the MRN complex—composed of MRE11, RAD50 and NBS1-, DNA-PKcs, ATM, or ATR. In telomeres, these proteins are required for efficient replication and proper trimming-processing of extremities after replication, and help in double-strand invasion to form the D-loop structure (Verdun and Karlseder, 2007; Gilson and Geli, 2007).

III. TELOMERE MAINTENANCE, CELL PROLIFERATION, AND TELOMERE UNCAPPING

A. Telomere Length Declines with Cell Divisions in Most Cells

Telomerase is a specialized ribonucleoprotein complex that includes two essential components: an RNA template (TR or TERC) and a reverse transcriptase catalytic subunit (TERT) (Greider and Blackburn, 1985, 1989; Lingner *et al.*, 1997). In addition to these essential components, a number of other telomerase-associated proteins have been identified, that is, dyskerin, NOP10, NHP2, GAR1, and TCAB1 (Meier, 2005; Venteicher and Artandi, 2009). These proteins are not indispensable for the catalytic activity of the enzyme *in vitro*; however, some likely regulate telomerase activity, assembly, and function *in vivo* (Artandi and DePinho, 2010). Elongation of telomeres by telomerase occurs during the S-phase of the cell cycle coincident with DNA replication. But both fork progression and accessibility of telomerase to the chromosome end are prevented by the presence of the t-loop structure. This topological barrier is remodeled upon the action of telomere binding proteins through co-factors that allow fork progression, control nuclease processing at the 5′ end needed for proper G-tail formation, facilitate telomere elongation when possible, and eventually recap the telomere (Gilson and Geli, 2007).

Considering that the expression of the telomerase enzyme is highly regulated and most somatic cells possess low levels or do not possess any detectable telomerase activity at all (Wright *et al.*, 1996), a reduction of the telomere track occurs at each cell division owing to the failure of conventional DNA polymerases to fully replicate the lagging strand terminus. Already predicted by Watson (1972) and Olovnikov (1973), this "end replication problem" causes a shortening in the newly synthesized chain of few base pairs/end/cell division. Besides this phenomenon, in humans and in mice, the active degradation of the 5′ chromosome ends necessary for proper folding to occur after replication supposes, in the absence of a compensatory mechanism, a loss of ~50–200 base pairs/end/cell division (Huffman *et al.*, 2000). Of note, telomere length may also decline through replication-independent pathways such as the presence of reactive oxygen species (Kawanishi and Oikawa, 2004).

B. Telomere Dysfunction Triggers the DNA Damage Response

Importantly, the physiological telomere shortening that occurs in many somatic cells with cell division can eventually halt cell proliferation. It appears that for proper telomere function a minimum length of telomere repeats is required. Sustained telomere shortening in proliferating cells produces critically short telomeres that activate the DDR rendering a permanent cellular arrest—senescence- or apoptosis, which prevents loss of genome integrity (Fig. 2). Evidence for a relationship between excessively short telomeres and replicative arrest comes from the observance that the "end-replication problem" can be circumvented by telomerase. Ectopic expression of hTERT arrests DDR signaling in senescent human cells and permits, and in some cases apparently promotes, the continued replication and expansion of cell populations (Bodnar *et al.*, 1998; Shay and Wright, 2007). Thus it appears that telomeres track the number of cell divisions through which a cell passes and control cell proliferation by

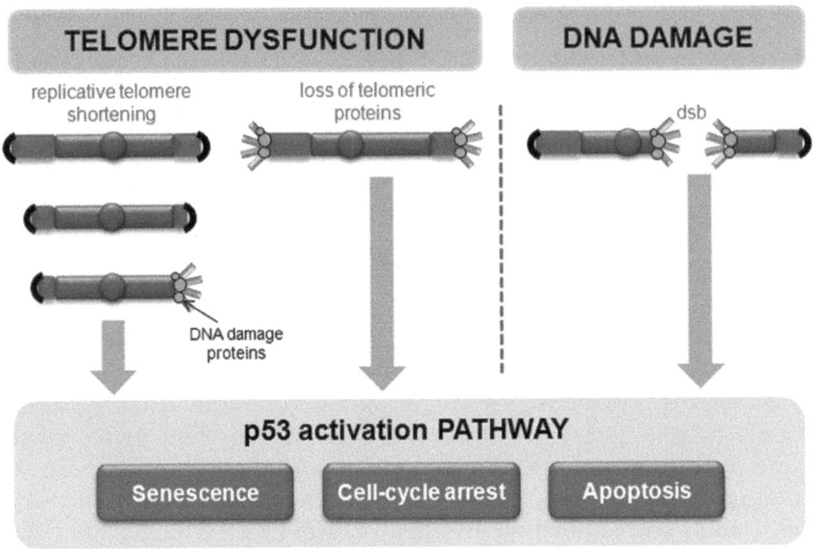

Fig. 2 **Cellular response to dysfunctional telomeres.** Telomere dysfunction, through either replicative shortening or disruption of telomeric proteins, elicits a DDR similar to that triggered after a DSB is produced. Activation of ATM/ATR at dysfunctional telomeres results in a p53-mediated cell-cycle arrest, replicative senescence, or apoptosis. (For color version of this figure, the reader is referred to the web version of this book.)

halting cell division once telomeres shorten at a certain predetermined length (Wright and Shay, 2001).

Another source of telomere damage that may lead to telomere uncapping is the sudden loss of telomeric proteins. Removal of TRF2 from telomeres, either by a dominant negative allele or by targeted mutation in mice, results in the loss of the 3′ overhang leading to acute telomere uncapping and, within a few divisions, all chromosome ends are detected as DNA breaks and then fuse with each other (Smogorzewska and de Lange, 2002; van Steensel *et al.*, 1998). The collapse of t-loops eliminates the discriminatory capacity that normal cells have to distinguish natural chromosome ends from double strand breaks (DSBs) even in the presence of long stretches of TTAGGG repeats. Accordingly, dysfunctional telomeres, either in cells where TRF2 function is disrupted or by excessive shortening, accumulate proteins involved in DNA damage recognition and/or repair distinctive of conventional DSBs, such as 53BP1, γH2AX, MRE11, and the phosphorylated forms of ATM and ATR (d'Adda di Fagagna *et al.*, 2003; Takai *et al.*, 2003) (Fig. 2). All these observations indicate that upon a DNA insult—either DSBs or uncapped telomeres—activation of the classic DDR occurs resulting in cell death via an apoptotic response or senescence.

In this context, it has been proposed that the slow accumulation of senescent cells with aging could eventually affect tissue fitness, thus contributing to age-related diseases. In humans, a decline in the telomere length correlates with a rise in the mortality rate from age-dependent diseases (Cawthon *et al.*, 2003). But also of relevance, limiting the proliferative potential of damaged cells would prevent the outgrowth of cell populations at risk of evolving into neoplasia. Consequently, the senescent phenotype has been suggested to be a potent tumor-suppressor mechanism.

IV. TELOMERES, CHECKPOINTS, AND CANCER

A. When Telomeres are Too Short, Lessons from Mice

The generation of mice deficient in the telomerase RNA gene, *mTerc*$^{-/-}$ (Blasco *et al.*, 1997) allowed the impact of specific reductions in telomere length to be studied at the organismal and cellular levels. These mice exhibit a progressive reduction in the telomere length throughout the cell divisions, as occurs in human somatic cells with age, and in subsequent generations. Early generations of these animals show no obvious pathological phenotype, probably because their cells still possess sufficiently long telomeres. In contrast, advanced generation *mTerc*$^{-/-}$ mice often show defects in highly proliferative tissues such as the hematopoietic and the reproductive system or the intestinal epithelium, in addition to presenting signs of premature aging (Lee *et al.*, 1998; Rudolph *et al.*, 1999). Overall, the decrease in tissue fitness was due to the loss of cell viability and an increase of apoptosis associated with telomere dysfunction, which in turn was manifested by the presence of signal free chromosome ends, end-to-end fusions and anaphase bridges. Both in mice and humans, cell protection from a variety of genotoxic insults is achieved by activating the p53 pathway, which prevents inappropriate proliferation of damaged cells. Accordingly, removal of the p53 function restored many of the cellular defects associated with the accumulation of short dysfunctional telomeres by attenuating their detrimental effect (Chin *et al.*, 1999). All these observations indicated that a minimum telomeric length is necessary to maintain the functions of tissues and fuelled speculation that short telomeres would dramatically limit tumor growth in the absence of telomerase when a fully functional p53 pathway is present. According to this view, a reduced incidence of spontaneous tumors was found in the *mTerc*$^{-/-}$ mouse model (Blasco *et al.*, 1997; Feldser and Greider, 2007).

To further examine the contribution of short dysfunctional telomeres in neoplasia, telomerase-knockout mice were mated with different mouse models bearing defective proteins related to cancer such as *Apc*$^{Min+/-}$—a model for multiple intestinal neoplasia (Rudolph *et al.*, 2001); *Ink4a*$^{-/-}$/*Arf*$^{-/-}$—mice prone to mesenchymal and lymphoid tumors (Greenberg *et al.*, 1999); *Atm*$^{-/-}$—mice that develop thymic lymphoma (Qi *et al.*, 2003; Wong *et al.*, 2003); or Alb-uPA$^{-/-}$—a hepatocellular carcinoma model (Farazi *et al.*, 2003) among others. In all these mouse models, the impact of telomere shortening also produced a reduction of tumor incidence. Nevertheless, these investigations importantly contributed to the knowledge of the operating pathways in cancer development. In particular, studies in the intestinal neoplasia *Apc*$^{Min+/-}$ model allowed telomere dysfunction to be associated with early stages of malignancy. While *mTerc*$^{-/-}$/*Apc*$^{Min+-}$- mice developed more early-stage microadenomas than *Apc*$^{Min+/-}$ mice with intact telomere function, growth and progression of adenomas was significantly suppressed in the double mutant because of the inability to activate telomerase (Rudolph *et al.*, 2001). These studies made it possible to establish that telomere dysfunction is necessary at first, to promote the CIN capable of generating preneoplastic lesions. But later, to allow for tumor progression, it is necessary to reactivate telomerase or alternative recombination mechanisms in order to reach the adequate telomere length compatible with cell survival. If this does not happen, excessive CIN becomes unsustainable and most cells die by apoptosis. Together these results show that dysfunctional telomeres inhibit tumor formation *in vivo* in the setting of an intact DNA damage p53 signaling pathway.

B. Cancer Arises When Cell-Cycle Checkpoint Defects Accompany Short Telomeres

Seminal studies proposing a model where telomere dysfunction together with impaired cell-cycle control promote cellular transformation come from crossing *mTerc*$^{-/-}$ mice with mice that were haploinsufficient in p53 (Artandi *et al.*, 2000). In this setting, loss of the p53 function accelerated tumorigenesis and produced a shift of the tumor spectrum. In contrast to initial generations of *mTerc*$^{-/-}$/*p53*$^{+/-}$ mice with intact telomere function, which normally succumb to lymphomas and sarcomas, the *mTerc*$^{-/-}$/*p53*$^{+/-}$ mice with dysfunctional telomeres showed a shift in their tumor spectra toward epithelial cancers, the tumor type prevalent in adult humans. In addition, the genomic profile of such murine carcinomas changed, showing complex structural chromosomal abnormalities similar to those found in human carcinomas, which are generally rare in mouse tumors. Similar results were obtained when telomere dysfunction was induced directly without telomere shortening but through a mutation of the shelterin protein TPP1. In a context of impaired p53 function,

Tpp1$^{acd/acd}$ mice became highly prone to epithelial carcinogenesis (Else *et al.*, 2009). In all, these findings demonstrate that telomere uncapping, either through loss of the G-rich overhang itself or by critically shortening the (TTAGGG)n tract of DNA, can trigger, in the setting of p53 deficiency, the genome instability that promotes the development of epithelial cancers.

V. TELOMERE UNCAPPING AS A SOURCE OF GENOMIC INSTABILITY

Given that loss of telomere repeats has been implicated in both tumorigenesis and the aging process, great efforts have been made to measure in human cells and model organisms the minimal length required for a functional telomere, as this could be a biomarker forecasting an individual's longevity and risk of disease (Cawthon *et al.*, 2003).

A. Telomere Lengths are Highly Heterogeneous within Somatic Cells

It has been thought that the uncapping signal is generated when the average telomere length reaches a threshold of about 2 kb (Levy *et al.*, 1992). However, examination of the telomere lengths of primary human fibroblasts at senescence by single telomere length analysis (STELA) has demonstrated the presence of very few telomeres that are virtually devoid of telomere repeats (Baird *et al.*, 2003). These results support the notion that it is not the average length of short telomeres but a few sentinel telomeres, those with the shortest lengths, that preferentially trigger cell cycle arrest (Zou *et al.*, 2004).

Various studies have paid attention to unraveling whether a specific telomere length at individual chromosome ends exists in cells and tissues. In humans, the distribution of telomere lengths among the different chromosome arms is heterogeneous, with some chromosome arms having longer telomeres than others. Importantly, this telomere signature is a heritable trait that is genetically determined at conception and maintained throughout life (Graakjaer *et al.*, 2004), and consequently it is also shared within families (Graakjaer *et al.*, 2006a, 2006b). So, when a telomere length distribution in a cell is fixed, the time point at which these particular chromosome arms become uncapped will depend on the specific telomere shortening rate occurring in each cell type or tissue. As a whole, given that each human being has a specific profile of telomere lengths, the risk of a particular chromosome being unstable also differs amongst individuals and will probably be reproduced in a particular organ for a given individual.

B. Initiation of Breakage-Fusion-Bridge Cycles by Dysfunctional Telomeres

Importantly, studies in mouse and human telomerase-deficient cells have demonstrated that the particular chromosomes with a critical telomere length in each animal or individual are those most frequently involved in end-to-end fusion events, namely ring or dicentric chromosomes (Deng *et al.*, 2003, 2004; der-Sarkissian *et al.*, 2004; Plug-DeMaggio *et al.*, 2004; Soler *et al.*, 2005). These unstable chromosome configurations can set in motion BFB cycles, as already noted in maize by McClintock (McClintock, 1939), a mechanism capable of producing rapid and widespread changes in gene dosage as well as complex chromosome rearrangements. When fused chromosomes—interchromosome dicentrics—or sister chromatids—isodicentric chromatids—are pulled to opposite poles during cell division, anaphase bridges are likely to occur (Fig. 3A and 3B). However, the probabilities of bridging and the chromosome aberrations eventually formed after bridge resolution are different depending on the initial substrate for joining. While isodicentric chromatids always bridge at anaphase as they proceed from a continuous DNA molecule, dicentric chromosomes may only contribute to genomic instability if a twist between the two centromeres occurs during division. Thus, dicentric chromosomes with a larger intercentromeric distance present higher probabilities of twisting and bridging.

Studies using real-time imaging have shown that anaphase bridges usually break during cell division (Hoffelder *et al.*, 2004; Pampalona *et al.*, 2010a; Shimizu *et al.*, 2005; Titen and Golic, 2008). Upon breakage of an isodicentric chromatid a partial duplication on a chromosome end in one daughter cell, and a terminal deletion on the same chromosome in the other daughter cell, are generated (reviewed in Murnane and Sabatier, 2004) (Fig. 3A1). Such new uncapped broken chromosomes may again fuse with their replica after DNA synthesis if no other substrate for joining exists, resulting in further amplification of the genes that are close to the break point or fusion point (Lo *et al.*, 2002; Tusell *et al.*, 2008). In contrast, resolution of interchromosome dicentrics results in partial losses—terminal deletions—and partial gains of the involved chromosomes that are also susceptible to be repaired by fusing with other unprotected extremities afterwards (Soler *et al.*, 2005) (Fig. 3B1). As a result, broken extremities can be healed in the form of stable chromosome aberrations such as nonreciprocal translocations or other nonbalanced rearrangements, diminishing CIN. But remarkably, chromosomes can also undergo prolonged periods of chromosome instability, involving repeated BFB cycles, if new dicentric chromosomes are formed. Thus, in the case of a telomere-dysfunction background, CIN is initiated by the formation of end-to-end fusions that can bridge and break during cell division. But relevantly, the continuous proliferation of cells showing increasing telomere dysfunction enhances the formation of new unstable chromosome aberrations, which can perpetuate BFB cycles eventually leading to massive structural CIN.

C. Missegregation of Anaphase-Bridged Chromatin Generates Aneuploid Cell Populations

A wide variety of cancer cells, besides anaphase bridges, also exhibit micronuclei (MN) and nuclear buds (Gisselsson *et al.*, 2001a). The presence of nuclear buds has been related to chromatin regression after anaphase bridge breakage (Pampalona *et al.*, 2010a) although this abnormal nuclear morphology has also been related to the budding, outside of the nucleus, of highly amplified DNA sequences (Shimizu *et al.*, 1998). Regarding micronuclei, these have always been considered to originate from improper segregation of chromosomes during cell division. At telophase, lagging whole chromosomes or chromosome fragments are eventually enclosed by a nuclear membrane separated from that of the daughter nuclei. Specifically, MN containing parts of chromosomes have been shown to originate from fragmented chromosome material when anaphase bridges are formed, stretched, and broken during telophase in carcinoma cell lines (Hoffelder *et al.*, 2004) or in human epithelial cells showing telomere dysfunction (Pampalona *et al.*, 2010a). But relevantly, a significant increase in centromere-labeled MN has also been observed along with telomere dysfunction, thus raising the possibility that bridged dicentric chromatids may also lead to migration defects at anaphase (Pampalona *et al.*, 2010b).

Pioneering work by Barbara McClintock in 1938 gave an early suggestion that anaphase bridges resulting from interlocked ring chromosomes can be excluded from the telophase nuclei, thus leading to whole chromosome losses (McClintock, 1938). In order to directly evaluate the contribution of telomere dysfunction in missegregation events we took advantage of the cytokinesis-blocked micronucleus assay (Pampalona *et al.*, 2010b). Segregation pattern analysis of specific chromosomes in binucleated human mammary epithelial cells demonstrated a preferential erroneous distribution of chromosomes with critically short telomeres. Abnormal segregation of bridged end-to-end fusions resulted in both nondisjuntion and anaphase loss, thus supporting the notion that the mechanical tension of bridged chromatin could result from defective

A

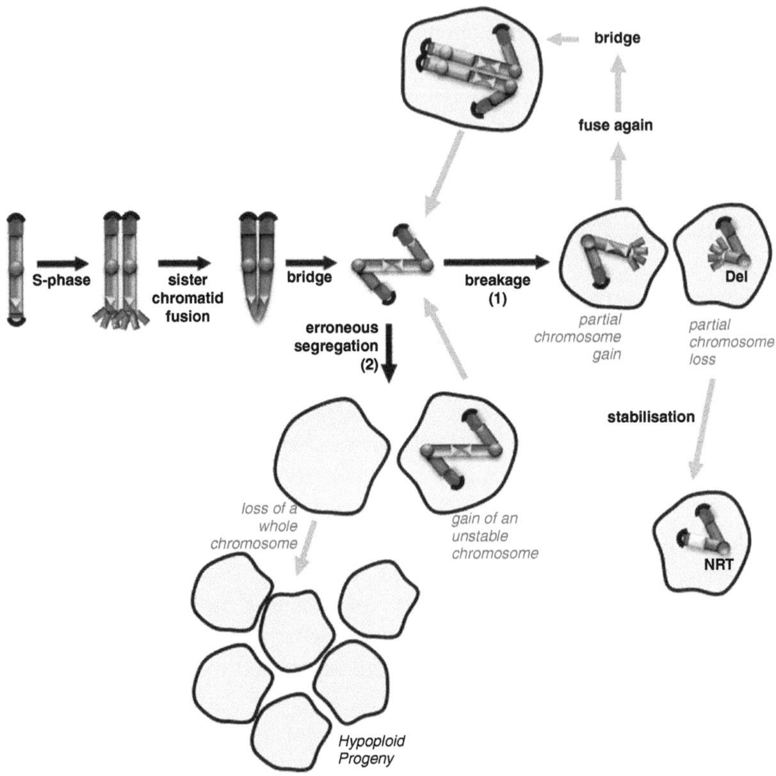

Fig. 3 Chromosome instability mediated by dysfunctional telomeres

(A) Sister chromatid fusion generates a continuous DNA molecule that always bridges at anaphase. (1) Resolution of the bridge through breakage would lead to inverted repeats (yellow triangles) on the end of a chromosome in a daughter cell and a deletion (Del) in the other. These chromosome aberrations are again susceptible to reorganize, leading to nonreciprocal translocations (NRT) or new dicentric chromosomes. While the formation of a NRT avoids further instability, the new dicentric may bridge again at anaphase, thus perpetuating chromosome instability. (2) Anaphase bridges may also erroneously segregate chromosomes between daughter cells. This results in the formation of aneuploid cells. Whereas the loss of a chromosome results in a real hypodiploid progeny, the gain of an unstable chromosome will produce a broad range of abnormalities since it can reorganize further.

(B) Fusion of chromosomes before the S-phase results in conventional interchromosome dicentrics, which will bridge at anaphase only if a twist between the two centromeres is produced. The two bridged dicentric chromatids can either break or missegregate. (1) Rather than amplification events, breakage of both chromatids generates partial gains and losses in the daughter cells. Again, the broken extremities can further reorganize in the form of stable or unstable aberrations, and consequently new BFB cycles can be initiated. (2) In the case that one dicentric chromatid breaks and the other missegregates, partial and whole chromosome losses and gains will be produced. Eventually, a hypodiploid clone of cells will emerge after proliferation of the monosomic cell, whereas the gained dicentric could re-enter into the BFB cycle.

B

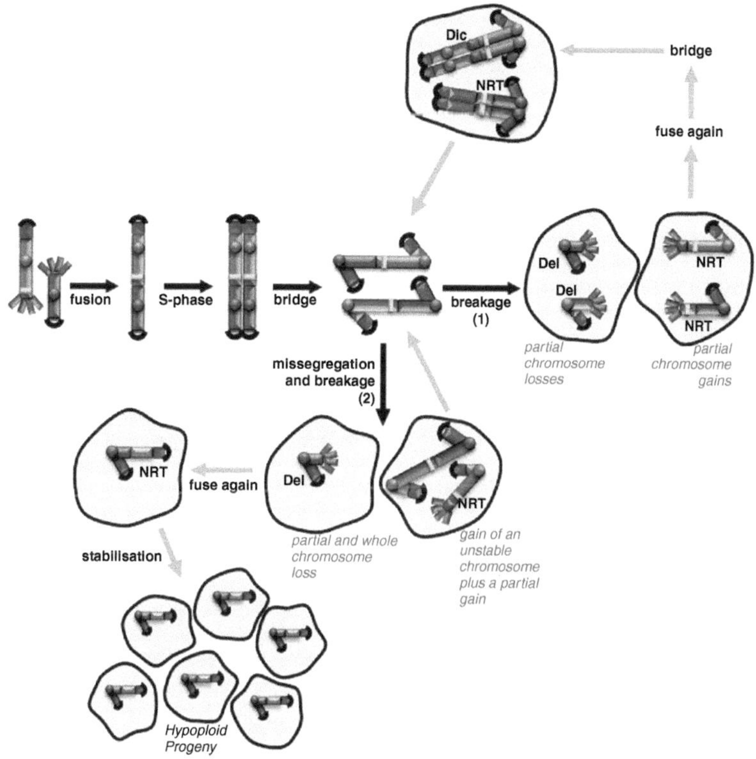

C

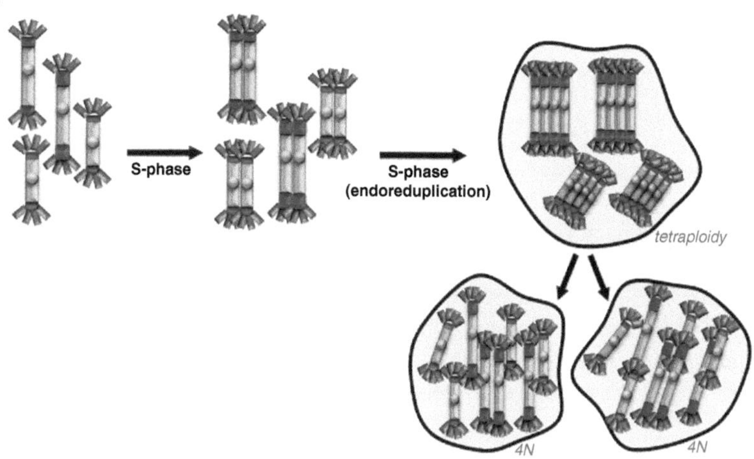

(C) Persistent telomere damage can induce bypass of mitosis and re-entry into S-phase, resulting in a tetraploid cell.

(D) Tetraploidization mediated by telomere dysfunction might also be produced by persistent chromatin bridges during mitosis. The presence of chromatin at the cleavage plane could induce furrow regression and cytokinesis failure, eventually leading to tetraploid cells.

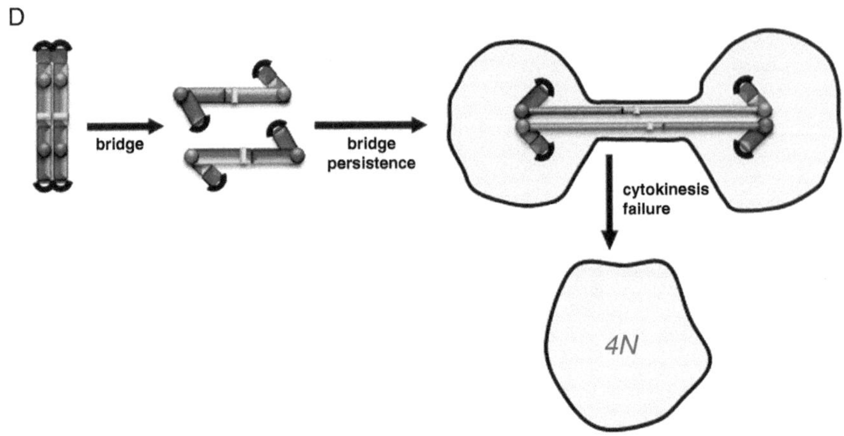

Fig. 3 *(Cont.)* (See Page 4 in Color Section at the back of the book.)

microtubule–kinetochore attachments that eventually end in an abnormal chromosome segregation (Pampalona *et al.*, 2010b). Consistent with these results, a strong correlation between telomere shortening and losses of both chromosome arms and centromeres was observed in biopsies of colon mucosa from ulcerative colitis progressors (O'Sullivan *et al.*, 2002). Also of note is the fact that the frequency of anaphase bridges in the epithelium of these individuals was highly significant. Finally, studies in colorectal cancer cell lines also indicated that detachment of anaphase bridges could contribute to aneuploidy by loss of chromosomes during cell division (Stewénius *et al.*, 2005).

All of this points to the notion that shortening of telomeric DNA repeats results in anaphase bridges that, through BFB cycles, are able to generate unbalanced structural chromosome rearrangements. But in addition, bridged chromatin may also unfaithfully segregate between the daughter cells resulting in aneuploid progeny (Fig. 3A2 and 3B2). According to an anaphase bridge-mediated increase in genome complexity, tumors with anaphase bridges typically had complex karyotypes with both numerical and structural changes. In contrast, tumors without anaphase bridges had either few (<3) cytogenetic changes or numerical changes only (Galli *et al.*, 2004).

D. Telomere Uncapping also Fuels Changes in Cell Ploidy

It is clear that anaphase bridges underlie CIN in the form of structural and numerical aberrations when telomeres become critically short and prone to end fusions. In addition, telomere uncapping has been recently associated with the emergence of whole genome duplication events (Davoli *et al.*, 2010), through anaphase-bridge-independent mechanisms. Depletion of the shelterin proteins POT1a and POT1b, in p53 defective mice, produced a generalized telomere dysfunction leading to the progressive accumulation of polyploid cells containing duplo- and quadruplo-chromosomes. Polyploidization in the form of endomitosis was also described in TRF2 depleted mouse fibroblasts deficient in Lig4 and p53 (Davoli *et al.*, 2010; Smogorzewska *et al.*, 2002). In these cases, polyploidization was induced by persistent DNA damage signaling of uncapped telomeres, as repair is repressed by TRF2 presence at telomeres or by blocking the nonhomologous end-joining repair pathway. These cells experienced a prolonged G2 arrest due to ATM- or ATR-pathways activation but eventually skip mitosis and re-entered cell cycle, thus duplicating the whole genome (Fig. 3C). In this setting, polyploidization was dependent on the DDR as inactivation of the DNA damage signaling cascade of ATM/ATR diminished the tetraploid phenotype.

It is clear that massive telomere damage can fuel polyploidization through endoreduplication; however, this does not exclude other possible causes of mitotic failure mediated by telomere dysfunction. In this sense, it can be envisaged than anaphase bridges resulting from end-to-end chromosome fusions may remain unbroken during mitosis, thus mechanically preventing cell abscission (McClintock, 1938; Mullins and Biesele, 1977) (Fig. 3D). The presence of chromatin occluding the cleavage plane could induce furrow regression giving rise to tetraploid cells. In yeast, the NoCut pathway monitors the clearance of chromatin from the midzone to ensure that cytokinesis completes only after all chromosomes have migrated to the poles (Mendoza *et al.*, 2009; Norden *et al.*, 2006). A similar mechanism has recently been observed in human cells, in which the protein Aurora B remains phosphorylated when chromatin occludes the cleavage plane, t allowing extra time for the chromatin to be released (Steigemann *et al.*, 2009). Although more studies are needed to support this possibility, it is tempting to speculate that uncleared bridged chromatin from the midzone could prevent cytokinesis and result in tetraploid cells.

In the context of neoplasia, different studies support the view that tetraploid cells act as a source of cancer cell precursors. Polyploid cells are often found in tumors of all stages (Storchova and Kuffer, 2008) and tetraploidy renders an increase in CIN (Fujiwara *et al.*, 2005; Mayer and Aguilera, 1990; Storchová *et al.*, 2006) that could constitute an important step in early cell transformation. It has been hypothesized that while diploid cells might die after losing multiple chromosomes, the presence of extra chromosome sets in polyploid cells would buffer the effects of deleterious mutations, allowing cells with DNA damage to survive longer until a crucial growth-enhancing or transforming mutation occurs (Ganem *et al.*, 2007).

The overall conclusion from these studies is that telomere-dependent instability could drive the formation of rearranged karyotypes showing both structural and numerical chromosome aberrations and also changes in ploidy levels, all of which are a hallmark of cancer cells. But more importantly, the inherent instability of these polyploid aberrant cells, through BFB-cycles and errors in merotelic kinetochore-microtubule attachments (Ganem *et al.*, 2009; Silkworth *et al.*, 2009), can result in additional mutations essential for progression toward malignant transformation.

VI. SHORTENING TELOMERES IN AGED HUMAN CELLS

There are a large number of studies demonstrating the harmful impact of relatively short telomeres in cells. Excessive telomere shortening is a relevant feature of both premature aging inherited diseases, often related to genetic defects in telomerase function (Monaghan, 2010), and cancer, a common age-related disease where initial stages have also been linked to eroded telomeres (Aubert and Lansdorp, 2008).

A. Telomere Attrition Occurs in Epithelial Compartments

Multicellular organisms with renewing epithelia need to maintain tissue fitness throughout the life of individuals through a cycle of stem cell proliferation and replacement. Nevertheless, telomerase expression in most adult stem cell compartments is not sufficient to maintain telomere homeostasis, and therefore telomere length of almost all epithelial tissues decreases with human aging (Takubo *et al.*, 2010). Coincident with these observations, most prevalent cancers in adult humans arise from epithelial compartments and their frequency increases exponentially with age.

Studies in mouse models have provided clues about the influence of excessive telomere shortening with organismal age and with cell divisions in culture (Fig. 4). When one or a few telomeres reach a critically short length, the DDR signals to cells to either senesce or die through apoptosis. Senescent cells remain viable but arrest growth and cannot initiate DNA replication in response to physiological mitogens (Cristofalo and Pignolo, 1993; Stanulis-Praeger, 1987). Because senescence stops incipient cancer cells from proliferating, it was proposed to be a powerful tumor-suppressive mechanism (Krtolica and Campisi, 2002). In this context, the senescence response is considered beneficial because it protects organisms from cancer,

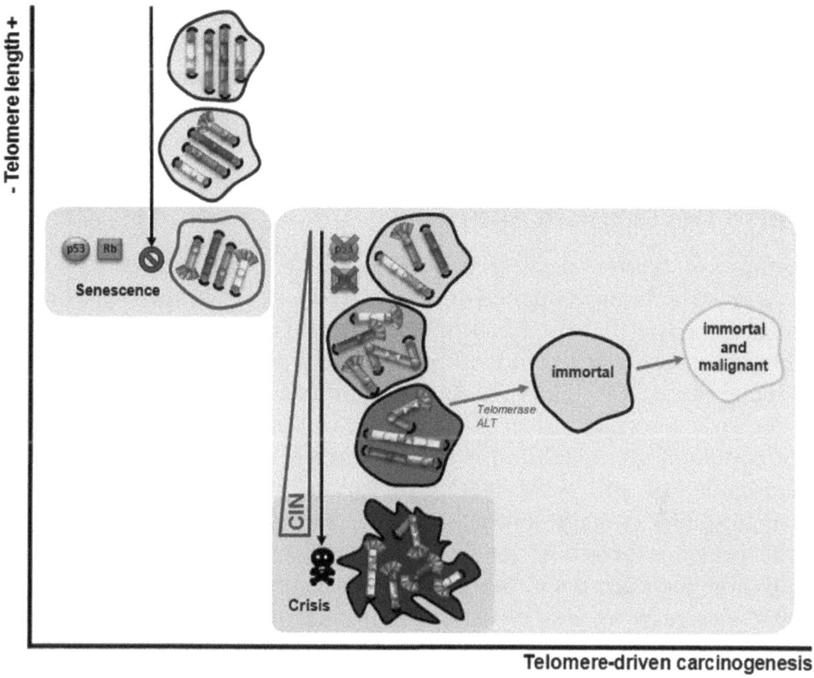

Fig. 4 Model for telomere-driven carcinogenesis. Telomere length diminishes during successive rounds of cell division in an environment where telomerase cannot compensate the end-replication problem. When a few sentinel chromosomes become uncapped, the p53- and Rb-dependent pathways induce a permanent cell cycle arrest known as senescence. Abrogation of these surveillance mechanisms would allow proliferation of cells carrying dysfunctional telomeres. The fusion of uncapped ends will fire the BFB cycles, which would later perpetuate, leading to structural, numerical, and ploidy chromosome abnormalities. Massive chromosome instability will frequently end in crisis, a phase of high rates of cell death. However, the high rate of mutation in these cells could also result in the activation of the mechanisms required to stabilize the chromosome ends. The reduction of chromosome instability to a permissible rate would fuel the genetic diversity needed for eventually acquiring a tumor phenotype. (For color version of this figure, the reader is referred to the web version of this book.)

a major life-threatening disease. However, it must be considered that aging itself could deteriorate the p53 and retinoblastoma (Rb) protein-p16 pathways that govern the senescent growth arrest. Abrogation of these pathways would allow cells, which are destined to senesce to propagate, thus further reducing the length of their telomeres and promoting CIN (Fig. 4). In preneoplastic lesions and cancer cells, components of the signaling cascade governed by p53 and Rb, such as the cyclin-dependent kinase inhibitors

$p21^{CIP1}$ and $p16^{INK4a}$, are frequently disrupted (Gemma, 2002). This highlights the possibility that many more mutations could contribute to faulty cell-cycle checkpoints with age. Or, in other words, that advancing age could contribute to an environment more permissive for carcinogenesis.

B. Does Tumor Suppressor Gene Expression also Decline with Age?

Aging is a natural process with gradual decline of many biological functions, which may enhance mutation frequency. Compelling evidence in mice indicates that the functions of critical proteins involved in DNA repair, cell-cycle control, and apoptosis decline in an age-dependent manner (Feng *et al.*, 2007). In virtually all tissues examined, p53 activity was significantly lower in old mice compared to young. Moreover, less efficient apoptosis was exhibited in the old tissues, thus indicating a reduced functionality of p53 with age. Importantly, the activity levels of the upstream DNA damage sensor kinase ATM, which phosphorylates p53 in the presence of DNA damage, was also found to be reduced after irradiation in older mice. Similarly, in humans, different reports have shown a progressive loss of the ability to repair DNA damage with age *in vitro* (Lambert *et al.*, 1979; Nette *et al.*, 1984). This reduced repair capacity with age correlated well with the increase of mutation rate *in vivo* with human aging (King *et al.*, 1994; Robinson *et al.*, 1994).

Yet, another condition leading to disabled cell-cycle checkpoints involves epigenetic changes in chromatin, such as DNA methylation or histone modification, which may modify gene expression patterns. There is now substantial evidence that aging also affects DNA methylation of specific loci, including cancer-related genes (reviewed in Fraga and Esteller, 2007). In this sense, silencing of tumor suppressor and tumor-related genes, such as *APC*, *CDH1*, and *CDKN2A*, by hypermethylation increases with age in nonneoplastic tissues of individuals (Waki *et al.*, 2003). But most importantly, hypermethylation of different tumor suppressor genes was more frequent in nonneoplastic epithelial tissue of cancer patients than in those individuals without cancer. In agreement with these observations, an age-related increased methylation level has been associated with cancer susceptibility in the elderly (So *et al.*, 2006).

All these studies indicate that the aging process is linked not only to telomere shortening and chromosome uncapping but also to harmful inactivation or defects in surveillance mechanisms that normally monitor genomic integrity, thus potentially increasing the possibilities for preneoplastic cells to proliferate indiscriminately and ultimately acquire a malignant phenotype.

C. BFB-Intermediates and Short Telomeres in Early Preneoplastic Lesions

In humans, excessive telomere shortening is observed in some chronic human diseases associated with high cell turnover and an increased risk of cancer, such as liver cirrhosis (Kitada *et al.*, 1995) or pancreatitis (van Heek *et al.*, 2002). In an analogous fashion, very short telomeres have been reported to be an early alteration in many human cancers, such as those from the urinary bladder, esophagus, large intestine, oral cavity, uterine cervix, and prostate (Meeker *et al.*, 2002, 2004). But more importantly, in Barrett's esophagus, the preneoplastic state that precedes esophageal adenocarcinoma, the samples with the shorter telomeres were also those with the highest degree of chromosomal abnormalities (Finley *et al.*, 2006). Similarly, in ulcerative colitis, a chronic inflammatory disease of the colon, a mechanistic link between telomere shortening and CIN was supported by a higher frequency of anaphase bridges in cancer precursors (O'Sullivan *et al.*, 2002). Collectively, such observations suggest that precancerous cells go through a period of excessive telomere erosion that may initiate CIN, thus increasing the mutability of the genome.

Therefore, as a general rule, it is expected that telomere dysfunction fires the initial BFB cycles that will go on to self-perpetuate through the formation of new unstable chromosome aberrations. Of course, the characteristic telomere length distribution of each individual chromosome could have a direct impact on the way chromosome instability is initiated. At first, only aberrations involving specific chromosomes with the shortest telomeres are evident, but later the random scrambling of the karyotype, through repeated BFB cycles, allows for the genetic divergence of the genome that fosters the characteristic karyotype heterogeneity within tumors. Strong evidence supporting the existence of a period of telomere-crisis before cancer onset comes from the evolution of the chromosome breakpoints during tumor progression in pancreatic carcinoma and osteosarcoma, two tumor types characterized by multiple complex aberrations. While cases with few chromosomal aberrations showed preferential clustering of breakpoints to the terminal bands, tumors harboring complex, reorganized genomes showed a gradual shift of breakpoints to interstitial and centromeric regions (Gisselsson *et al.*, 2001b). The high rate of terminal breakpoints in low rearranged tumors suggested an early disruption of the telomere integrity, and was consistent with persistent BFB cycles gradually eroding the chromosome arms (McClintock, 1941) during tumor evolution. Accordingly, both chromosomes lacking telomeric FISH signals and anaphase bridges were observed in cells from low-grade tumor biopsies. Together, this study further strengthened the possible relationship of telomere dysfunction with cancer initiation in human cells.

D. Toward a Malignant Phenotype

While genomic instability generated through telomere dysfunction might be necessary for tumor development, too much instability can be detrimental to tumor growth. *In vitro* models have shown that rampant telomere-driven instability threatens cell viability as cells accumulate very short telomeres, telomeric fusions, and anaphase bridges, and display a high degree of apoptosis (Wright and Shay, 1992) (Fig. 4). Most cells that enter this second proliferative block, termed crisis, die. That is why this state, along with replicative senescence, is also considered a barrier for dangerous cell proliferation. However, the constant reshuffling of the karyotype during telomere-dependent instability could provide the genetic diversity needed for some cells to acquire certain genetic alterations, which subvert the proliferation barriers to continue growth and eventually transform (Shay and Wright, 2005).

Overcoming the second proliferative barrier is mostly achieved by upregulating telomerase expression or, less frequently, via an alternative recombination-based telomere maintenance mechanism (Hanahan and Weinberg, 2011), as stabilization of chromosome ends, at least to a permissible level, avoids widespread cell death. Telomerase is expressed in the majority of advanced human carcinomas (Kim *et al.*, 1994; Shay and Bacchetti, 1997), suggesting that telomerase reactivation is a critically important step for initiated lesions to progress to malignancy, since it removes telomere shortening's inhibitory barrier to tumor progression. Of note, studies in human cell lines have established that karyotype evolution correlates with the lifetime of the cells prior to telomerase reactivation. Cultures immortalized after a short period of telomere-dependent CIN had very few abnormalities whereas cell lines immortalized after a long instability period were characterized by having multiple gross chromosomal abnormalities (der-Sarkissian *et al.*, 2004). All of this indicates that a delayed acquisition of telomerase function could be a mandatory step to boost malignancy (Hanahan and Weinberg, 2011), since acquiring much more CIN would enhance the probabilities of generating tumor-promoting mutations. Telomerase reactivation *per se* does not cause cells to transform into a tumorigenic state and, at least *in vitro*, conversion to a fully tumorigenic state occurs when normal cells inactivate the p53 and Rb pathways, abrogate PP2A (protein phosphatase 2A) function and express oncogenic H-Ras and hTERT (Hahn, 2002; Hahn *et al.*, 1999). Therefore, in order to become tumorigenic, cells with telomerase must acquire additional mutations to evolve into malignancy.

Importantly, after subsequent upregulation of telomerase expression, telomere-instability does not disappear but decreases to a point where it affects cellular fitness to a sufficiently small extent to allow tumor development. In many tumor types, this chromosome instability is translated into a stepwise accumulation of cytogenetic changes during tumor progression (Al-Mulla *et al.*, 1999; Mitelman *et al.*, 1997; Nishizaki *et al.*, 1997), through a continuous cycle of DNA DSB formation and repair. This is evidenced from the presence of DSB markers, such as phosphorylated histone H2AX and 53BP1 nuclear foci (Bartkova *et al.*, 2005), as well as anaphase bridges and/or mitotically unstable chromosomes (Gisselsson *et al.*, 2004; Rampazzo *et al.*, 2010) in cancer cells. At this stage, both residual traces of the initial telomere-dependent instability and newly attained CIN mechanisms could be responsible for the accumulation of cytogenetic changes. In this context, even though telomerase is expressed, recurrent instability allows for evolution and adaptation of the cancerous cells to the environment as well as the development of the cancer hallmarks that eventually lead to metastatic spread.

VII. SUMMARY

Accumulated data support the notion that loss of telomere repeats contributes to human carcinogenesis; however, the nature of this link remains elusive. It is not an easy issue to demonstrate, since telomere-driven instability should occur early in the neoplastic process and the subsequent derivation of the genome would mask telomere-dysfunctional traces. However, evidence from model organisms, human *in vitro* cell systems, and tissue samples has revealed the existence of scars that denote the flow through a state of telomere instability.

The physiological shortening of telomeres with age in human tissues may detonate CIN through anaphase bridge intermediates when uncapped chromosomes fuse with each other. Massive remodeling and scrambling of the genome, in proliferating cells, is then achieved through repeated BFB cycles of unstable chromosomes resulting from the joining of both initially broken chromosomes and those that later became unprotected. As a result, telomere dysfunction serves as a generator of genomic instability where an arbitrary pattern of structural, numerical and ploidy abnormalities can occur. Adding more complexity to the ongoing telomere-driven instability, the presence of merotelic attachments in polyploid cells, may further increase chromosome missegregation in an already reorganized genome. Thus, telomere-driven genome instability could well be one of the sources leading to a hypermutable heterogeneous genome prone to accumulating the genetic changes that drive cells into malignity.

ACKNOWLEDGMENTS

Research in the Tusell lab is supported by grants from the *Consejo de Seguridad Nuclear*; *Instituto de Salud Carlos III* (RD06/0020/1020) and the *Generalitat de Catalunya* (2009SGR-282).

REFERENCES

Al-Mulla, F., Keith, W. N., Pickford, I. R., Going, J. J., and Birnie, G. D. (1999). Comparative genomic hybridization analysis of primary colorectal carcinomas and their synchronous metastases. *Genes Chromosomes Cancer* **24**, 306–314.

Amiard, S., Doudeau, M., Pinte, S., Poulet, A., Lenain, C., Faivre-Moskalenko, C., Angelov, D., Hug, N., Vindigni, A., Bouvet, P., Paoletti, J., Gilson, E., et al., (2007). A topological mechanism for TRF2-enhanced strand invasion. *Nat. Struct. Mol. Biol.* **14**, 147–154.

Artandi, S. E., Chang, S., Lee, S. L., Alson, S., Gottlieb, G. J., Chin, L., and DePinho, R. A. (2000). Telomere dysfunction promotes non-reciprocal translocations and epithelial cancers in mice. *Nature* **406**, 641–645.

Artandi, S. E., and DePinho, R. A. (2010). Telomeres and telomerase in cancer. *Carcinogenesis* **31**, 9–18.

Aubert, G., and Lansdorp, P. (2008). Telomeres and aging. *Physiol. Rev.* **88**, 557–579.

Baird, D. M., Rowson, J., Wynford-Thomas, D., and Kipling, D. (2003). Extensive allelic variation and ultrashort telomeres in senescent human cells. *Nat. Genet.* **33**, 203–207.

Bartkova, J., Horejsíí, Z., Koed, K., Krämer, A., Tort, F., Zieger, K., Guldberg, P., Sehested, M., Nesland, J. M., Lukas, C., Ørntoft, T., Lukas, J., et al., (2005). DNA damage response as a candidate anti-cancer barrier in early human tumorigenesis. *Nature* **434**, 864–870.

Bayani, J., Selvarajah, S., Maire, G., Vukovic, B., Al-Romaih, K., Zielenska, M., and Squire, J. A. (2007). Genomic mechanisms and measurement of structural and numerical instability in cancer cells. *Semin. Cancer Biol.* **17**, 5–18.

Blackburn, E. H. (1984). The molecular structure of centromeres and telomeres. *Annu. Rev. Biochem* **53**, 163–194.

Blasco, M., Lee, H., Hande, M., Samper, E., Lansdorp, P., DePinho, R., and Greider, C. (1997). Telomere shortening and tumor formation by mouse cells lacking telomerase RNA. *Cell* **91**, 25–34.

Blegen, H., Will, J. S., Ghadimi, B. M., Nash, H. P., Zetterberg, A., Auer, G., and Ried, T. (2003). DNA amplifications and aneuploidy, high proliferative activity and impaired cell cycle control characterize breast carcinomas with poor prognosis. *Anal. Cell Pathol.* **25**, 103–114.

Bodnar, A. G., Ouellette, M., Frolkis, M., Holt, S. E., Chiu, C. P., Morin, G. B., Harley, C. B., Shay, J. W., Lichtsteiner, S., and Wright, W. E. (1998). Extension of life-span by introduction of telomerase into normal human cells. *Science* **279**, 349–352.

Camps, J., Ponsa, I., Ribas, M., Prat, E., Egozcue, J., Peinado, M., and Miró, R. (2005). Comprehensive measurement of chromosomal instability in cancer cells: combination of fluorescence *in situ* hybridization and cytokinesis-block micronucleus assay. *FASEB J.* **19**, 828–830.

Carter, S. L., Eklund, A. C., Kohane, I. S., Harris, L. N., and Szallasi, Z. (2006). A signature of chromosomal instability inferred from gene expression profiles predicts clinical outcome in multiple human cancers. *Nat. Genet.* **38**, 1043–1048.

Cawthon, R. M., Smith, K. R., O'Brien, E., Sivatchenko, A., and Kerber, R. A. (2003). Association between telomere length in blood and mortality in people aged 60 years or older. *Lancet* **361**, 393–395.

Chin, L., Artandi, S., Shen, Q., Tam, A., Lee, S., Gottlieb, G., Greider, C., and DePinho, R. (1999). p53 deficiency rescues the adverse effects of telomere loss and cooperates with telomere dysfunction to accelerate carcinogenesis. *Cell* **97**, 527–538.

Cristofalo, V. J., and Pignolo, R. J. (1993). Replicative senescence of human fibroblast-like cells in culture. *Physiol. Rev.* **73**, 617–638.

d'Adda di Fagagna, F., Reaper, P., Clay-Farrace, L., Fiegler, H., Carr, P., Von Zglinicki, T., Saretzki, G., Carter, N., and Jackson, S. (2003). A DNA damage checkpoint response in telomere-initiated senescence. *Nature* **426**, 194–198.

Davoli, T., Denchi, E., and de Lange, T. (2010). Persistent telomere damage induces bypass of mitosis and tetraploidy. *Cell* **141**, 81–93.

de Lange, T. (2005). Shelterin: the protein complex that shapes and safeguards human telomeres. *Genes Dev.* **19**, 2100–2110.

Dejardin, J., and Kingston, R. E. (2009). Purification of proteins associated with specific genomic Loci. *Cell* **136**, 175–186.

Deng, W., Tsao, S., Guan, X., Lucas, J., and Cheung, A. (2003). Role of short telomeres in inducing preferential chromosomal aberrations in human ovarian surface epithelial cells: a combined telomere quantitative fluorescence *in situ* hybridization and whole-chromosome painting study. *Genes Chromosomes Cancer* **37**, 92–97.

Deng, W., Tsao, S., Guan, X., Lucas, J., Si, H., Leung, C., Mak, P., Wang, L., and Cheung, A. (2004). Distinct profiles of critically short telomeres are a key determinant of different chromosome aberrations in immortalized human cells: whole-genome evidence from multiple cell lines. *Oncogene* **23**, 9090–9101.

der-Sarkissian, H., Bacchetti, S., Cazes, L., and Londoño-Vallejo, J. (2004). The shortest telomeres drive karyotype evolution in transformed cells. *Oncogene* **23**, 1221–1228.

Else, T., Trovato, A., Kim, A. C., Wu, Y., Ferguson, D. O., Kuick, R. D., Lucas, P. C., and Hammer, G. D. (2009). Genetic p53 deficiency partially rescues the adrenocortical dysplasia phenotype at the expense of increased tumorigenesis. *Cancer Cell* **6**, 465–476.

Farazi, P. A., Glickman, J., Jiang, S., Yu, A., Rudolph, K. L., and DePinho, R. A. (2003). Differential impact of telomere dysfunction on initiation and progression of hepatocellular carcinoma. *Cancer Res.* **63**, 5021–5027.

Feldser, D. M., and Greider, C. W. (2007). Short telomeres limit tumor progression *in vivo* by inducing senescence. *Cancer Cell* **11**, 461–469.

Feng, Z., Hu, W., Teresky, A. K., Hernando, E., Cordon-Cardo, C., and Levine, A. J. (2007). Declining p53 function in the aging process: a possible mechanism for the increased tumor incidence in older populations. *Proc. Natl Acad. Sci. U.S.A.* **104**, 16633–16638.

Finley, J., Reid, B., Odze, R., Sanchez, C., Galipeau, P., Li, X., Self, S., Gollahon, K., Blount, P., and Rabinovitch, P. (2006). Chromosomal instability in Barrett's esophagus is related to telomere shortening. *Cancer Epidemiol. Biomarkers Prev.* **15**, 1451–1457.

Fraga, M. F., and Esteller, M. (2007). Epigenetics and aging: the targets and the marks. *Trends Genet.* **23**, 413–418.

Fujiwara, T., Bandi, M., Nitta, M., Ivanova, E. V., Bronson, R. T., and Pellman, D. (2005). Cytokinesis failure generating tetraploids promotes tumorigenesis in p53-null cells. *Nature* **437**, 1043–1047.

Galli, R., Binda, E., Orfanelli, U., Cipelletti, B., Gritti, A., De Vitis, S., Fiocco, R., Foroni, C., Dimeco, F., and Vescovi, A. (2004). Isolation and characterization of tumorigenic, stem-like neural precursors from human glioblastoma. *Cancer Res.* **64**, 7011–7021.

Ganem, N., Storchova, Z., and Pellman, D. (2007). Tetraploidy, aneuploidy and cancer. *Curr. Opin. Genet. Dev.* **17**, 157–162.

Ganem, N., Godinho, S., and Pellman, D. (2009). A mechanism linking extra centrosomes to chromosomal instability. *Nature* **460**, 278–282.

Gemma, A. (2002). Genomic alterations in preneoplastic lesions. *Gan To Kagaku Ryoho* **29**, 2451–2457.

Gilson, E., and Geli, V. (2007). How telomeres are replicated. *Nat. Rev. Mol. Cell. Biol.* **8**, 825–838.

Gisselsson, D., Björk, J., Höglund, M., Mertens, F., Dal Cin, P., Akerman, M., and Mandahl, N. (2001 a). Abnormal nuclear shape in solid tumors reflects mitotic instability. *Am. J. Pathol* **158**, 199–206.

Gisselsson, D., Jonson, T., Petersén, A., Strömbeck, B., Dal Cin, P., Höglund, M., Mitelman, F., Mertens, F., and Mandahl, N. (2001 b). Telomere dysfunction triggers extensive DNA fragmentation and evolution of complex chromosome abnormalities in human malignant tumors. *Proc. Natl Acad. Sci. U.S.A.* **98**, 12683–12688.

Gisselsson, D., Gorunova, L., Höglund, M., Mandahl, N., and Elfving, P. (2004). Telomere shortening and mitotic dysfunction generate cytogenetic heterogeneity in a subgroup of renal cell carcinomas. *Br. J. Cancer* **91**, 327–332.

Gorgoulis, V. G., Vassiliou, L. V., Karakaidos, P., Zacharatos, P., Kotsinas, A., Liloglou, T., Venere, M., Ditullio, R. A., Kastrinakis, N. G., Levy, B., Kletsas, D., Yoneta, A., et al., (2005). Activation of the DNA damage checkpoint and genomic instability in human precancerous lesions. *Nature* **434**, 907–913.

Graakjaer, J., Pascoe, L., Der-Sarkissian, H., Thomas, G., Kolvraa, S., Christensen, K., and Londoño-Vallejo, J. (2004). The relative lengths of individual telomeres are defined in the zygote and strictly maintained during life. *Aging Cell* **3**, 97–102.

Graakjaer, J., Der-Sarkissian, H., Schmitz, A., Bayer, J., Thomas, G., Kolvraa, S., and Londono-Vallejo, J. A. (2006 a). Allele-specific relative telomere lengths are inherited. *Hum. Genet.* **119**, 344–350.

Graakjaer, J., Londono-Vallejo, J., Christensen, K., and Kølvraa, S. (2006 b). The pattern of chromosome-specific variations in telomere length in humans shows signs of heritability and is maintained through life. *Ann. NY Acad. Sci.* **1067**, 311–316.

Greenberg, R. A., Chin, L., Femino, A., Lee, K. H., Gottlieb, G. J., Singer, R. H., Greider, C. W., and DePinho, R. A. (1999). Short dysfunctional telomeres impair tumorigenesis in the INK4a(delta2/3) cancer-prone mouse. *Cell* **97**, 515–525.

Greider, C. W., and Blackburn, E. H. (1985). Identification of a specific telomere terminal transferase activity in Tetrahymena extracts. *Cell*405–413.

Greider, C. W., and Blackburn, E. H. (1989). A telomeric sequence in the RNA of Tetrahymena telomerase required for telomere repeat synthesis. *Nature* **337**, 331–337.

Griffith, J. D., Comeau, L., Rosenfield, S., Stansel, R. M., Bianchi, A., Moss, H., and de Lange, T. (1999). Mammalian telomeres end in a large duplex loop. *Cell* **97**, 503–514.

Habermann, J. K., Paulsen, U., Roblick, U. J., Upender, M. B., McShane, L. M., Korn, E. L., Wangsa, D., Kruger, S., Duchrow, M., Bruch, H. P., Auer, G., and Ried, T. (2007). Stage-specific alterations of the genome, transcriptome, and proteome during colorectal carcinogenesis. *Genes Chromosomes Cancer* **46**, 10–26.

Hahn, W. C., Counter, C. M., Lundberg, A. S., Beijersbergen, R. L., Brooks, M. W., and Weinberg, R. A. (1999). Creation of human tumour cells with defined genetic elements. *Nature* **400**, 464–468.

Hahn, W. C. (2002). Immortalization and transformation of human cells. *Mol. Cells* **13**, 351–361.

Hanahan, D., and Weinberg, R. A. (2000). The hallmarks of cancer. *Cell* **100**, 57–70.

Hanahan, D., and Weinberg, R. A. (2011). Hallmarks of cancer: the next generation. *Cell* **144**, 646–674.

Heilig, C. E., Loffler, H., Mahlknecht, U., Janssen, J. W., Ho, A. D., Jauch, A., and Kramer, A. (2010). Chromosomal instability correlates with poor outcome in patients with myelodysplastic syndromes irrespectively of the cytogenetic risk group. *J. Cell Mol. Med.* **14**, 895–902.

Henderson, E. R., and Blackburn, E. H. (1989). An overhanging 3′ terminus is a conserved feature of telomeres. *Mol. Cell Biol.* **9**, 345–348.

Hoffelder, D., Luo, L., Burke, N., Watkins, S., Gollin, S., and Saunders, W. (2004). Resolution of anaphase bridges in cancer cells. *Chromosoma* **112**, 389–397.

Huffman, K. E., Levene, S. D., Tesmer, V. M., Shay, J. W., and Wright, W. E. (2000). Telomere shortening is proportional to the size of the G-rich telomeric 3′-overhang. *J. Biol. Chem.* **275**, 19719–19722.

Kawanishi, S., and Oikawa, S. (2004). Mechanism of telomere shortening by oxidative stress. *Ann. NY Acad. Sci.* **1019**, 278–284.

Kim, N. W., Piatyszek, M. A., Prowse, K. R., Harley, C. B., West, M. D., Ho, P. L., Coviello, G. M., Wright, W. E., Weinrich, S. L., and Shay, J. W. (1994). Specific association of human telomerase activity with immortal cells and cancer. *Science* **266**, 2011–2015.

King, C. M., Gillespie, E. S., McKenna, P. G., and Barnett, Y. A. (1994). An investigation of mutation as a function of age in humans. *Mutat. Res.* **316**, 79–90.

Kitada, T., Seki, S., Kawakita, N., Kuroki, T., and Monna, T. (1995). Telomere shortening in chronic liver diseases. *Biochem. Biophys. Res. Commun.* **211**, 33–39.

Krtolica, A., and Campisi, J. (2002). Cancer and aging: a model for the cancer promoting effects of the aging stroma. *Int. J. Biochem. Cell Biol.* **34**, 1401–1414.

Lambert, B., Ringborg, U., and Skoog, L. (1979). Age-related decrease of ultraviolet light-induced DNA repair synthesis in human peripheral leukocytes. *Cancer Res.* **39**, 2792–2795.

Lee, H., Blasco, M., Gottlieb, G., Horner, J. N., Greider, C., and DePinho, R. (1998). Essential role of mouse telomerase in highly proliferative organs. *Nature* **392**, 569–574.

Lei, M., Podell, E. R., and Cech, T. R. (2004). Structure of human POT1 bound to telomeric single-stranded DNA provides a model for chromosome end-protection. *Nat. Struct. Mol. Biol.* **11**, 1223–1229.

Lejnine, S., Makarov, V. L., and Langmore, J. P. (1995). Conserved nucleoprotein structure at the ends of vertebrate and invertebrate chromosomes. *Proc. Natl Acad. Sci. U.S.A.* **92**, 2393–2397.

Lengauer, C., Kinzler, K. W., and Vogelstein, B. (1997). Genetic instability in colorectal cancers. *Nature* **386**, 623–627.

Lengauer, C., Kinzler, K., and Vogelstein, B. (1998). Genetic instabilities in human cancers. *Nature* **396**, 643–649.

Levy, M. Z., Allsopp, R. C., Futcher, A. B., Greider, C. W., and Harley, C. B. (1992). Telomere end-replication problem and cell aging. *J. Mol. Biol.* **225**, 951–960.

Lingner, J., Hughes, T. R., Shevchenko, A., Mann, M., Lundblad, V., and Cech, T. R. (1997). Reverse transcriptase motifs in the catalytic subunit of telomerase. *Science* **276**, 561–567.

Liu, D., O'Connor, M. S., Qin, J., and Songyang, Z. (2004). Telosome, a mammalian telomere-associated complex formed by multiple telomeric proteins. *J. Biol. Chem.* **279**, 51338–51342.

Lo, A., Sabatier, L., Fouladi, B., Pottier, G., Ricoul, M., and Murnane, J. (2002). DNA amplification by breakage/fusion/bridge cycles initiated by spontaneous telomere loss in a human cancer cell line. *Neoplasia* **4**, 531–538.

Loayza, D., Parsons, H., Donigian, J., Hoke, K., and de Lange, T. (2004). DNA binding features of human POT1: a nonamer 5′-TAGGGTTAG-3′ minimal binding site, sequence specificity, and internal binding to multimeric sites. *J. Biol. Chem.* **279**, 13241–13248.

Maser, R., and DePinho, R. (2002). Connecting chromosomes, crisis, and cancer. *Science* **297**, 565–569.

Mayer, V., and Aguilera, A. (1990). High levels of chromosome instability in polyploids of *Saccharomyces cerevisiae. Mutat. Res.* **231**, 177–186.

McClintock, B. (1938). The production of homozygous deficient tissues with mutant characteristics by means of the aberrant mitotic behavior of ring-shaped chromosomes. *Genetics* **23**, 315–376.

McClintock, B. (1939). The Behavior in successive nuclear divisions of a chromosome broken at meiosis. *Proc. Natl Acad. Sci. U.S.A.* **25**, 405–416.

McClintock, B. (1941). The stability of broken ends of chromosomes in zea mays. *Genetics* **26**, 234–282.

Meeker, A. K., Hicks, J. L., Platz, E. A., March, G. E., Bennett, C. J., Delannoy, M. J., and De Marzo, A. M. (2002). Telomere shortening is an early somatic DNA alteration in human prostate tumorigenesis. *Cancer Res.* **62**, 6405–6409.

Meeker, A., Hicks, J., Iacobuzio-Donahue, C., Montgomery, E., Westra, W., Chan, T., Ronnett, B., and De Marzo, A. (2004). Telomere length abnormalities occur early in the initiation of epithelial carcinogenesis. *Clin. Cancer Res.* **10**, 3317–3326.

Meier, U. T. (2005). The many facets of H/ACA ribonucleoproteins. *Chromosoma* **114**, 1–14.

Mendoza, M., Norden, C., Durrer, K., Rauter, H., Uhlmann, F., and Barral, Y. (2009). A mechanism for chromosome segregation sensing by the NoCut checkpoint. *Nat. Cell. Biol.* **11**, 477–483.

Meyne, J., Ratliff, R. L., and Moyzis, R. K. (1989). Conservation of the human telomere sequence (TTAGGG)n among vertebrates. *Proc. Natl Acad. Sci. U.S.A.* **86**, 7049–7053.

Mitelman, F., Johansson, B., Mandahl, N., and Mertens, F. (1997). Clinical significance of cytogenetic findings in solid tumors. *Cancer Genet. Cytogenet.* **95**, 1–8.

Monaghan, P. (2010). Telomeres and life histories: the long and the short of it. *Ann. NY Acad. Sci.* **1206**, 130–142.

Mullins, J., and Biesele, J. (1977). Terminal phase of cytokinesis in D-98s cells. *J. Cell Biol.* **73**, 672–684.

Murnane, J., and Sabatier, L. (2004). Chromosome rearrangements resulting from telomere dysfunction and their role in cancer. *Bioessays* **26**, 1164–1174.

Negrini, S., Gorgoulis, V. G., and Halazonetis, T. D. (2010). Genomic instability—an evolving hallmark of cancer. *Nat. Rev. Mol. Cell. Biol.* **11**, 220–228.

Nette, E. G., Xi, Y. P., Sun, Y. K., Andrews, A. D., and King, D. W. (1984). A correlation between aging and DNA repair in human epidermal cells. *Mech. Ageing Dev.* **24**, 283–292.

Nikitina, T., and Woodcock, C. L. (2004). Closed chromatin loops at the ends of chromosomes. *J. Cell Biol.* **166**, 161–165.

Nishizaki, T., DeVries, S., Chew, K., Goodson 3rd, W. H., Ljung, B. M., Thor, A., and Waldman, F. M. (1997). Genetic alterations in primary breast cancers and their metastases: direct comparison using modified comparative genomic hybridization. *Genes Chromosomes Cancer* **19**, 267–272.

Nishizaki, T., Harada, K., Kubota, H., Furuya, T., Suzuki, M., and Sasaki, K. (2002). Chromosome instability in malignant astrocytic tumors detected by fluorescence *in situ* hybridization. *J. Neurooncol.* **56**, 159–165.

Norden, C., Mendoza, M., Dobbelaere, J., Kotwaliwale, C., Biggins, S., and Barral, Y. (2006). The NoCut pathway links completion of cytokinesis to spindle midzone function to prevent chromosome breakage. *Cell* **125**, 85–98.

Nowell, P. C. (1976). The clonal evolution of tumor cell populations. *Science* **194**, 23–28.

O'Sullivan, J., Bronner, M., Brentnall, T., Finley, J., Shen, W., Emerson, S., Emond, M., Gollahon, K., Moskovitz, A., Crispin, D., Potter, J., and Rabinovitch, P. (2002). Chromosomal instability in ulcerative colitis is related to telomere shortening. *Nat. Genet.* **32**, 280–284.

Olovnikov, A. M. (1973). A theory of marginotomy. The incomplete copying of template margin in enzymic synthesis of polynucleotides and biological significance of the phenom enon. *J. Theor. Biol.* **41**, 181–190.

Palm, W., and de Lange, T. (2008). How shelterin protects mammalian telomeres. *Annu. Rev. Genet.* **42**, 301–334.

Pampalona, J., Soler, D., Genescà, A., and Tusell, L. (2010 a). Telomere dysfunction and chromosome structure modulate the contribution of individual chromosomes in abnormal nuclear morphologies. *Mutat. Res* **683**, 16–22.

Pampalona, J., Soler, D., Genescà, A., and Tusell, L. (2010 b). Whole chromosome loss is promoted by telomere dysfunction in primary cells. *Genes Chromosomes Cancer* **49**, 368–378.

Plug-DeMaggio, A., Sundsvold, T., Wurscher, M., Koop, J., Klingelhutz, A., and McDougall, J. (2004). Telomere erosion and chromosomal instability in cells expressing the HPV oncogene 16E6. *Oncogene* **23**, 3561–3571.

Qi, L., Strong, M. A., Karim, B. O., Armanios, M., Huso, D. L., and Greider, C. W. (2003). Short telomeres and ataxia-telangiectasia mutated deficiency cooperatively increase telomere dysfunction and suppress tumorigenesis. *Cancer Res.* **63**, 8188–8196.

Rampazzo, E., Bertorelle, R., Serra, L., Terrin, L., Candiotto, C., Pucciarelli, S., Del Bianco, P., Nitti, D., and De Rossi, A. (2010). Relationship between telomere shortening, genetic instability, and site of tumour origin in colorectal cancers. *Br. J. Cancer* **102**, 1300–1305.

Reshmi, S. C., and Gollin, S. M. (2005). Chromosomal instability in oral cancer cells. *J. Dent. Res.* **84**, 107–117.

Ried, T., Heselmeyer-Haddad, K., Blegen, H., Schrock, E., and Auer, G. (1999). Genomic changes defining the genesis, progression, and malignancy potential in solid human tumors: a phenotype/genotype correlation. *Genes Chromosomes Cancer* **25**, 195–204.

Risques, R. A., Moreno, V., Ribas, M., Marcuello, E., Capella, G., and Peinado, M. A. (2003). Genetic pathways and genome-wide determinants of clinical outcome in colorectal cancer. *Cancer Res.* **63**, 7206–7214.

Robinson, D. R., Goodall, K., Albertini, R. J., O'Neill, J. P., Finette, B., Sala-Trepat, M., Moustacchi, E., Tates, A. D., Beare, D. M., Green, M. H., et al., (1994). An analysis of *in vivo* hprt mutant frequency in circulating T-lymphocytes in the normal human population: a comparison of four datasets. *Mutat. Res.* **313**, 227–247.

Roschke, A. V., Stover, K., Tonon, G., Schaffer, A. A., and Kirsch, I. R. (2002). Stable karyotypes in epithelial cancer cell lines despite high rates of ongoing structural and numerical chromosomal instability. *Neoplasia* **4**, 19–31.

Roschke, A. V., and Kirsch, I. R. (2010). Targeting karyotypic complexity and chromosomal instability of cancer cells. *Curr. Drug Targets* **11**, 1341–1350.

Rudolph, K., Millard, M., Bosenberg, M., and DePinho, R. (2001). Telomere dysfunction and evolution of intestinal carcinoma in mice and humans. *Nat. Genet.* **28**, 155–159.

Rudolph, K. L., Chang, S., Lee, H. -L., Blasco, M., Gottlieb, G. J., Greider, C., and DePinho, R. A. (1999). Longevity, stress response, and cancer in aging Telomerase-deficient mice. *Cell* **98**, 701–712.

Schvartzman, J. M., Sotillo, R., and Benezra, R. (2010). Mitotic chromosomal instability and cancer: mouse modelling of the human disease. *Nat. Rev. Cancer* **10**, 102–115.

Selvarajah, S., Yoshimoto, M., Ludkovski, O., Park, P. C., Bayani, J., Thorner, P., Maire, G., Squire, J. A., and Zielenska, M. (2008). Genomic signatures of chromosomal instability and osteosarcoma progression detected by high resolution array CGH and interphase FISH. *Cytogenet. Genome Res.* **122**, 5–15.

Shay, J. W., and Bacchetti, S. (1997). A survey of telomerase activity in human cancer. *Eur. J. Cancer* **33**, 787–791.

Shay, J. W., and Wright, W. E. (2005). Senescence and immortalization: role of telomeres and telomerase. *Carcinogenesis* **26**, 867–874.

Shay, J. W., and Wright, W. E. (2007). Hallmarks of telomeres in ageing research. *J. Pathol.* **211**, 114–123.

Shimizu, N., Itoh, N., Utiyama, H., and Wahl, G. (1998). Selective entrapment of extrachromosomally amplified DNA by nuclear budding and micronucleation during S phase. *J. Cell Biol.* **140**, 1307–1320.

Shimizu, N., Shingaki, K., Kaneko-Sasaguri, Y., Hashizume, T., and Kanda, T. (2005). When, where and how the bridge breaks: anaphase bridge breakage plays a crucial role in gene amplification and HSR generation. *Exp. Cell Res.* **302**, 233–243.

Silkworth, W., Nardi, I., Scholl, L., and Cimini, D. (2009). Multipolar spindle pole coalescence is a major source of kinetochore mis-attachment and chromosome mis-segregation in cancer cells. *PLoS One* **4**, e6564.

Smogorzewska, A., and de Lange, T. (2002). Different telomere damage signaling pathways in human and mouse cells. *EMBO J.* **21**, 4338–4348.

Smogorzewska, A., Karlseder, J., Holtgreve-Grez, H., Jauch, A., and de Lange, T. (2002). DNA ligase IV-dependent NHEJ of deprotected mammalian telomeres in G1 and G2. *Curr. Biol.* **12**, 1635–1644.

So, K., Tamura, G., Honda, T., Homma, N., Waki, T., Togawa, N., Nishizuka, S., and Motoyama, T. (2006). Multiple tumor suppressor genes are increasingly methylated with age in non-neoplastic gastric epithelia. *Cancer Sci.* **97**, 1155–1158.

Soler, D., Genescà, A., Arnedo, G., Egozcue, J., and Tusell, L. (2005). Telomere dysfunction drives chromosomal instability in human mammary epithelial cells. *Genes Chromosomes Cancer* **44**, 339–350.

Stanulis-Praeger, B. M. (1987). Cellular senescence revisited: a review. *Mech. Ageing Dev.* **38**, 1–48.

Steigemann, P., Wurzenberger, C., Schmitz, M., Held, M., Guizetti, J., Maar, S., and Gerlich, D. (2009). Aurora B-mediated abscission checkpoint protects against tetraploidization. *Cell* **136**, 473–484.

Stewénius, Y., Gorunova, L., Jonson, T., Larsson, N., Höglund, M., Mandahl, N., Mertens, F., Mitelman, F., and Gisselsson, D. (2005). Structural and numerical chromosome changes in colon cancer develop through telomere-mediated anaphase bridges, not through mitotic multipolarity. *Proc. Natl Acad. Sci. U.S.A.* **102**, 5541–5546.

Storchová, Z., Breneman, A., Cande, J., Dunn, J., Burbank, K., O'Toole, E., and Pellman, D. (2006). Genome-wide genetic analysis of polyploidy in yeast. *Nature* **443**, 541–547.

Storchová, Z., and Kuffer, C. (2008). The consequences of tetraploidy and aneuploidy. *J. Cell Sci.* **121**, 3859–3866.

Takai, H., Smogorzewska, A., and de Lange, T. (2003). DNA damage foci at dysfunctional telomeres. *Curr. Biol.* **13**, 1549–1556.

Takubo, K., Aida, J., Izumiyama-Shimomura, N., Ishikawa, N., Sawabe, M., Kurabayashi, R., Shiraishi, H., Arai, T., and Nakamura, K. (2010). Changes of telomere length with aging. *Geriatr. Gerontol. Int.* **10**(suppl 1), S197–S206.

Teixeira, M. T., Arneric, M., Sperisen, P., and Lingner, J. (2004). Telomere length homeostasis is achieved via a switch between telomerase-extendible and -nonextendible states. *Cell* **117**, 323–335.

Titen, S., and Golic, K. (2008). Telomere loss provokes multiple pathways to apoptosis and produces genomic instability in *Drosophila melanogaster*. *Genetics* **180**, 1821–1832.

Tusell, L., Soler, D., Agostini, M., Pampalona, J., and Genescà, A. (2008). The number of dysfunctional telomeres in a cell: one amplifies; more than one translocate. *Cytogenet. Genome Res.* **122**, 315–325.
van Heek, N. T., Meeker, A. K., Kern, S. E., Yeo, C. J., Lillemoe, K. D., Cameron, J. L., Offerhaus, G. J., Hicks, J. L., Wilentz, R. E., Goggins, M. G., De Marzo, A. M., Hruban, R. H., et al., (2002). Telomere shortening is nearly universal in pancreatic intraepithelial neoplasia. *Am. J. Pathol.* **161**, 1541–1547.
van Steensel, B., Smogorzewska, A., and de Lange, T. (1998). TRF2 protects human telomeres from end-to-end fusions. *Cell* **92**, 401–413.
Venteicher, A. S., and Artandi, S. E. (2009). TCAB1: driving telomerase to Cajal bodies. *Cell Cycle* **8**, 1329–1331.
Verdun, R. E., and Karlseder, J. (2007). Replication and protection of telomeres. *Nature* **447**, 924–931.
Waki, T., Tamura, G., Sato, M., and Motoyama, T. (2003). Age-related methylation of tumor suppressor and tumor-related genes: an analysis of autopsy samples. *Oncogene* **22**, 4128–4133.
Watson, J. D. (1972). Origin of concatemeric T7 DNA. *Nat. New Biol.* **239**, 197–201.
Willeit, P., Willeit, J., Mayr, A., Weger, S., Oberhollenzer, F., Brandstatter, A., Kronenberg, F., and Kiechl, S. (2010). Telomere length and risk of incident cancer and cancer mortality. *JAMA* **304**, 69–75.
Wong, K. K., Maser, R. S., Bachoo, R. M., Menon, J., Carrasco, D. R., Gu, Y., Alt, F. W., and DePinho, R. A. (2003). Telomere dysfunction and Atm deficiency compromises organ homeostasis and accelerates ageing. *Nature* **421**, 643–648.
Wright, W. E., and Shay, J. W. (1992). The two-stage mechanism controlling cellular senescence and immortalization. *Exp. Gerontol.* **27**, 383–389.
Wright, W. E., Piatyszek, M. A., Rainey, W. E., Byrd, W., and Shay, J. W. (1996). Telomerase activity in human germline and embryonic tissues and cells. *Dev. Genet.* **18**, 173–179.
Wright, W. E., and Shay, J. W. (2001). Cellular senescence as a tumor-protection mechanism: the essential role of counting. *Curr. Opin. Genet. Dev.* **11**, 98–103.
Zijlmans, J. M., Martens, U. M., Poon, S. S., Raap, A. K., Tanke, H. J., Ward, R. K., and Lansdorp, P. M. (1997). Telomeres in the mouse have large inter-chromosomal variations in the number of T2AG3 repeats. *Proc. Natl Acad. Sci. U.S.A.* **94**, 7423–7428.
Zou, Y., Sfeir, A., Gryaznov, S. M., Shay, J. W., and Wright, W. E. (2004). Does a sentinel or a subset of short telomeres determine replicative senescence? *Mol. Biol. Cell* **15**, 3709–3718.

How Mitotic Errors Contribute to Karyotypic Diversity in Cancer

Joshua M. Nicholson and Daniela Cimini

Department of Biological Sciences, Virginia Tech, Blacksburg, VA, USA

Aneuploidy is a common feature of cancer cells, and is believed to play a critical role in tumorigenesis and cancer progression. Most cancer cells also exhibit high rates of mitotic chromosome mis-segregation, a phenomenon known as chromosomal instability, which leads to high variability of the karyotype. Here, we describe the nature, nuances, and implications of cancer karyotypic diversity. Moreover, we summarize recent studies aimed at identifying the mitotic defects that may be responsible for inducing chromosome mis-segregation in cancer cells. These include kinetochore attachment errors, spindle assembly checkpoint dysfunction, mitotic spindle defects, and other cell division inaccuracies. Finally, we discuss how such mitotic errors generate karyotypic diversity in cancer cells. © 2011 Elsevier Inc.

I. INTRODUCTION

At the crux of carcinogenesis lies complexity. Amongst a single tumor, cell-to-cell variability is manifest in many different properties: karyotype, morphology, enzyme receptors, ability to metastasize, mutations, and

Advances in CANCER RESEARCH, Volume 112
Copyright 2011, Elsevier Inc. All right reserved.
0065-230X/10 $35.00
DOI: 10.1016/B978-0-12-387688-1.00003-X

drug resistance. While seemingly chaotic per cell, the tumor as a whole is much like a mosaic displaying interdependency and organization. For this reason it has been called a species (Duesberg and Rasnick, 2000; Duesberg *et al.*, 2011; Huxley, 1956; Klein *et al.*, 2010; Van Valen and Maiorana, 1991; Vincent, 2010), a society of cells (Heppner, 1984), and a complex ecosystem (Heppner, 1984; Merlo *et al.*, 2006; Sachs and Hlatky, 2010). Whereas it is likely that a myriad of mechanisms contribute to cancer complexity, in this chapter we specifically focus on how mitotic defects contribute to the complexity of the cancer cell karyotype. It has long been known that cancer cells are characterized by aneuploid karyotypes (Bayreuther and Klein, 1958; Chu and Giles, 1958; Hauschka and Levan, 1958), and an increasing body of work has unveiled a causal relationship between aneuploidy and tumorigenesis [reviewed in (Cimini, 2008; Pavelka *et al.*, 2010a; Sen, 2000)]. A key role of aneuploidy in tumorigenesis and tumor progression is also indicated by the fact that measuring DNA indices or ploidy is clinically informative as a prognostic indicator in various cancers (Atkin and Kay, 1979). For instance, ploidy measurements in different types of cancer (Frankfurt *et al.*, 1985; Grote *et al.*, 2001; Jakobsen, 1984; Jakobsen *et al.*, 1988; Kallioniemi *et al.*, 1987; Susini *et al.*, 2011) are as, if not more, accurate when predicting survival than other widely used measures such as the prostate-specific antigen (PSA) test (Stamey, 2004; Vickers *et al.*, 2011).

Karyotypic analysis also indicates that there is wide variability in the chromosome number within the same cancer cell population, suggesting that errors in mitotic chromosome segregation are recurrent in cancer cells. In the remaining part of this chapter, we will start by reviewing what is known about intratumor karyotypic diversity. Next, we will describe the mitotic mechanisms that promote chromosome mis-segregation in cancer cells, and thus can produce karyotypic diversity.

II. INTRATUMOR DIVERSITY AT THE KARYOTYPIC LEVEL

A. Preneoplastic Aneuploidy

Aneuploidy is ubiquitous in cancer (Gebhart and Liehr, 2000; Mertens *et al.*, 1997; Mitelman *et al.*, 2011; Weaver and Cleveland, 2006), and was proposed to have a causal role in cancer already over a century ago (Boveri, 1902, 1914). Over the years, the idea that aneuploidy plays a causal role in the origin of cancer has been supported by substantial experimental evidence. For example, random aneuploidy appears after carcinogen application but before transformation in Chinese hamster (Duesberg *et al.*, 2000a), mouse (Conti *et al.*, 1986), rat (Sudilovsky and Hei, 1991), and human cells (Hittelman, 2001; Hittelman *et al.*, 1996). Furthermore, although the results were occasionally ambiguous (see below for further discussion), a number of studies in mouse models carrying mutations/deletions that cause aneuploidy showed a correlation between aneuploidy and tumorigenesis (Babu *et al.*, 2003; Diaz-Rodriguez *et al.*, 2008; Iwanaga *et al.*, 2007; Michel *et al.*, 2001; Sotillo *et al.*, 2007; Weaver *et al.*, 2007). For aneuploidy to play a causal role in tumorigenesis, it would be expected to occur in undifferentiated dividing cells within the organism's tissues. Unfortunately, the rates of chromosome mis-segregation and the frequencies of aneuploidy *in situ* are very difficult to assess due to the inaccessibility of primary tissues. One exception in humans is represented by peripheral blood lymphocytes, which can be easily isolated. Isolated lymphocytes can be either analyzed directly by chromosome- or centromere-specific fluorescence *in situ* hybridization (FISH) or used to establish short-term cultures. Cultured lymphocytes can be used for metaphase spread analysis (Minissi *et al.*, 1999), micronucleus assay (Countryman and Heddle, 1976), or cytokinesis-block assay (Fenech, 1993) to determine spontaneous rates of chromosome mis-segregation and aneuploidy. Studies using these approaches have yielded a large amount of information, and indicate that the basal frequencies of aneuploidy for individual chromosomes in human lymphocytes range between 0.03% and 10% (Carere *et al.*, 1998; Catalan *et al.*, 2000; Hayes *et al.*, 2000; Minissi *et al.*, 1999; Ramirez *et al.*, 2007; Rehen *et al.*, 2005; Zhang *et al.*, 1998;). Furthermore, studies in brain cells have found aneuploidy for individual chromosomes to be ~4% (Kingsbury *et al.*, 2006; Rehen *et al.*, 2005). Such frequencies are surprisingly high. Indeed, if the same aneuploidy rates were occurring in all dividing cells, hundreds of thousands aneuploid cells would be produced every day in each individual, given that $\sim 2.5 \times 10^8$ cells are dividing at any given time (Alberts *et al.*, 1994). If such frequencies were realistic, cancer would be expected to occur at much higher rates in the population. So, how can these high frequencies of aneuploidy be reconciled with the relatively low incidence of cancer if indeed aneuploidy is a cause of cancer? One possibility is that the rates of aneuploidy observed in peripheral blood lymphocytes and neurons are not representative of all cell types. However, it is noteworthy that the frequency of aneuploidy in peripheral blood lymphocytes of healthy individuals correlates with cancer risk (Hagmar *et al.*, 1994; Olaharski *et al.*, 2006; Rossner *et al.*, 2005). In addition, aneuploidy in human peripheral blood lymphocytes is increased by various factors typically associated with cancer, such as smoking and age (Brogger *et al.*, 1990; Hittelman, 2001), suggesting that aneuploidy in lymphocytes may be, if not equivalent, at least proportional to that of other cell types. The other possible explanation for such high frequencies of spontaneous aneuploidy resulting in a relatively low incidence of cancer is that not all aneuploidies are equal and the majority of aneuploid cells die (Li *et al.*, 2009; Zhivotovsky and Kroemer, 2004). We favor this latter explanation given that high rates of aneuploidy are found in a number of premalignant lesions, such as head and neck amongst others (Hittelman, 2001; Hittelman *et al.*, 1996; Ried *et al.*, 1999).

B. The Effects of Aneuploidy

In aneuploid cells, large numbers of genes are either lacking or present in excess. This may lead to underproduction or excessive production of both mRNA and proteins. Indeed, numerous studies have found chromosome copy number, including subchromosomal derivative chromosomes, to directly affect global transcript levels [Table I (FitzPatrick *et al.*, 2002; Gao *et al.*, 2007; Hyman *et al.*, 2002; Phillips *et al.*, 2001; Pollack *et al.*, 2002; Upender *et al.*, 2004; Virtaneva *et al.*, 2001; Williams *et al.*, 2008)]. Moreover, recent studies in aneuploid budding yeast show a clear copy number transcriptome–proteome relationship (Pavelka *et al.*, 2010b). Despite this connection, the overall relationship between karyotype and phenotype cannot be considered linear. For instance, single trisomies, arising spontaneously or produced artificially, not only raise the gene expression of the trisomic chromosome, but alter hundreds of other genes on different chromosomes (FitzPatrick, 2005; FitzPatrick *et al.*, 2002; Saran *et al.*, 2003; Upender *et al.*, 2004;). Thus, chromosomal gains or losses exhibit pleiotropic effects across the whole genome. The complex effects of aneuploidy can alter gene expression both negatively and positively (underexpression and overexpression, respectively), and can act either directly or indirectly (Table I and Fig. 1).

Studies on the effects of aneuploidy on the cell phenotype have yielded contrasting results. For instance, experiments with trisomic mouse embryonic fibroblasts show that aneuploidy impairs proliferation, alters metabolic properties, and negatively affects immortalization (Williams *et al.*, 2008). Yet, aneuploidy remains ubiquitous in cancer. A recent study identified "aneuploidy tolerating" mutations capable of attenuating the changes in intracellular protein composition caused by aneuploidy (Torres *et al.*, 2010). These findings could reconcile the prevalence of

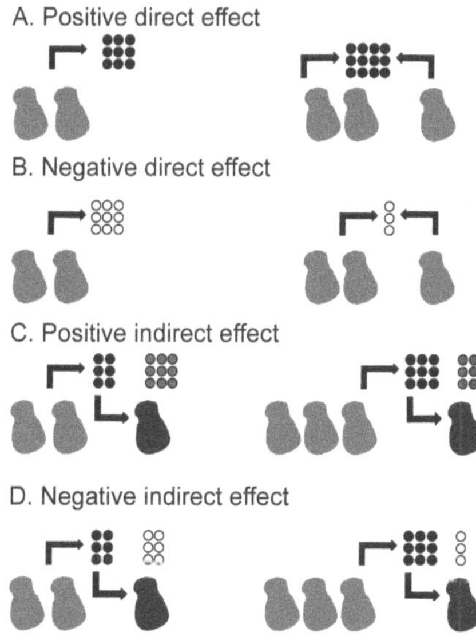

Fig. 1 Pleiotropic effects of aneuploidy. Diagrammatic representation of the effects of trisomy on gene expression. The small circles in the figure represent gene products. The diagrams on the left show gene regulation under disomic conditions, whereas the diagrams on the right show possible effects of trisomy on gene regulation. Trisomy of an individual chromosome produces a 1.5-fold increase in gene copy number of genes on that chromosome. This would be expected to induce a 1.5-fold increase in the amount of gene products. However, the effect seen is often more complex and pleiotropic. For example, an individual trisomic chromosome may increase (A) or decrease (B) the expression of various genes on that chromosome. Additionally, a trisomic chromosome may also increase or decrease (C and D, respectively) the gene expression on one or more other chromosomes.

aneuploidy in cancer with the detrimental effects caused by it. However, another study found no evidence of benefit from stress response gene enrichment and/or no stress response gene enrichment at all in various aneuploid strains of budding yeast (Pavelka *et al.*, 2010b). Importantly, there is no universal gene expression signature specific to aneuploidy (Pavelka *et al.*, 2010b; Upender *et al.*, 2004). Therefore, it appears that whether aneuploidy promotes or inhibits tumorigenesis largely depends on the context and karyotype. Accordingly, aneuploidy is detrimental to cell growth in most normal conditions; however, under certain stressful conditions (cytotoxic drugs, low pH, high salt, etc.) it is beneficial to cells (Ly *et al.*, 2011; Pavelka *et al.*, 2010b; Tang *et al.*, 2011).

Table I Effect of Aneuploidy on Global Gene Expression

			Over-expressed genes		Under-expressed genes			
Cell Type	Aneusomy	Genes on chr.	# (on chr.)	Fold (or %) increase	# (on chr.)	Fold (or %) decrease	Method	References
DLD1	Trisomy 3	1924	81 (17 on chr3)	>1.7-fold	67 (0 on chr3)	>1.7-fold	Micro-cell mediated chr. transfer	Upender et al., 2004
	Trisomy 7	1880	155 (32 on chr7)	>1.7-fold	47 (3 on ch7)	>1.7-fold		
	Trisomy 13	703	92 (10 on chr13)	>1.7-fold	72(0 on chr13)	>1.7-fold		
HME	Trisomy 3	1924	91 (21 on chr3)	>1.7-fold	49 (2 on chr3)	>1.7-fold	Same as above	
Amniocyte	Trisomy 21	449	54 (3 on chr21)	>1.7-fold	54 (0 on chr21)	>1.7-fold	Spontaneous trisomy	Fitzpatrick *et al.*, 2002
	Trisomy 13	703	273 (7 on chr13)	>1.7-fold	86 (0 on chr13)	>1.7-fold		
MEF	Trisomy 19	1037	10(2); 64(10)	2-fold; 1.5-fold	8(0); 75(0)	2-fold; 1.5-fold	Selective breeding	Williams *et al.*, 2008
	Trisomy 1	1944	18(4); 139(29)	2-fold; 1.5-fold	33(0); 144(2)	2-fold; 1.5-fold		
	Trisomy 13	1497	6(3); 76(10)	2-fold; 1.5-fold	6(0); 118(1)	2-fold; 1.5-fold		
	Trisomy 16	1025	2(1); 50(11)	2-fold; 1.5-fold	4(0); 45(1)	2-fold; 1.5-fold		
DBA-2/DB-P	No change (avg)	NA	NA	Log_2 ratio (0.02)	NA	NA	Subclone of established glioblastoma cell line versus parental gliobastoma line	Gao *et al.*, 2007
	Chr2 (4somy/3somy)	2346	NA	Log_2 ratio (0.18)	NA	NA		
	Chr7 (6somy/5somy)	1880	NA	Log_2 ratio (0.16)	NA	NA		

	Chr8 (4somy/3somy)	1317	NA	Log_2 ratio (0.30)	NA	NA		
	Chr14 (2somy/3somy)	1526	NA	NA	NA	Log_2 ratio (-0.33)		
	Chr15 (4somy/5somy)	1272	NA	NA	NA	Log_2 ratio (-0.25)		
	Chr20 (4somy/5somy)	884	NA	NA	NA	Log_2 ratio (-0.32)		
	Chr21 (4somy/7somy)	449	NA	NA	NA	Log_2 ratio (−0.54)		
AML	Trisomy 8	1317	NA	32% up compared to other chr.	NA	NA	Natural	Virtaneva *et al.*, 2001
Prostate cancers	Various	NA	Up-regulated genes on chr. with increased copy number	42%	Down-regulated genes on chr. with decreased copy number	51%	Natural	Phillips *et al.*, 2001

Chr., chromosome(s); avg, average; NA, not analyzed.

C. Chromosome Instability and the Cancer Karyotype: between Order and Disorder

In addition to being aneuploid, cancer cells commonly exhibit high rates of chromosome mis-segregation, a phenomenon termed chromosomal instability (CIN). One consequence of CIN is the generation of karyotypic diversity in cancer cells. Such karyotypic variability within a cancer cell population typically ranges at any one time ±20% around the modal chromosome number (Beheshti *et al.*, 2001; Li *et al.*, 2009; Nicholson and Duesberg, 2009). For example, the modal chromosome number of the breast cancer cell line Ma6 is 83 and can fluctuate within the population by ±13% (Li *et al.*, 2009). In other words, many cells possess 83 chromosomes, but cells with as few as 72, and as many as 94 chromosomes can be found within the population (one cell with 83 vs. another with 72 chromosomes can be observed in the example shown in Fig. 2). Polyploidization can also occur causing a 100% change, but these cells only represent a minority of the entire cell population. Moreover, polyploidization may occur sporadically rather than recurrently in the cancer cell population (see following section for further discussion). Thus, the majority of karyotypic variations within a cancer cell population generally result from gains or losses of individual chromosomes (Furuya *et al.*, 2000; Haruki *et al.*, 2001a; Li *et al.*, 2009; Miyazaki *et al.*, 1999; Nicholson and Duesberg, 2009; Yoon *et al.*, 2002). Indeed, 30% of cells in Ma6 deviate from the modal copy number of chromosome 13 ($n = 4$) by ±1–2 chromosomes whereas only 5% of cells deviate from the modal copy number of chromosome 6 by ±1–2 chromosomes ($n = 3$) (Fig. 2).

While the karyotypes of cancer are heterogeneous, they are not totally random. Indeed, the karyotypes of cancer cells are, over time (68 cell generations), relatively stable, varying only ±6% despite a variability of ±22% at any one time (Nicholson and Duesberg, 2009). The idea of a characteristic karyotype, termed the "stemline concept," was first described in the analysis of tumor ascites by Winge (Winge, 1930), and later supported with additional cytological evidence by Hauschka (Hauschka, 1953), Levan (Hauschka and Levan, 1958; Levan, 1956), and Makino (Makino, 1956). In short, the stemline concept holds that within each tumor, out of the heterogeneity of cancer cells is an underlying karyotypic pattern or karyotypic stemline. Analysis of thousands of tumors from different tissues of origin using array comparative genomic hybridization (aCGH) showed these karyotypic patterns to be strongly tissue dependent (Beroukhim *et al.*, 2010; Gebhart and Liehr, 2000; Mertens *et al.*, 1997). A possible explanation for this is that aneuploidies for specific chromosomes undergo negative selection in certain tissues but not others. Karyotypic stemlines in cancers have also been recently

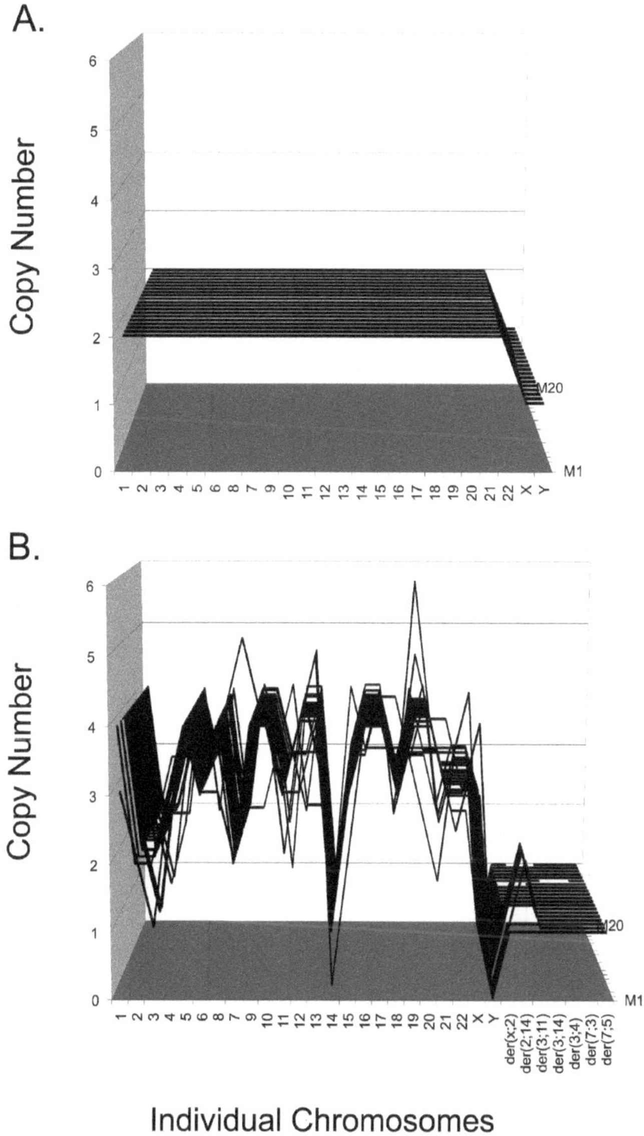

Fig. 2 Visualization of the karyotypic stemline in a karyograph. The three-dimensional graphs (karyographs) show the copy number of individual chromosomes per cell for a population of cells. (A) Karyograph of normal male diploid cells. (B) Karyograph of the transformed mammary epithelial cell line Ma6 (Kendall *et al.*, 2005). Each karyograph shows the results obtained by m-FISH analysis of 20 different cells. "M" denotes metaphase; M1–M20 are displayed along the *y*-axis. Note, both the karyotypic stemline and the karyotypic heterogeneity of the cancer cell population (B) are immediately apparent when the karyotypes are displayed in a karyograph.

identified using multicolor fluorescence *in situ* hybridization (mFISH), corroborating the initial identification by Winge and leading to the conclusion that the cancer karyotype is "selected for oncogenic function" (Nicholson and Duesberg, 2009) or that "a dynamic equilibrium is maintained" (Heim and Mitelman, 2010). This stemline or equilibrium can be visualized graphically in a karyograph (Fig. 2), which plots the copy number for each chromosome in individual cells and in the cell population.

These stemline patterns have led some to propose that a specific karyotype may be the result of selective gain of chromosomes containing activated oncogenes and/or loss of chromosomes carrying tumor-suppressor genes (Lengauer, Kinzler, and Vogelstein, 1997). In support of this idea, Baker *et al.* found that whole chromosome mis-segregation results in loss of heterozygosity for the tumor suppressor gene p53 (Baker *et al.*, 2009). However, it is unclear in this study why cancer cells gained an extra copy of the chromosome carrying the mutant gene rather than simply losing the chromosome harboring the wild-type copy of p53. Moreover, the hypothesis of preferential gain/loss of chromosomes carrying oncogenes/tumor suppressor genes predicts that specific karyotypes should be expected because of the specific location of certain gene loci. However, such specific karyotypes have yet to be identified. Accordingly, clones with identical oncogenic mutations display distinct karyotypes (Duelli *et al.*, 2007; Nicholson and Duesberg, 2009) and breast cancers sharing karyotypic patterns (Beroukhim *et al.*, 2010; Gebhart and Liehr, 2000; Mertens *et al.*, 1997) do not share mutations (Wadman, 2011). Finally, both transgenic and transfected oncogenes have been shown to induce cancers that become oncogene independent over time (Gilbert and Harris, 1988; Klein *et al.*, 2010).

An alternative theory, explaining why cancer karyotypes are not completely random, proposes that the selection of the karyotype depends on the fact that the karyotype as a whole is responsible for carcinogenesis (Nicholson and Duesberg, 2009). Equivalently, just as a species cannot be defined by a singular gene or a singular chromosome, nor can cancer (Heng, 2009).

Identifying specific contributions of individual chromosomes or individual genes in cancer will be exceedingly difficult given the complexity of aneuploid karyotypes and the continual generation of new karyotypes (CIN). For example, 5^{23} or $\sim 10^{16}$ different karyotypes are possible if we assume five copy number states (monosomy, disomy, trisomy, etc.) per chromosome. Moreover, the difficulty may be exacerbated by the dependence of outcome on the karyotypic contexts. For instance, activated oncogenes of the RAS family have different effects in different karyotypic contexts, ranging from loss of tumorigenicity to enhanced tumorigenicity (Geiser, Anderson, and Stanbridge, 1989). Additionally, the insertion of specific chromosomes into different cancer cell lines via microcell-mediated chromosome transfer shows different effects on tumorigenicity, further demonstrating dependency on the karyotypic context (Cao *et al.*, 2001; Negrini *et al.*, 1994; Rimessi *et al.*, 1994).

In conclusion, although analysis of cancer cell populations reveals some kind of equilibrium (order), the CIN of cancer cells generates ever-changing karyotypes (disorder). Regardless of how and why cancer karyotypic diversity originates, it yields phenotypic differences per cell, conferring flexibility to the tumor as a whole. Such flexibility is apparent in numerous properties of cancer, wherein CIN is associated with poor patient prognosis (Carter *et al.*, 2006; Sheffer *et al.*, 2009; Walther, Houlston, and Tomlinson, 2008), and is responsible for cancer drug resistance (Duesberg *et al.*, 2000b; McClelland, Burrell, and Swanton, 2009; Swanton *et al.*, 2009).

III. HOW MITOTIC ERRORS CONTRIBUTE TO THE KARYOTYPIC DIVERSITY OF CANCER CELLS

The goal of mitosis is to accurately segregate the replicated chromosomes into two daughter cells. This is achieved through the interaction between the sister chromatids of each replicated chromosome with microtubules from opposite poles of the mitotic spindle (Fig. 3). This interaction occurs at the kinetochore, a specialized protein structure that assembles at the centromeric region of each chromatid [for review see (Cheeseman and Desai, 2008; Maiato *et al.*, 2004; Santaguida and Musacchio, 2009)], and leads to the alignment of chromosomes at the metaphase plate with the two sister chromatids oriented back to back (Fig. 3). When the cohesion between the two sisters is "dissolved" at anaphase onset, the two chromatids (now chromosomes) are pulled poleward by their associated kinetochore microtubules (Fig. 3). A mitotic checkpoint (also known as the spindle assembly checkpoint or SAC) delays anaphase onset until amphitelic attachment has been achieved by all chromosomes [reviewed in (Musacchio and Salmon, 2007)], thus increasing the fidelity of mitosis. It is easy to infer that the chromosome number instability associated with cancer must be a result of inaccurate mitotic chromosome segregation in cancer cells. Indeed, a number of recurrent errors have been observed in cancer cells, and we will dedicate the remaining part of this section to the discussion of mitotic errors typical of cancer cells and how they contribute to cancer karyotypic diversity.

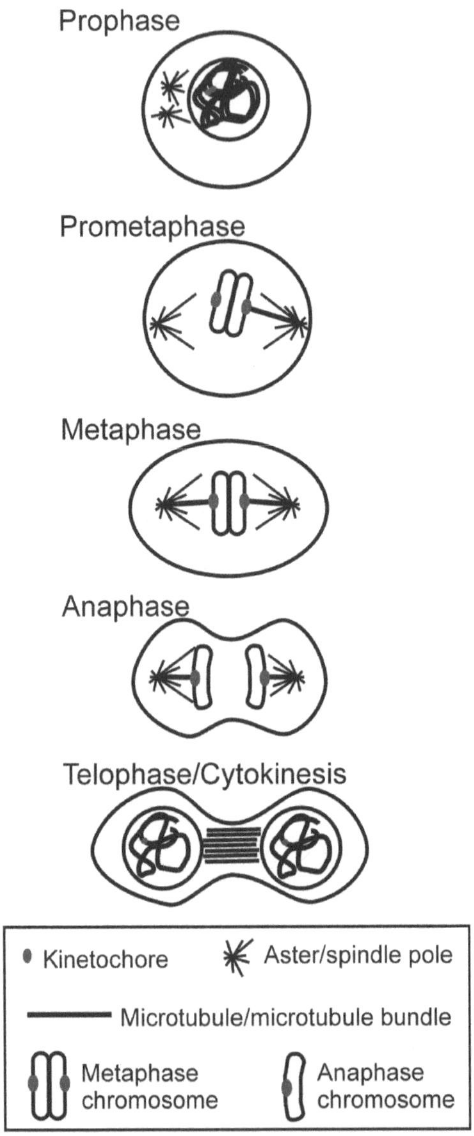

Fig. 3 Diagrammatic representation of the stages of mitosis. A simplified view of the five stages of mitosis is illustrated in this figure. For simplicity, only one chromosome is diagrammed. Upon mitotic entry, each chromosome is constituted of two sister chromatids, which represent the products of DNA replication from the preceding S-phase. *Prophase* is characterized by chromosome condensation and separation of the duplicated centrosomes (spindle poles). Nuclear envelope breakdown marks the beginning of *prometaphase*. During prometaphase, microtubules emanating from the spindle poles establish attachment with the kinetochores. Typically, one sister kinetochore binds microtubules first (monotelic attachment). When the

A. Kinetochore–Microtubule Attachments

Accurate chromosome segregation is ensured by the attachment of the two sister chromatids to microtubules from opposite spindle poles (amphitelic attachment; Fig. 4A). However, other types of attachments can and do occur in mitosis. These include monotelic, syntelic, and merotelic attachment (Fig. 4B–D).

Monotelic attachment occurs when one of the two sister kinetochores of an individual chromosome is attached to one spindle pole whereas the other sister is unattached (Fig. 4B). This type of attachment is very common in early mitosis, and it is known to trigger a checkpoint-dependent mitotic delay (Rieder *et al.*, 1994, 1995). Thus, if the mitotic checkpoint is functional (see below for a more specific discussion of mitotic checkpoint and cancer), monotelic kinetochore attachment will not contribute to cancer karyotypic diversity.

Syntelic attachment occurs when the two sister kinetochores of an individual chromosome bind microtubules from the same spindle pole (Fig. 4C). Syntelic attachments are not commonly observed in untreated cells (Hauf *et al.*, 2003), and there are no reports of increased frequencies of syntelic attachments in cancer cells, suggesting that syntelic attachment does not play an important role in cancer cell chromosome mis-segregation.

Finally, merotelic attachment occurs when a single kinetochore binds microtubules from two poles rather than just one [Fig. 4D; (Cimini *et al.*, 2001; Ladrach and LaFountain, 1986)]. Merotelic attachments occur frequently in early mitosis (Cimini *et al.*, 2003) and can be corrected via an Aurora B-dependent mechanism prior to anaphase onset (Cimini, 2007; Cimini *et al.*, 2006; DeLuca *et al.*, 2006;). However, merotelic attachments do not induce a SAC-dependent mitotic delay (Cimini *et al.*, 2002, 2004; Khodjakov *et al.*, 1997; Wise and Brinkley, 1997; Yu and Dawe, 2000), and thus they can persist into anaphase, at

---

other sister kinetochore binds microtubules, the chromosome congresses to the spindle equator. When all the chromosomes are aligned at the spindle equator, the cell is said to be in *metaphase*. During metaphase, the aligned chromosomes, forming the so-called metaphase plate, undergo oscillatory movements about the spindle equator. Once all the kinetochores are bound to microtubules and all chromosomes are aligned at the metaphase plate, the sister chromatids separate, marking the metaphase-to-anaphase transition. During *anaphase*, chromosome segregation is ensured by two mechanisms. First, the sister chromatids are pulled apart toward opposite spindle poles (anaphase A). Second, the spindle poles move apart causing the spindle to elongate (anaphase B, not depicted here). When chromosome segregation is complete, the cell is said to be in *telophase*, at which time the nuclear envelope reforms around the two groups of chromosomes. During the latest stages of mitosis, an actin–myosin ring assembles at the cell equator and pinches the cell into two in a process termed cytokinesis, which ensures cytoplasmic division and formation of two individual daughter cells.

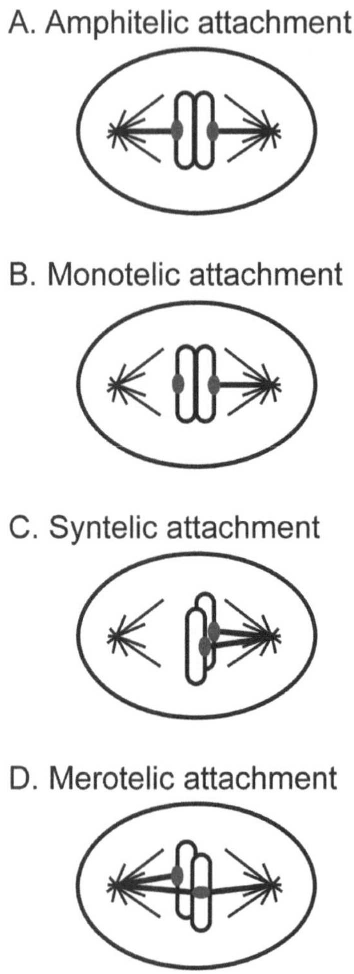

Fig. 4 Types of kinetochore–microtubule attachments. Four different types of attachments can be formed during the early stages of mitosis. These include amphitelic, monotelic, syntelic, and merotelic attachments. (A) Amphitelic attachments are defined by the attachment of the two sister kinetochores to microtubules from opposite spindle poles. This type of attachment is necessary for accurate chromosome segregation. (B) Monotelic attachment occurs when one of the two sister kinetochores of an individual chromosome is attached to one spindle pole whereas the other sister is unattached. (C) Syntelic attachment occurs when the two sister kinetochores of an individual chromosome bind microtubules from the same spindle pole. (D) Merotelic attachment occurs when a single kinetochore binds microtubules from two poles rather than just one. See the text for further details.

which time their behavior will depend on the relative size of the microtubule bundles bound to the individual kinetochore [Fig. 5; (Cimini *et al.*, 2004) reviewed in (Cimini, 2008; Gregan *et al.*, 2011)]. If the two microtubule bundles attached to a merotelic kinetochore are different in size, during anaphase the chromosome will be shifted to the side attached to the thicker bundle [Fig. 5A; (Cimini *et al.*, 2004)]. Conversely, if the two microtubule bundles are comparable in size, the merotelically attached chromosome will lag behind at the spindle equator while all the other chromosomes segregate to opposite spindle poles [Fig. 5B; (Cimini *et al.*, 2001, 2004)]. Although the experimental observations on the fate of these anaphase lagging chromosomes are relatively limited (Cimini

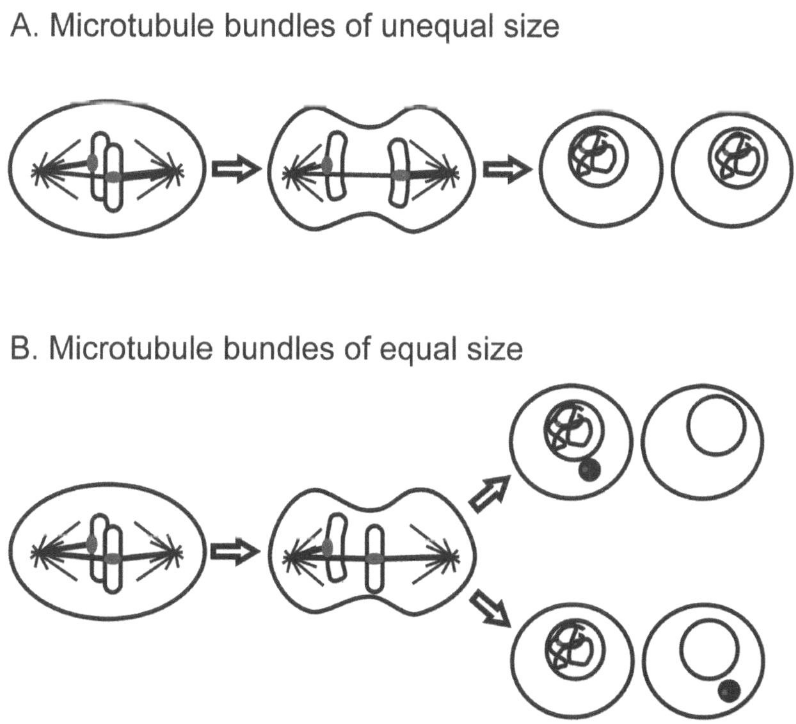

Fig. 5 Possible fates of merotelically attached anaphase lagging chromosomes. Merotelic attachments do not induce a SAC-dependent mitotic delay, and thus they can persist into anaphase, at which time their behavior will depend on the relative size of the microtubule bundles bound to the individual kinetochore. If the two microtubule bundles attached to a merotelic kinetochore are different in size (A), during anaphase the chromosome will be shifted to the side attached to the thicker bundle. Conversely, if the two microtubule bundles are comparable in size (B), the merotelically attached chromosome will lag behind at the spindle equator while all the other chromosomes segregate to opposite spindle poles. Upon mitotic exit, the lagging chromosome forms a micronucleus. See the text for further details.

et al., 2002; Utani *et al.*, 2010), they seem to suggest that the lagging chromosome can end up in either one of the two daughter cells, depending on where the cleavage furrow ingresses with respect to the lagging chromosome (Fig. 5B, middle and right panels). These experiments (Cimini *et al.*, 2002; Utani *et al.*, 2010) also suggest that anaphase lagging chromosomes form micronuclei upon mitotic exit (Fig. 5B, right panels). Interestingly, micronuclei are frequently observed in cancer cells (Brogger *et al.*, 1990) and represent a biomarker for increased risk of cancer [(reviewed in (Fenech, 2002; Majer *et al.*, 2001)]. Although some micronuclei may contain chromosome fragments rather than entire chromosomes, it is noteworthy that a number of studies have shown that anaphase lagging chromosomes are frequently observed in cancer cells (Ganem *et al.*, 2009; Gisselsson *et al.*, 2005; Reing *et al.*, 2004; Saunders *et al.*, 2000; Silkworth *et al.*, 2009; Thompson and Compton, 2008;), and can possibly represent the most common type of chromosome mis-segregation in CIN cancer cells [reviewed in (Cimini and Degrassi, 2005; Compton, 2011)]. Notably, mutations in (or depletion of) certain genes traditionally classified as oncogenes or tumor suppressor genes cause defects that lead to merotelic attachments and anaphase lagging chromosomes. For example, truncation of the adenomatous polyposis coli (APC) gene or depletion of the APC protein causes anaphase lagging chromosomes (Draviam *et al.*, 2006; Green *et al.*, 2005; Green and Kaplan, 2003). Similarly, a recent series of studies showed that the retinoblastoma protein (pRb) plays an important role in mitosis and its depletion results in aberrant chromosome condensation (Coschi *et al.*, 2010), centromere dysfunction (Manning *et al.*, 2010), chromatid breaks, and cohesion defects (van Harn *et al.*, 2010), which can in turn lead to merotelic attachment and anaphase lagging chromosomes (Manning *et al.*, 2010; Sage and Straight, 2010).

B. Spindle Assembly Checkpoint (SAC)

The spindle assembly checkpoint (SAC) is a biochemical pathway that can delay the metaphase to anaphase transition in the presence of unattached kinetochores and possibly lack of tension across sister kinetochores (Li and Nicklas, 1995; Musacchio and Salmon, 2007; Rieder *et al.*, 1995; Rudner and Murray, 1996). The various genes and corresponding proteins that act within the SAC pathway were initially identified in genetic screens of yeast. These screens revealed six genes necessary for the SAC function, *MAD1-3* (mitotic arrest defective), *BUB1*, *BUB3* (budding uninhibited by benomyl), and *MPS1* (monopolar spindles) (Hardwick and Murray, 1995; Hoyt *et al.*, 1991; Li and Murray, 1991; Roberts *et al.*, 1994; Weiss and Winey, 1996; Winey *et al.*, 1991). Subsequent studies have identified homologs in higher eukaryotes, wherein *MAD3* is termed *BUBR1*.

Upon mitotic entry, the SAC is active (Khodjakov and Rieder, 2009), and it is maintained in an active state as long as unattached kinetochores are present. Both *Mad1* and *Mad2* localize at unattached kinetochores (Howell *et al.*, 2004; Shah *et al.*, 2004; Waters *et al.*, 1998), and the interaction between *Mad1* and *Mad2* confers a conformational change in *Mad2* (DeAntoni *et al.*, 2005), driving it to bind to and sequester Cdc20 (Fang *et al.*, 1998; Hwang *et al.*, 1998; Kim *et al.*, 1998), an activator of the anaphase promoting complex, or cyclosome (APC/C) (Hwang *et al.*, 1998; Kim *et al.*, 1998; Li *et al.*, 1997). Additionally, Bub1 and *BubR1* (*MAD3* in yeast) are recruited to tensionless kinetochores (Skoufias *et al.*, 2001), where they further interact with *Mad2* forming the mitotic checkpoint complex (Sudakin *et al.*, 2001; Tang *et al.*, 2001), which also inhibits APC/C. APC/C is an E3 ubiquitin ligase that tags specific substrates for degradation by the proteasome. Two key substrates are the proteins Securin and Cyclin B, both necessary for the completion of mitosis (Cohen-Fix *et al.*, 1996; Funabiki *et al.*, 1996; Minshull *et al.*, 1990; Peters, 2006; Whitfield *et al.*, 1990; Zou *et al.*, 1999). Securin inhibits the enzyme separase, which is necessary to cleave the cohesin proteins holding the sister chromatids together (Cohen-Fix *et al.*, 1996; Funabiki *et al.*, 1996; Uhlmann *et al.*, 1999). Thus, degradation of securin leads to activation of separase, degradation of cohesin, and sister chromatid separation, which marks anaphase onset (Cohen-Fix *et al.*, 1996; Funabiki *et al.*, 1996; Zou *et al.*, 1999). Cyclin B degradation will then ensure mitotic exit and completion of cell division.

SAC dysfunction invariably leads to chromosome mis-segregation because it causes cells to enter anaphase before amphitelic attachment can be achieved by all chromosomes either by accelerating mitotic progression or by rendering the cells unable to delay anaphase onset in the presence of unattached kinetochores (Meraldi *et al.*, 2004). As a consequence, precocious anaphase onset is associated with a number of mitotic defects, including anaphase lagging chromosomes and segregation of both sister chromatids to the same daughter cell that generate aneuploidy in the progeny. Thus, mutations in SAC genes have the potential to play a role in tumorigenesis and cancer cell karyotypic diversity. Indeed, Cahill and coworkers identified a mutation in the SAC gene *BUB1* in a CIN colorectal cell line (Cahill *et al.*, 1998) and postulated that this could explain the CIN that had been previously observed in colorectal cancers (Lengauer *et al.*, 1997). Interestingly, *BUB1* and *BUBR1* mutations have been identified in some individuals affected by mosaic variegated aneuploidy (MVA) (Hanks *et al.*, 2004; Suijkerbuijk *et al.*, 2010), a syndrome associated with increased risk of cancer, and in which the patients' somatic cells exhibit high levels of trisomies and monosomies (Kajii *et al.*, 1998, 2001).

The idea that SAC dysfunction may play a role in tumorigenesis prompted a number of studies in mouse models. Because SAC-null mice are usually embryonic lethal, these studies used mice that either were haploinsufficient or carried hypomorphic mutations in specific SAC genes. The results varied somewhat and did not reveal a clear correlation between mutations in SAC genes, aneuploidy, and tumorigenesis. However, in many cases the haploinsufficient or mutant animals were more prone to develop either spontaneous (Iwanaga *et al.*, 2007; Michel *et al.*, 2001) or carcinogen-induced (Babu *et al.*, 2003; Dai *et al.*, 2004; Iwanaga *et al.*, 2007; Jeganathan *et al.*, 2007) tumors compared to wild-type mice. One exception is represented by *BubR1*, whose mutations do not result in increased tumor incidence but rather in senescence phenotypes (Baker *et al.*, 2004). Interestingly, overexpression of various SAC genes is found in cancer cells (Yuan *et al.*, 2006) and functional tests of *Mad2* overexpression in mice result in 40–55% aneuploid MEFs (mouse embryonic fibroblasts) and a 50% tumor incidence (Sotillo *et al.*, 2007). These findings suggest that the SAC pathway must be finely balanced to prevent aneuploidy. Alternatively, this effect may be specific to *Mad2* overexpression and may be due to other SAC-independent function(s) of the protein (Weaver *et al.*, 2008).

Because of the initial identification of a checkpoint gene mutation in colorectal cancer cells (Cahill *et al.*, 1998) and the impact that SAC gene mutations have on tumorigenesis in mouse models, many researchers embarked in a number of mutational analyses of cancer cells of different origin with the idea that most cancers would display mutations in SAC genes. Surprisingly, many of these studies found no mutations in SAC genes (Myrie *et al.*, 2000; Saeki *et al.*, 2002; Sato *et al.*, 2000; Yamaguchi *et al.*, 1999) and only a few identified mutations in a minor fraction of the cell lines analyzed (Haruki *et al.*, 2001b; Sato *et al.*, 2000), pointing to the idea that mutations in SAC genes may result in rates of chromosome mis-segregation that are too high to be compatible with cell survival. Thus, although SAC mutations can cause aneuploidy and in principle induce tumors (see above), the fact that SAC gene mutations are largely absent in cancer indicates that they do not significantly contribute to the CIN phenotype and the karyotypic diversity of cancer cells.

C. Multipolarity and Spindle Geometry

It has long been known that centrosome amplification and multipolar mitotic spindles are common features of cancer cells (Ganem *et al.*, 2009; Ghadimi *et al.*, 2000; Gisselsson, 2005; Jin *et al.*, 2007; Kaplan *et al.*, 2001; Lingle *et al.*, 2002; Lingle and Salisbury, 1999; Pihan *et al.*, 1998; Reing *et al.*, 2004; Sato *et al.*, 2001; Saunders *et al.*, 2000; Silkworth *et al.*, 2009; Stewenius *et al.*, 2005, 2007). Centrosome amplification can arise through different mechanisms, including cytokinesis failure (Fujiwara *et al.*, 2005), mitotic slippage (Dai *et al.*, 2004), centriole overduplication (Balczon *et al.*, 1995; Basto *et al.*, 2008; Habedanck *et al.*, 2005), centriole fragmentation (Peloponese *et al.*, 2005), aberrant centriole cohesion (Wang *et al.*, 2008), and cell fusion (Duelli and Lazebnik, 2003), although it is not clear how prevalent these mechanisms are in cancer. The presence of extra centrosomes leads to the assembly of multipolar mitotic spindles (Fig. 6A), which may in turn lead to the

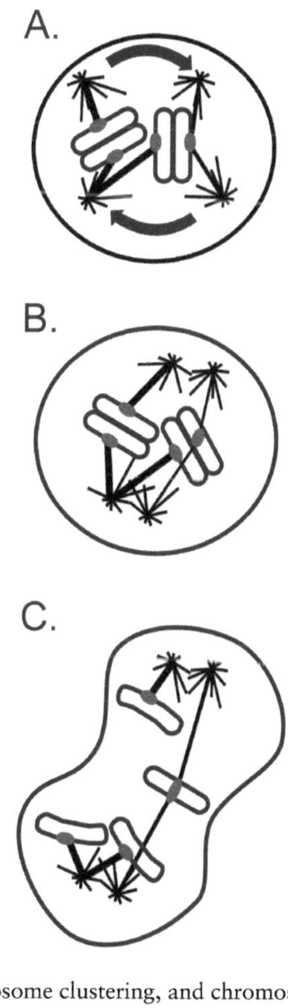

Fig. 6 Multipolarity, centrosome clustering, and chromosome mis-segregation. Within a multipolar spindle, a single kinetochore is more likely to face and establish attachment with two spindle poles than it would be in a bipolar spindle (A). Most cancer cells possessing extra centrosomes undergo spindle bipolarization by centrosome clustering before anaphase onset (B), but not all merotelic attachments are corrected before anaphase onset. Those that persist into anaphase can give rise to a lagging chromosome (C).

segregation of chromosomes into more than two daughter cells, thus yielding high rates of chromosome mis-segregation and aneuploidy. Nevertheless, whereas multipolar spindles are frequently observed in prometaphase, they are not very common in anaphase in a majority of cancer cell types. This observation led to the finding that some cancer cell types can cluster the extra centrosomes to produce a bipolar spindle before anaphase onset [Fig. 6B; (Brinkley, 2001; Ganem *et al.*, 2009; Kwon *et al.*, 2008; Quintyne *et al.*, 2005; Silkworth *et al.*, 2009)]. Indeed, this appears to be the case for most cancer cells displaying centrosome amplification and multipolar spindles. One exception is represented by Wilms tumor cells, which frequently exhibit tripolar chromosome segregation followed by cleavage of the cell in two instead of three, so that approximately two-thirds of the chromosomes will end up in one daughter cell and one-third in the other daughter cell (Gisselsson *et al.*, 2010). In most other cancer cells studied, the initial assembly of a multipolar spindle is followed by centrosome clustering and bipolarization, so that chromosome segregation occurs in a bipolar fashion [Fig. 6; (Brinkley, 2001; Ganem *et al.*, 2009; Kwon *et al.*, 2008; Quintyne *et al.*, 2005; Silkworth *et al.*, 2009)]. In fact, those cells that undergo multipolar division produce daughter cells that exhibit either cell cycle arrest or postmitotic cell death (Ganem *et al.*, 2009). Despite the ability of multipolar cells to undergo bipolar cell division, the transient assembly of a multipolar spindle was recently shown to be strictly linked to the high rates of chromosome mis-segregation in CIN cancer cells (Ganem *et al.*, 2009; Silkworth *et al.*, 2009). Indeed, the transient multipolar prometaphase promotes the formation of large numbers of kinetochore misattachments, particularly merotelic [Fig. 6A; (Ganem *et al.*, 2009; Silkworth *et al.*, 2009)]. Because merotelic attachments do not trigger a SAC-dependent mitotic delay, such high rates of merotelic attachments result in high rates of anaphase lagging chromosomes [Fig. 6C; (Ganem *et al.*, 2009; Silkworth *et al.*, 2009; Thompson and Compton, 2008)].

Interestingly, other factors previously believed to possess strictly intrinsic oncogenic properties, such as proteins from different viruses implicated in cancer, have been shown to affect spindle assembly and geometry. For instance, the protein E7 of human papillomavirus 16 (HPV-16) induces rapid centrosome duplication (Duensing *et al.*, 2000), and transfection of viral genes encoding viral proteins E7 and E6 results in 50% of cells assembling multipolar spindles (Duensing *et al.*, 2000). Similarly, expression of the viral protein HBX of hepatitis B virus also induces centrosome amplification (Fujii *et al.*, 2006; Yun *et al.*, 2004). These observations, plus the fact that systemic infection results in monoclonal cancers, may point to viruses as indirect carcinogens that would induce cancer by inducing chromosome mis-segregation and karyotypic evolution. This would explain the long latency between infection by oncogenic viruses and cancer development, as cancer would result from the emergence of a specific karyotype within the infected cell population.

In conclusion, multipolarity and aberrant spindle geometry in cancer cells can be caused by a number of different mechanisms, which make it a widespread feature of cancer cells and a major mechanism of CIN. The extent to which abnormal spindle geometry and multipolarity contribute to cancer progression is likely a function of the variability they produce. Indeed, centrosomal abnormalities parallel CIN in numerous cancers (D'Assoro *et al.*, 2002; Ghadimi *et al.*, 2000; Pihan *et al.*, 2001; Sato *et al.*, 2001; Skyldberg *et al.*, 2001).

D. Microtubule Dynamics

As described above, merotelic kinetochore attachment can persist into anaphase and induce chromosome mis-segregation in the form of lagging chromosomes. Most cancer cells analyzed so far exhibit significantly higher rates of merotelically attached anaphase lagging chromosomes than nontumor cells (Ganem *et al.*, 2009; Silkworth *et al.*, 2009; Thompson and Compton, 2008). Increased rates of merotelic attachments can result either from increased rates of merotelic kinetochore formation, or from decreased rates of correction of merotelic attachments. Recent studies have shown that cancer cells incur both these problems. Indeed, correction of kinetochore mis-attachments relies on the ability of incorrectly attached kinetochore microtubules to detach from their attachment site and leave an empty site to which microtubules from the correct spindle pole can attach. This is possible thanks to the intrinsic properties of kinetochore microtubules, which turn over with half-lives on the order of a few minutes during the early stages of mitosis (Zhai *et al.*, 1995) to allow for such correction to occur (Cimini *et al.*, 2006; DeLuca *et al.*, 2006). A recent study showed that microtubule dynamics must be finely controlled, and that experimentally induced minor reductions in kinetochore microtubule turnover are associated with high rates of chromosome mis-segregation resulting from the inability to correct merotelic attachments (Bakhoum *et al.*, 2009b). Interestingly, many different cancer cell lines were found to exhibit kinetochore microtubule turnover rates falling within those ranges known to prevent efficient correction of kinetochore mis-attachments (Bakhoum *et al.*, 2009a). Therefore, chromosome mis-segregation in cancer cells appears to depend on both increased rates of formation (see previous section) and decreased rates of correction of merotelic (and possibly other) kinetochore mis-attachments.

E. Polyploidization

Polyploidy is the condition of a cell or organism possessing extra haploid sets of chromosomes. Although species with up to eight sets of chromosomes (octoploidy) are known in plants, the most typical polyploidies observed in humans are triploidy and tetraploidy (Comai, 2005; Leitch and Leitch, 2008). The addition of a complete set of chromosomes, while seemingly a large genetic perturbation, is better tolerated than certain trisomies in humans, with reported cases of both triploid and tetraploid live births surviving up to 1 year (Golbus *et al.*, 1976; Pitt *et al.*, 1981; Russell *et al.*, 1981; Scarbrough *et al.*, 1984; Sherard *et al.*, 1986). Moreover, polyploid cells are found in some normal tissues, including the liver and aorta (Barrett *et al.*, 1983; Biesterfeld *et al.*, 1994; Celton-Morizur and Desdouets, 2010; Goldberg *et al.*, 1984). Such tolerance of polyploidy has been suggested to depend on the balance in genomic content being maintained with polyploidy as opposed to aneuploidy (Papp *et al.*, 2003; Tal, 1979; Torres *et al.*, 2008; Veitia, 2002). Indeed, in a comparison between mouse hepatocytes with different polyploidies the gene transcriptome was largely unchanged (Lu *et al.*, 2007).

While the karyotypes of most cancer cells are not perfect multiples of the haploid karyotype many are found to be near-triploid or near-tetraploid with chromosome number frequently ranging between ~60 and ~100 (Mitelman *et al.*, 2011).

Polyploidization can be caused by several different mechanisms, including cytokinesis failure, mitotic slippage, and endoreduplication. Cytokinesis failure can occur due to a failure of the actin–myosin cytokinetic ring to assemble or to initial furrow ingression followed by cleavage furrow regression. Typically, this would produce a single binucleate daughter cell. If such a cell progressed through the cell cycle, the following mitosis would be characterized by a tetraploid karyotype and four centrosomes. This could lead to high rates of chromosome mis-segregation due to transient multipolarity (Fig. 6) or multipolar cell division (see above for a detailed discussion on multipolarity). Cytokinesis failure is unlikely to be recurrent in cancer cells because this would lead to an ever-increasing chromosome number. However, it is possible that occasional failure of cytokinesis may be a source of karyotype diversity in cancer. Moreover, it is possible that polyploidization may represent an early step in carcinogenesis (Boveri and Harris, 2008; Fujiwara *et al.*, 2005; Ganem *et al.*, 2007). Indeed, tetraploid cells are found in a number of premalignant lesions, including those of the cervix and esophagus (Galipeau *et al.*, 1996; Olaharski *et al.*, 2006). In addition, experimentally induced cytokinesis failure can promote tumorigenesis when the tetraploid cells are injected in nude mice (Fujiwara *et al.*, 2005). Interestingly, the carcinogenic effect of asbestos fibers has been associated with their ability to physically block cytokinesis and induce furrow regression (Jensen *et al.*, 1996). Similarly, chromatin bridges spanning the spindle midzone can induce furrow regression (Mullins and Biesele, 1977; Steigemann *et al.*, 2009).

Polyploidization can also occur by mitotic slippage, a phenomenon by which cells exit mitosis in the absence of chromosome segregation. Mitotic slippage typically occurs when cells are kept in the presence of microtubule poisons for prolonged periods of time (Brito and Rieder, 2006; Di Leonardo *et al.*, 1997; Rieder and Palazzo, 1992). This adaptation has also been observed *in vivo* with the polyploidization of liver cells, a mechanism proposed to confer resistance to different cellular stresses (Duncan *et al.*, 2010). Whereas mitotic slippage has not been documented as a common phenomenon in cancer cells, it may occur during chemotherapeutic treatment (e.g., paclitaxel treatment), and may play a role in development of drug resistance.

Finally, polyploidization can be generated by endoreduplication, a phenomenon by which the DNA replicates without intervening cell division (Edgar and Orr-Weaver, 2001). Although endoreduplication has been suggested to be associated with tumorigenesis (Lee *et al.*, 2009; Storchova and Pellman, 2004), to date, there are no reports of experimental data linking endoreduplication to tumorigenesis or CIN.

Other mechanisms capable of inducing polyploidization are cellular fusion and entosis, both generating polyploid cells starting from two different nuclei. In the process of cellular fusion, two cells fuse to become one (Duelli and Lazebnik, 2003; Lu and Kang, 2009). Cellular fusion can arise spontaneously or as an effect of different viruses (Duelli and Lazebnik, 2003). For example, otherwise harmless viruses contribute to CIN independently of the expression of viral proteins by inducing cell fusion (Duelli *et al.*, 2007), which has been shown to induce destabilization of the genome (possibly due to the presence of extra centrosomes) and cause global gene expression changes (Duelli and Lazebnik, 2007). In the process of entosis one cell envelopes an entire separate cell (Krajcovic *et al.*, 2011). The consequences of entosis mirror that of other polyploidization events, ultimately leading to chromosome mis-segregation and intratumor heterogeneity.

In summary, polyploidization contributes to karyotypic diversity not only at the time it arises (by inducing a sudden change in the chromosome number), but also over subsequent cell cycles by triggering a chain of chromosome mis-segregation events. This is because, first, many of the polyploidization pathways will generate cells with multiple centrosomes, which are a major source of chromosome instability (Fig. 6) in cancer cells (see above), and second, polyploid cells may be more tolerant to chromosome mis-segregation.

IV. CONCLUSION

We summarized in this chapter how errors in mitotic chromosome segregation produce karyotypic diversity in cancer cells. Although a recent study identified a phenomenon, termed chromothripsis, by which a massive genomic rearrangement occurs in a single step (Stephens *et al.*, 2011), this only appears to occur in a minor fraction of all cancers. Instead, the mitotic errors most commonly observed in cancer cells do not cause massive chromosome mis-segregation, rather yield small changes in the chromosome number. Although small in terms of the karyotype, such changes reflect a large genomic change compared to single mutations. These karyotypic changes reflect the punctuated evolution of cancer cells recently identified in single-cell analysis (Navin *et al.*, 2011). Moreover, by occurring repetitively, such small changes in chromosome number generate enough variability to make the cancer cell population plastic and adaptable. For instance, CIN is responsible for intrinsic resistance to numerous drugs with many different mechanisms of action (Lee *et al.*, 2011). Indeed, when a panel of cancer cell lines were challenged with inhibitors of kinases involved in different pathways, a strong correlation was observed between CIN and intrinsic drug resistance, independent of somatic mutation and proliferation rate (Lee *et al.*, 2011). These results provide further support to previous studies suggesting that karyotype alterations are a route to drug resistance in cancer cells (Baroja *et al.*, 1998; Duesberg *et al.*, 2001, 2007) and act as a mechanism to evade cellular stress in both budding yeast (Pavelka *et al.*, 2010b; Rancati *et al.*, 2008) and mouse embryonic fibroblasts (Tang *et al.*, 2011).

In conclusion, we highlight here the central role of the karyotype in cancer, the mechanisms through which it arises and contributes to diversity in cancer cells, and the problems it presents to clinicians. To this end, chromosome mis-segregation, while considered a defect in normal cells, is a generator of phenotypic diversity and flexibility in cancer cells. Thus, treatments aimed at reducing chromosome segregation errors may prove successful as a coadjuvant to traditional treatments by delaying tumor progression.

ACKNOWLEDGMENTS

We would like to thank the members of the Cimini Lab for helpful discussion, and P. Duesberg (UC Berkeley) for sharing unpublished data. Work in the Cimini lab is supported by NSF Grant MCB-0842551 and HFSP Grant RGY0069/2010.

REFERENCES

Alberts, B., Bray, D., Lewis, J., Raff, M., Roberts, K., and Watson, J. D. (1994). Molecular Biology of the Cell. Garland Publishing, New York.

Atkin, N. B., and Kay, R. (1979). *Br. J. Cancer* **40**, 210.

Babu, J. R., Jeganathan, K. B., Baker, D. J., Wu, X., Kang-Decker, N., and van Deursen, J. M. (2003). *J. Cell Biol.* **160**, 341–353.

Baker, D. J., Jeganathan, K. B., Cameron, J. D., Thompson, M., Juneja, S., Kopecka, A., Kumar, R., Jenkins, R. B., de Groen, P. C., Roche, P., and van Deursen, J. M. (2004). *Nat. Genet.* **36**, 744–749.

Baker, D. J., Jin, F., Jeganathan, K. B., and van Deursen, J. M. (2009). *Cancer Cell* **16**, 475–486.

Bakhoum, S. F., Genovese, G., and Compton, D. A. (2009 a). *Curr. Biol.* **19**, 1937–19342.

Bakhoum, S. F., Thompson, S. L., Manning, A. L., and Compton, D. A. (2009 b). *Nat. Cell Biol.* **11**, 27–35.

Balczon, R., Bao, L., Zimmer, W. E., Brown, K., Zinkowski, R. P., and Brinkley, B. R. (1995). *J. Cell Biol.* **130**, 105–115.

Baroja, A., de la Hoz, C., Alvarez, A., Vielba, R., Sarrat, R., Arechaga, J., and de Gandarias, J. M. (1998). *Life Sci.* **62**, 2275–2282.

Barrett, T. B., Sampson, P., Owens, G. K., Schwartz, S. M., and Benditt, E. P. (1983). *Proc. Natl. Acad. Sci. U.S.A.* **80**, 882–885.

Basto, R., Brunk, K., Vinadogrova, T., Peel, N., Franz, A., Khodjakov, A., and Raff, J. W. (2008). *Cell* **133**, 1032–1042.

Bayreuther, K., and Klein, E. (1958). *J. Natl. Cancer Inst.* **21**, 885–923.

Beheshti, B., Park, P. C., Sweet, J. M., Trachtenberg, J., Jewett, M. A., and Squire, J. A. (2001). *Neoplasia* **3**, 62–69.

Beroukhim, R., Mcrmel, C. H., Porter, D., Wei, G., Raychaudhuri, S., Donovan, J., Barretina, J., Boehm, J. S., Dobson, J., Urashima, M., Mc Henry, K. T., Pinchback, R. M., Ligon, A. H., Cho, Y. -J., Haery, L., Greulich, H., Reich, M., Winckler, W., Lawrence, M. S., Weir, B. A., Tanaka, K. E., Chiang, D. Y., Bass, A. J., Loo, A., Hoffman, C., Prensner, J., Liefeld, T., Gao, Q., Yecies, D., Signoretti, S., Maher, E., Kaye, F. J., Sasaki, H., Tepper, J. E., Fletcher, J. A., Tabernero, J., Baselga, J., Tsao, M. -S., Demichelis, F., Rubin, M. A., Janne, P. A., Daly, M. J., Nucera, C., Levine, R. L., Ebert, B. L., Gabriel, S., Rustgi, A. K., Antonescu, C. R., Ladanyi, M., Letai, A., Garraway, L. A., Loda, M., Beer, D. G., True, L. D., Okamoto, A., Pomeroy, S. L., Singer, S., Golub, T. R., Lander, E. S., Getz, G., Sellers, W. R., and Meyerson, M. (2010). *Nature* **463**, 899–905.

Biesterfeld, S., Gerres, K., Fischer-Wein, G., and Bocking, A. (1994). *J. Clin. Pathol.* **47**, 38–42.

Boveri, T. (1902). Über Mehrpolige Mitosen als Mittel Zur Analyse Des Zellkerns. Stuber, Würzburg.

Boveri, T. (1914). Zur Frage Der Entstehung Maligner Tumoren. Gustav Fischer Verlag, Jena, Germany.

Boveri, T., and Harris, H. (2008). *J. Cell Sci.* **121**(Suppl. 1), 1–84.

Brinkley, B. R. (2001). *Trends Cell Biol.* **11**, 18–21.

Brito, D. A., and Rieder, C. L. (2006). *Curr. Biol.* **16**, 1194–1200.

Brogger, A., Hagmar, L., Hansteen, I. -L., Heim, S., Hogstedt, B., Knudsen, L., Lambert, B., Linnainmaa, K., Mitelman, F., Nordenson, I., Reuterwall, C., Salomaa, S., Skerfving, S., and Sorsa, M. (1990). *Mutat. Res.* **241**, 325–337.

Cahill, D. P., Lengauer, C., Yu, J., Riggins, G. J., Willson, J. K. V., Markowitz, S. D., Kinzler, K. W., and Vogelstein, B. (1998). *Nature* **392**, 300–303.
Cao, Q., Abeysinghe, H., Chow, O., Xu, J., Kaung, H., Fong, C., Keng, P., Insel, R. A., Lee, W. M., Barrett, J. C., and Wang, N. (2001). *Cancer Genet. Cytogenet.* **129**, 131–137.
Carere, A., Antoccia, A., Cimini, D., Crebelli, R., Degrassi, F., Leopardi, P., Marcon, F., Sgura, A., Tanzarella, C., and Zijno, A. (1998). *Environ. Mol. Mutagen.* **32**, 130–138.
Carter, S. L., Eklund, A. C., Kohane, I. S., Harris, L. N., and Szallasi, Z. (2006). *Nat. Genet.* **38**, 1043–1048.
Catalan, J., Surralles, J., Falck, G. C., Autio, K., and Norppa, H. (2000). *Mutagenesis* **15**, 251–255.
Celton-Morizur, S., and Desdouets, C. (2010). *Adv. Exp. Med. Biol.* **676**, 123–135.
Cheeseman, I. M., and Desai, A. (2008). *Nat. Rev. Mol. Cell Biol.* **9**, 33–46.
Chu, E. H. Y., and Giles, N. H. (1958). *J. Natl. Cancer. Inst.* **20**, 383–401.
Cimini, D. (2007). *Cell Cycle* **6**, 1558–1564.
Cimini, D. (2008). *Biochim. Biophys. Acta.* **1786**, 32–40.
Cimini, D., Cameron, L. A., and Salmon, E. D. (2004). *Curr. Biol.* **14**, 2149–2155.
Cimini, D., and Degrassi, F. (2005). *Trends Cell Biol.* **15**, 442–451.
Cimini, D., Fioravanti, D., Salmon, E. D., and Degrassi, F. (2002). *J. Cell Sci.* **115**, 507–515.
Cimini, D., Howell, B., Maddox, P., Khodjakov, A., Degrassi, F., and Salmon, E. D. (2001). *J. Cell Biol.* **153**, 517–527.
Cimini, D., Moree, B., Canman, J. C., and Salmon, E. D. (2003). *J. Cell Sci.* **116**, 4213–4225.
Cimini, D., Wan, X., Hirel, C. B., and Salmon, E. D. (2006). *Curr. Biol.* **16**, 1711–1718.
Cohen-Fix, O., Peters, J. M., Kirschner, M. W., and Koshland, D. (1996). *Genes Dev.* **10**, 3081–3093.
Comai, L. (2005). *Nat. Rev. Genet.* **6**, 836–846.
Compton, D. A. (2011). *Curr. Opin. Cell Biol.* **23**, 109–113.
Conti, J. C., Aldaz, C. M., O'Connell, J., Klen-Szanto, A. J. -P., and Slaga, T. J. (1986). *Carcinogenesis* **7**, 1845–1848.
Coschi, C. H., Martens, A. L., Ritchie, K., Francis, S. M., Chakrabarti, S., Berube, N. G., and Dick, F. A. (2010). *Genes Dev.* **24**, 1351–1363.
Countryman, P. I., and Heddle, J. A. (1976). *Mutat. Res.* **41**, 321–332.
D'Assoro, A. B., Barrett, S. L., Folk, C., Negron, V. C., Boeneman, K., Busby, R., Whitehead, C., Stivala, F., Lingle, W. L., and Salisbury, J. L. (2002). *Breast Cancer Res. Treat* **75**, 25–34.
Dai, W., Wang, Q., Liu, T., Swamy, M., Fang, Y., Xie, S., Mahmood, R., Yang, Y. M., Xu, M., and Rao, C. V. (2004). *Cancer Res.* **64**, 440–445.
DeAntoni, A., Sala, V., and Musacchio, A. (2005). *Philos. Trans. R. Soc. Lond. Biol. Sci.* **360**, 637–647 discussion 447–448.
DeLuca, J. G., Gall, W. E., Ciferri, C., Cimini, D., Musacchio, A., and Salmon, E. D. (2006). *Cell* **127**, 969–982.
Di Leonardo, A., Khan, S. H., Linke, S. P., Greco, V., Seidita, G., and Wahl, G. M. (1997). *Cancer Res.* **57**, 1013–1019.
Diaz-Rodriguez, E., Sotillo, R., Schvartzman, J. M., and Benezra, R. (2008). *Proc. Natl. Acad. Sci. U.S.A.* **105**, 16719–16724.
Draviam, V. M., Shapiro, I., Aldridge, B., and Sorger, P. K. (2006). *J. Eur. Mol. Biol. Org.* **25**, 2814–2827.
Duelli, D., and Lazebnik, Y. (2003). *Cancer Cell* **3**, 445–448.
Duelli, D., and Lazebnik, Y. (2007). *Nat. Rev. Cancer* **7**, 968–976.
Duelli, D. M., Padilla-Nash, H. M., Berman, D., Murphy, K. M., Ried, T., and Lazebnik, Y. (2007). *Curr. Biol.* **17**, 431–437.

Duensing, S., Lee, L. Y., Duensing, A., Basile, J., Piboonniyom, S., Gonzalez, S., Crum, C. P., and Munger, K. (2000). *Proc. Natl. Acad. Sci. U.S.A.* **97**, 10002–10007.

Duesberg, P., Li, R., Rasnick, D., Rausch, C., Willer, A., Kraemer, A., Yerganian, G., and Hehlmann, R. (2000 a). *Cancer Genet. Cytogenet.* **119**, 83–93.

Duesberg, P., Li, R., Sachs, R., Fabarius, A., Upender, M. B., and Hehlmann, R. (2007). *Drug Resist. Updat.* **10**, 51–58.

Duesberg, P., Mandrioli, D., McCormack, A., and Nicholson, J. M. (2011). *Cell Cycle* **10**, 2100–2114.

Duesberg, P., and Rasnick, D. (2000). *Cell Motil. Cytoskeleton* **47**, 81–107.

Duesberg, P., Stindl, R., and Hehlmann, R. (2000 b). *Proc. Natl. Acad. Sci. U.S.A.* **97**, 14295–14300.

Duesberg, P., Stindl, R., and Hehlmann, R. (2001). *Proc. Natl. Acad. Sci. U.S.A.* **98**, 11283–11288.

Duncan, A. W., Taylor, M. H., Hickey, R. D., Hanlon Newell, A. E., Lenzi, M. L., Olson, S. B., Finegold, M. J., and Grompe, M. (2010). *Nature* **467**, 707–710.

Edgar, B. A., and Orr-Weaver, T. L. (2001). *Cell* **105**, 297–306.

Fang, G., Yu, H., and Kirschner, M. W. (1998). *Genes Dev.* **12**, 1871–1883.

Fenech, M. (1993). *Mutat Res.* **285**, 35–44.

Fenech, M. (2002). *Toxicology* **181–182**, 411–416.

FitzPatrick, D. R. (2005). *Trends Genet.* **21**, 249–253.

FitzPatrick, D. R., Ramsay, J., McGill, N. I., Shade, M., Carothers, A. D., and Hastie, N. D. (2002). *Hum. Mol. Genet* **11**, 3249–3256.

Frankfurt, O. S., Chin, J. L., Englander, L. S., Greco, W. R., Pontes, J. E., and Rustum, Y. M. (1985). *Cancer Res.* **45**, 1418–1423.

Fujii, R., Zhu, C., Wen, Y., Marusawa, H., Bailly-Maitre, B., Matsuzawa, S. -i., Zhang, H., Kim, Y., Bennett, C. F., Jiang, W., and Reed, J. C. (2006). *Cancer Res.* **66**, 9099–9107.

Fujiwara, T., Bandi, M., Nitta, M., Ivanova, E. V., Bronson, R. T., and Pellman, D. (2005). *Nature* **437**, 1043–1047.

Funabiki, H., Yamano, H., Kumada, K., Nagao, K., Hunt, T., and Yanagida, M. (1996). *Nature* **381**, 438–441.

Furuya, T., Uchiyama, T., Murakami, T., Adachi, A., Kawauchi, S., Oga, A., Hirano, T., and Sasaki, K. (2000). *Clin. Cancer Res.* **6**, 2815–2820.

Galipeau, P. C., Cowan, D. S., Sanchez, C. A., Barrett, M. T., Emond, M. J., Levine, D. S., Rabinovitch, P. S., and Reid, B. J. (1996). *Proc. Natl. Acad. Sci. U.S.A.* **93**, 7081–7084.

Ganem, N. J., Godinho, S. A., and Pellman, D. (2009). *Nature* **460**, 278–282.

Ganem, N. J., Storchova, Z., and Pellman, D. (2007). *Curr. Opin. Genet. Dev.* **17**, 157–162.

Gao, C., Furge, K., Koeman, J., Dykema, K., Su, Y., Cutler, M. L., Werts, A., Haak, P., and Vande Woude, G. F. (2007). *Proc. Natl. Acad. Sci. U.S.A.* **104**, 8995–9000.

Gebhart, E., and Liehr, T. (2000). *Int. J. Oncol.* **16**, 383–399.

Geiser, A. G., Anderson, M. J., and Stanbridge, E. J. (1989). *Cancer Res.* **49**, 1572–1577.

Ghadimi, B. M., Sackett, D. L., Difilippantonio, M. J., Schrock, E., Neumann, T., Jauho, A., Auer, G., and Ried, T. (2000). *Gen. Chrom. Canc.* **27**, 183–190.

Gilbert, P. X., and Harris, H. (1988). *J. Cell Sci.* **90**(Pt 3), 433–446.

Gisselsson, D. (2005). *Cell Cycle* **4**, 1007–1010.

Gisselsson, D., Jin, Y., Lindgren, D., Persson, J., Gisselsson, L., Hanks, S., Sehic, D., Mengelbier, L. H., Ora, I., Rahman, N., Mertens, F., Mitelman, F., and Mandahl, N. (2010). *Proc. Natl. Acad. Sci. U.S.A.* **107**, 20489–20493.

Gisselsson, D., Lv, M., Tsao, S. W., Man, C., Jin, C., Hoglund, M., Kwong, Y. L., and Jin, Y. (2005). *Genes Chromosomes Cancer* **42**, 22–33.

Golbus, M. S., Bachman, R., Wiltse, S., and Hall, B. D. (1976). *J. Med. Genet.* **13**, 329–332.

Goldberg, I. D., Rosen, E. M., Shapiro, H. M., Zoller, L. C., Myrick, K., Levenson, S. E., and Christenson, L. (1984). *Science* **226**, 559–561.
Green, R. A., and Kaplan, K. B. (2003). *J. Cell Biol.* **163**, 949–961.
Green, R. A., Wollman, R., and Kaplan, K. B. (2005). *Mol. Biol. Cell* **16**, 4609–4622.
Gregan, J., Polakova, S., Zhang, L., Tolic-Norrelykke, I. M., and Cimini, D. (2011). *Trends Cell Biol.* **21**, 374–381.
Grote, H. J., Friedrichs, N., Pomjanski, N., Guhde, H. F., Reich, O., and Bocking, A. (2001). *Anal. Cell Pathol.* **23**, 97–105.
Habedanck, R., Stierhof, Y. D., Wilkinson, C. J., and Nigg, E. A. (2005). *Nat. Cell Biol.* **7**, 1140–1146.
Hagmar, L., Brogger, A., Hansteen, I. L., Heim, S., Hogstedt, B., Knudsen, L., Lambert, B., Linnainmaa, K., Mitelman, F., Nordenson, I., et al., (1994). *Cancer Res.* **54**, 2919–2922.
Hanks, S., Coleman, K., Reid, S., Plaja, A., Firth, H., Fitzpatrick, D., Kidd, A., Mehes, K., Nash, R., Robin, N., Shannon, N., Tolmie, J., Swansbury, J., Irrthum, A., Douglas, J., and Rahman, N. (2004). *Nat. Genet.* **36**, 1159–1161.
Hardwick, K. G., and Murray, A. W. (1995). *J. Cell Biol.* **131**, 709–720.
Haruki, N., Harano, T., Masuda, A., Kiyono, T., Takahashi, T., Tatematsu, Y., Shimizu, S., Mitsudomi, T., Konishi, H., Osada, H., and Fujii, Y. (2001a). *Am. J. Pathol.* **159**, 1345–1352.
Haruki, N., Saito, H., Harano, T., Nomoto, S., Takahashi, T., Osada, H., and Fujii, Y. (2001b). *Cancer Lett.* **162**, 201–205.
Hauf, S., Cole, R. W., LaTerra, S., Zimmer, C., Schnapp, G., Walter, R., Heckel, A., van Meel, J., Rieder, C. L., and Peters, J. M. (2003). *J. Cell Biol.* **161**, 281–294.
Hauschka, T. S. (1953). *Trans. N Y Acad. Sci.* **16**, 64–73.
Hauschka, T. S., and Levan, A. (1958). *J. Natl. Cancer Inst.* **21**, 77–135.
Hayes, R. B., Zhang, L., Yin, S., Swenberg, J. A., Xi, L., Wiencke, J., Bechtold, W. E., Yao, M., Rothman, N., Haas, R., O'Neill, J. P., Zhang, D., Wiemels, J., Dosemeci, M., Li, G., and Smith, M. T. (2000). *Carcinogenesis* **21**, 55–62.
Heim, S., and Mitelman, F. (2010). A New Approach to an Old Problem. John Wiley & Sons, Inc, Hoboken,NJ.
Heng, H. H. (2009). *BioEssays* **31**, 512–525.
Heppner, G. H. (1984). *Cancer Res.* **44**, 2259–2265.
Hittelman, W. N. (2001). *Ann. N Y Acad. Sci.* **952**, 1–12.
Hittelman, W. N., Kim, H. J., Lee, J. S., Shin, D. M., Lippman, S. M., Kim, J., Ro, J. Y., and Hong, W. K. (1996). *J. Cell. Biochem. Suppl.* **25**, 57–62.
Howell, B. J., Moree, B., Farrar, E. M., Stewart, S., Fang, G., and Salmon, E. D. (2004). *Curr. Biol.* **14**, 953–964.
Hoyt, M. A., Totis, L., and Roberts, B. T. (1991). *Cell* **66**, 507–517.
Huxley, J. (1956). *Biol. Rev.* **31**, 474–513.
Hwang, L. H., Lau, L. F., Smith, D. L., Mistrot, C. A., Hardwick, K. G., Hwang, E. S., Amon, A., and Murray, A. W. (1998). *Science* **279**, 1041–1044.
Hyman, E., Kauraniemi, P., Hautaniemi, S., Wolf, M., Mousses, S., Rozenblum, E., Ringner, M., Sauter, G., Monni, O., Elkahloun, A., Kallioniemi, O. P., and Kallioniemi, A. (2002). *Cancer Res.* **62**, 6240–6245.
Iwanaga, Y., Chi, Y. H., Miyazato, A., Sheleg, S., Haller, K., Peloponese Jr., J. M., Li, Y., Ward, J. M., Benezra, R., and Jeang, K. T. (2007). *Cancer Res.* **67**, 160–166.
Jakobsen, A. (1984). *Am. J. Clin. Oncol.* **7**, 475–480.
Jakobsen, A., Bichel, P., Kristensen, G. B., and Nyland, M. (1988). *Eur. J. Cancer Clin. Oncol.* **24**, 969–972.
Jeganathan, K., Malureanu, L., Baker, D. J., Abraham, S. C., and van Deursen, J. M. (2007). *J. Cell Biol.* **179**, 255–267.

Jensen, C. G., Jensen, L. C., Rieder, C. L., Cole, R. W., and Ault, J. G. (1996). *Carcinogenesis* **17**, 2013–2021.
Jin, Y., Stewenius, Y., Lindgren, D., Frigyesi, A., Calcagnile, O., Jonson, T., Edqvist, A., Larsson, N., Lundberg, L. M., Chebil, G., Liedberg, F., Gudjonsson, S., Mansson, W., Hoglund, M., and Gisselsson, D. (2007). *Clin. Cancer Res.* **13**, 1703–1712.
Kajii, T., Ikeuchi, T., Yang, Z. Q., Nakamura, Y., Tsuji, Y., Yokomori, K., Kawamura, M., Fukuda, S., Horita, S., and Asamoto, A. (2001). *Am. J. Med. Genet.* **104**, 57–64.
Kajii, T., Kawai, T., Takumi, T., Misu, H., Mabuchi, O., Takahashi, Y., Tachino, M., Nihei, F., and Ikeuchi, T. (1998). *Am. J. Med. Genet.* **78**, 245–249.
Kallioniemi, O. P., Blanco, G., Alvaikko, M., Hietanen, T., Mattila, J., Lauslahti, K., and Koivula, T. (1987). *Br. J. Cancer* **5**, 637–642.
Kaplan, K. B., Burds, A. A., Swedlow, J. R., Bekir, S. S., Sorger, P. K., and Nathke, I. S. (2001). *Nat. Cell. Bio.l* **3**, 429–432.
Kendall, S. D., Linardic, C. M., Adam, S. J., and Counter, C. M. (2005). *Cancer Res.* **65**, 9824–9828.
Khodjakov, A., Cole, R. W., McEwen, B. F., Buttle, K. F., and Rieder, C. L. (1997). *J. Cell Biol.* **136**, 229–240.
Khodjakov, A., and Rieder, C. L. (2009). *J. Biol.* **8**, 88.
Kim, S. H., Lin, D. P., Matsumoto, S., Kitazono, A., and Matsumoto, T. (1998). *Science* **279**, 1045–1047.
Kingsbury, M. A., Yung, Y. C., Peterson, S. E., Westra, J. W., and Chun, J. (2006). *Cell Mol. Life Sci.* **63**, 2626–2641.
Klein, A., Li, N., Nicholson, J. M., McCormack, A. A., Graessmann, A., and Duesberg, P. (2010). *Cancer Gen. Cytogen.* **200**, 79–99.
Krajcovic, M., Johnson, N. B., Sun, Q., Normand, G., Hoover, N., Yao, E., Richardson, A. L., King, R. W., Cibas, E. S., Schnitt, S. J., Brugge, J. S., and Overholtzer, M. (2011). *Nat. Cell Biol.* **13**, 324–330.
Kwon, M., Godinho, S. A., Chandhok, N. S., Ganem, N. J., Azioune, A., Thery, M., and Pellman, D. (2008). *Genes Dev.* **22**, 2189–21203.
Ladrach, K. S., and LaFountain Jr., J. R. (1986). *Cell Motil. Cytoskeleton* **6**, 419–427.
Lee, A. J., Endesfelder, D., Rowan, A. J., Walther, A., Birkbak, N. J., Futreal, P. A., Downward, J., Szallasi, Z., Tomlinson, I. P., Howell, M., Kschischo, M., and Swanton, C. (2011). *Cancer Res.* **71**, 1858–1870.
Lee, H. O., Davidson, J. M., and Duronio, R. J. (2009). *Genes Dev.* **23**, 2461–2477.
Leitch, A. R., and Leitch, I. J. (2008). *Science* **320**, 481–483.
Lengauer, C., Kinzler, K. W., and Vogelstein, B. (1997). *Nature* **386**, 623–627.
Levan, A. (1956). *Cancer* **9**, 648–663.
Li, L., McCormack, A. A., Nicholson, J. M., Fabarius, A., Hehlmann, R., Sachs, R. K., and Duesberg, P. H. (2009). *Cancer Genet. Cytogenet.* **188**, 1–25.
Li, R., and Murray, A. W. (1991). *Cell* **66**, 519–531.
Li, X., and Nicklas, R. B. (1995). *Nature* **373**, 630–632.
Li, Y., Gorbea, C., Mahaffey, D., Rechsteiner, M., and Benezra, R. (1997). *Proc. Natl. Acad. Sci. U.S.A.* **94**, 12431–12436.
Lingle, W. L., Barrett, S. L., Negron, V. C., D'Assoro, A. B., Boeneman, K., Liu, W., Whitehead, C. M., Reynolds, C., and Salisbury, J. L. (2002). *Proc. Natl. Acad. Sci. U.S. A.* **99**, 1978–1983.
Lingle, W. L., and Salisbury, J. L. (1999). *Am. J. Pathol.* **155**, 1941–1951.
Lu, P., Prost, S., Caldwell, H., Tugwood, J. D., Betton, G. R., and Harrison, D. J. (2007). *Mamm. Genome.* **18**, 617–626.
Lu, X., and Kang, Y. (2009). *Cancer Res.* **69**, 8536–8539.

Ly, P., Eskiocak, U., Kim, S. B., Roig, A. I., Hight, S. K., Lulla, D. R., Zou, Y. S., Batten, K., Wright, W. E., and Shay, J. W. (2011). *Neoplasia* **13**, 348–357.
Maiato, H., DeLuca, J., Salmon, E. D., and Earnshaw, W. C. (2004). *J. Cell Sci.* **117**, 5461–5477.
Majer, B. J., Laky, B., Knasmuller, S., and Kassie, F. (2001). *Mutat. Res.* **489**, 147–172.
Makino, S. (1956). *Ann. N Y Acad. Sci.* **63**, 818–830.
Manning, A. L., Longworth, M. S., and Dyson, N. J. (2010). *Genes Dev.* **24**, 1364–1376.
McClelland, S. E., Burrell, R. A., and Swanton, C. (2009). *Cell Cycle* **8**, 3262–3266.
Meraldi, P., Draviam, V. M., and Sorger, P. K. (2004). *Dev. Cell* **7**, 45–60.
Merlo, L. M., Pepper, J. W., Reid, B. J., and Maley, C. C. (2006). *Nat. Rev. Cancer* **6**, 924–935.
Mertens, F., Johansson, B., Hoeglund, M., and Mitelman, F. (1997). *Cancer Res.* **57**, 2765–2780.
Michel, L. S., Liberal, V., Chatterjee, A., Kirchwegger, R., Pasche, B., Gerald, W., Dobles, M., Sorger, P. K., Murty, V. V., and Benezra, R. (2001). *Nature* **409**, 355–359.
Minissi, S., Degrassi, F., Tanzarella, C., and Gustavino, B. (1999). *Mutagenesis* **14**, 557–562.
Minshull, J., Golsteyn, R., Hill, C. S., and Hunt, T. (1990). *J. Eur. Mol. Biol. Org.* **9**, 2865–2875.
Mitelman, F., Johansson, B., and Mertens, F. (2011). http://cgap.nci.nih.gov/Chromosomes/Mitelman.
Miyazaki, M., Furuya, T., Shiraki, A., Sato, T., Oga, A., and Sasaki, K. (1999). *Cancer Res.* **59**, 5283–5285.
Mullins, J. M., and Biesele, J. J. (1977). *J. Cell Biol.* **73**, 672–684.
Musacchio, A., and Salmon, E. D. (2007). *Nat. Rev. Mol. Cell Biol.* **8**, 379–393.
Myrie, K. A., Percy, M. J., Azim, J. N., Neeley, C. K., and Petty, E. M. (2000). *Cancer Lett.* **152**, 193–199.
Navin, N., Kendall, J., Troge, J., Andrews, P., Rodgers, L., McIndoo, J., Cook, K., Stepansky, A., Levy, D., Esposito, D., Muthuswamy, L., Krasnitz, A., McCombie, W. R., Hicks, J., and Wigler, M. (2011). *Nature* **472**, 90–94.
Negrini, M., Sabbioni, S., Possati, L., Rattan, S., Corallini, A., Barbanti-Brodano, G., and Croce, C. M. (1994). *Cancer Res.* **54**, 1331–1336.
Nicholson, J. M., and Duesberg, P. (2009). *Cancer Genet. Cytogenet.* **194**, 96–110.
Olaharski, A. J., Sotelo, R., Solorza-Luna, G., Gonsebatt, M. E., Guzman, P., Mohar, A., and Eastmond, D. A. (2006). *Carcinogenesis* **27**, 337–343.
Papp, B., Pal, C., and Hurst, L. D. (2003). *Nature* **424**, 194–197.
Pavelka, N., Rancati, G., and Li, R. (2010a). *Curr. Opin. Cell Biol.* **22**, 809–815.
Pavelka, N., Rancati, G., Zhu, J., Bradford, W. D., Saraf, A., Florens, L., Sanderson, B. W., Hattem, G. L., and Li, R. (2010 b). *Nature* **468**, 321–325.
Peloponese Jr., J. M., Haller, K., Miyazato, A., and Jeang, K. T. (2005). *Proc. Natl. Acad. Sci. U.S.A.* **102**, 18974–18979.
Peters, J. M. (2006). *Nat. Rev. Mol. Cell Biol.* **7**, 644–656.
Phillips, J. L., Hayward, S. W., Wang, Y., Vasselli, J., Pavlovich, C., Padilla-Nash, H., Pezullo, J. R., Ghadimi, B. M., Grossfeld, G. D., Rivera, A., Linehan, W. M., Cunha, G. R., and Ried, T. (2001). *Cancer Res.* **61**, 8143–8149.
Pihan, G. A., Purohit, A., Wallace, J., Knecht, H., Woda, B., Quesenberry, P., and Doxsey, S. J. (1998). *Cancer Res.* **58**, 3974–3985.
Pihan, G. A., Purohit, A., Wallace, J., Malhotra, R., Liotta, L., and Doxsey, S. J. (2001). *Cancer Res.* **61**, 2212–2219.
Pitt, D., Leversha, M., Sinfield, C., Campbell, P., Anderson, R., Bryan, D., and Rogers, J. (1981). *J. Med. Genet.* **18**, 309–311.

Pollack, J. R., Sorlie, T., Perou, C. M., Rees, C. A., Jeffrey, S. S., Lonning, P. E., Tibshirani, R., Botstein, D., Borresen-Dale, A. L., and Brown, P. O. (2002). *Proc. Natl. Acad. Sci. U.S. A.* **99**, 12963–12968.
Quintyne, N. J., Reing, J. E., Hoffelder, D. R., Gollin, S. M., and Saunders, W. S. (2005). *Science* **307**, 127–129.
Ramirez, T., Stopper, H., Hock, R., and Herrera, L. A. (2007). *Mutat. Res.* **617**, 16–22.
Rancati, G., Pavelka, N., Fleharty, B., Noll, A., Trimble, R., Walton, K., Perera, A., Staehling-Hampton, K., Seidel, C. W., and Li, R. (2008). *Cell* **135**, 879–893.
Rehen, S. K., Yung, Y. C., McCreight, M. P., Kaushal, D., Yang, A. H., Almeida, B. S., Kingsbury, M. A., Cabral, K. M., McConnell, M. J., Anliker, B., Fontanoz, M., and Chun, J. (2005). *J. Neurosci.* **25**, 2176–2180.
Reing, J. E., Gollin, S. M., and Saunders, W. S. (2004). *Cancer Genet. Cytogenet.* **150**, 57–61.
Ried, T., Heselmeyer-Haddad, K., Blegen, H., Schrock, E., and Auer, G. (1999). *Genes Chromosomes Cancer* **25**, 195–204.
Rieder, C. L., Cole, R. W., Khodjakov, A., and Sluder, G. (1995). *J. Cell Biol.* **130**, 941–948.
Rieder, C. L., and Palazzo, R. E. (1992). *J. Cell Sci.* **102**(Pt 3), 387–392.
Rieder, C. L., Schultz, A., Cole, R., and Sluder, G. (1994). *J. Cell Biol.* **127**, 1301–1310.
Rimessi, P., Gualandi, F., Morelli, C., Trabanelli, C., Wu, Q., Possati, L., Montesi, M., Barrett, J. C., and Barbanti-Brodano, G. (1994). *Oncogene* **9**, 3467–3474.
Roberts, B. T., Farr, K. A., and Hoyt, M. A. (1994). *Mol. Cell Biol.* **14**, 8282–8291.
Rossner, P., Boffetta, P., Ceppi, M., Bonassi, S., Smerhovsky, Z., Landa, K., Juzova, D., and Sram, R. J. (2005). *Environ. Health Perspect.* **113**, 517–520.
Rudner, A. D., and Murray, A. W. (1996). *Curr. Opin. Cell Biol.* **8**, 773–780.
Russell, M. H., Shenai, J. P., Niblack, G. D., and Kasselberg, A. G. (1981). *Clin. Pediatr. (Phila)* **20**, 806–807.
Sachs, R. K., and Hlatky, L. (2010). *Bull. Math. Biol.* **72**, 359–374.
Saeki, A., Tamura, S., Ito, N., Kiso, S., Matsuda, Y., Yabuuchi, I., Kawata, S., and Matsuzawa, Y. (2002). *Cancer* **94**, 2047–2054.
Sage, J., and Straight, A. F. (2010). *Genes Dev.* **24**, 1329–1333.
Santaguida, S., and Musacchio, A. (2009). *J. Eur. Mol. Biol. Org.* **28**, 2511–2531.
Saran, N. G., Pletcher, M. T., Natale, J. E., Cheng, Y., and Reeves, R. H. (2003). *Hum. Mol. Genet.* **12**, 2013–2019.
Sato, M., Sekido, Y., Horio, Y., Takahashi, M., Saito, H., Minna, J. D., Shimokata, K., and Hasegawa, Y. (2000). *Jpn. J. Cancer Res.* **91**, 504–509.
Sato, N., Mizumoto, K., Nakamura, M., Maehara, N., Minamishima, Y. A., Nishio, S., Nagai, E., and Tanaka, M. (2001). *Cancer Genet. Cytogenet.* **126**, 13–19.
Saunders, W. S., Shuster, M., Huang, X., Gharaibeh, B., Enyenihi, A. H., Petersen, I., and Gollin, S. M. (2000). *Proc. Natl. Acad. Sci. U.S.A.* **97**, 303–308.
Scarbrough, P. R., Hersh, J., Kukolich, M. K., Carroll, A. J., Finley, S. C., Hochberger, R., Wilkerson, S., Yen, F. F., and Althaus, B. W. (1984). *Am. J. Med. Genet.* **19**, 29–37.
Sen, S. (2000). *Curr. Opin. Oncol.* **12**, 82–88.
Shah, J. V., Botvinick, E., Bonday, Z., Furnari, F., Berns, M., and Cleveland, D. W. (2004). *Curr. Biol.* **14**, 942–952.
Sheffer, M., Bacolod, M. D., Zuk, O., Giardina, S. F., Pincas, H., Barany, F., Paty, P. B., Gerald, W. L., Notterman, D. A., and Domany, E. (2009). *Proc. Natl. Acad. Sci. U.S.A.* **106**, 7131–71316.
Sherard, J., Bean, C., Bove, B., DelDuca, V., Esterly, K. L., Karcsh, H. J., Munshi, G., Reamer, J. F., Suazo, G., Wilmoth, D., Dahlke, M. B., Weiss, C., Borgaonkar, D. S., and Reynolds, J. F. (1986). *Am. J. Med. Genet.* **25**, 307–312.
Silkworth, W. T., Nardi, I. K., Scholl, L. M., and Cimini, D. (2009). *PLoS One* **4**, e6564.

Skoufias, D. A., Andreassen, P. R., Lacroix, F. B., Wilson, L., and Margolis, R. L. (2001). *Proc. Natl. Acad. Sci. U.S.A.* **98**, 4492–4497.

Skyldberg, B., Fujioka, K., Hellstrom, A. C., Sylven, L., Moberger, B., and Auer, G. (2001). *Mod. Pathol.* **14**, 279–284.

Sotillo, R., Hernando, E., Diaz-Rodriguez, E., Teruya-Feldstein, J., Cordon-Cardo, C., Lowe, S. W., and Benezra, R. (2007). *Cancer Cell* **11**, 9–23.

Stamey, T. A. (2004). *Br. J. Urol. Int.* **94**, 963–964.

Steigemann, P., Wurzenberger, C., Schmitz, M. H. A., Held, M., Guizetti, J., Maar, S., and Gerlich, D. W. (2009). *Cell* **136**, 473–484.

Stephens, P. J., Greenman, C. D., Fu, B., Yang, F., Bignell, G. R., Mudie, L. J., Pleasance, E. D., Lau, K. W., Beare, D., Stebbings, L. A., McLaren, S., Lin, M. L., McBride, D. J., Varela, I., Nik-Zainal, S., Leroy, C., Jia, M., Menzies, A., Butler, A. P., Teague, J. W., Quail, M. A., Burton, J., Swerdlow, H., Carter, N. P., Morsberger, L. A., Iacobuzio-Donahue, C., Follows, G. A., Green, A. R., Flanagan, A. M., Stratton, M. R., Futreal, P. A., and Campbell, P. J. (2011). *Cell* **144**, 27–40.

Stewenius, Y., Gorunova, L., Jonson, T., Larsson, N., Hoglund, M., Mandahl, N., Mertens, F., Mitelman, F., and Gisselsson, D. (2005). *Proc. Natl. Acad. Sci. U.S.A.* **102**, 5541–5546.

Stewenius, Y., Jin, Y., Ora, I., de Kraker, J., Bras, J., Frigyesi, A., Alumets, J., Sandstedt, B., Meeker, A. K., and Gisselsson, D. (2007). *Clin. Cancer Res.* **13**, 6593–6602.

Storchova, Z., and Pellman, D. (2004). *Nat. Rev. Mol. Cell Biol.* **5**, 45–54.

Sudakin, V., Chan, G. K., and Yen, T. J. (2001). *J. Cell Biol.* **154**, 925–936.

Sudilovsky, O., and Hei, T. K. (1991). *Cancer Lett.* **56**, 131–135.

Suijkerbuijk, S. J., van Osch, M. H., Bos, F. L., Hanks, S., Rahman, N., and Kops, G. J. (2010). *Cancer Res.* **70**, 4891–4900.

Susini, T., Olivieri, S., Molino, C., Amunni, G., Rapi, S., Taddei, G., and Scarselli, G. (2011). *Int. J. Gynecol. Cancer* **21**, 678–684.

Swanton, C., Nicke, B., Schuett, M., Eklund, A. C., Ng, C., Li, Q., Hardcastle, T., Lee, A., Roy, R., East, P., Kschischo, M., Endesfelder, D., Wylie, P., Kim, S. N., Chen, J. G., Howell, M., Ried, T., Habermann, J. K., Auer, G., Brenton, J. D., Szallasi, Z., and Downward, J. (2009). *Proc. Natl. Acad. Sci. U.S.A.* **106**, 8671–8676.

Tal, M. (1979). *Basic Life Sci.* **13**, 61–75.

Tang, Y. C., Williams, B. R., Siegel, J. J., and Amon, A. (2011). *Cell* **144**, 499–512.

Tang, Z., Bharadwaj, R., Li, B., and Yu, H. (2001). *Dev. Cell* **1**, 227–237.

Thompson, S. L., and Compton, D. A. (2008). *J. Cell Biol.* **180**, 665–672.

Torres, E. M., Dephoure, N., Panneerselvam, A., Tucker, C. M., Whittaker, C. A., Gygi, S. P., Dunham, M. J., and Amon, A. (2010). *Cell* **143**, 71–83.

Torres, E. M., Williams, B. R., and Amon, A. (2008). *Genetics* **179**, 737–746.

Uhlmann, F., Lottspeich, F., and Nasmyth, K. (1999). *Nature* **400**, 37–42.

Upender, M. B., Habermann, J. K., McShane, L. M., Korn, E. L., Barrett, J. C., Difilippantonio, M. J., and Ried, T. (2004). *Cancer Res.* **64**, 6941–6949.

Utani, K., Kohno, Y., Okamoto, A., and Shimizu, N. (2010). *PLoS One* **5**, e10089.

van Harn, T., Foijer, F., van Vugt, M., Banerjee, R., Yang, F., Oostra, A., Joenje, H., and te Riele, H. (2010). *Genes. Dev.* **24**, 1377–1388.

Van Valen, L. M., and Maiorana, V. C. (1991). *Evol. Theory*71–74.

Veitia, R. A. (2002). *BioEssays* **24**, 175–184.

Vickers, A. J., Till, C., Tangen, C. M., Lilja, H., and Thompson, I. M. (2011). *J. Nat. Cancer Inst* **103**, 462–469.

Vincent, M. D. (2010). *Evolution* **64**, 1173–1183.

Virtaneva, K., Wright, F. A., Tanner, S. M., Yuan, B., Lemon, W. J., Caligiuri, M. A., Bloomfield, C. D., de La Chapelle, A., and Krahe, R. (2001). *Proc. Natl. Acad. Sci. U.S. A.* **98**, 1124–1129.

Wadman, M. (2011). *In* "Nature News." Nature, http://www.nature.com/news/2011/110402/full/news.2011.203.html.

Walther, A., Houlston, R., and Tomlinson, I. (2008). *Gut* **57**, 941–950.

Wang, X., Yang, Y., Duan, Q., Jiang, N., Huang, Y., Darzynkiewicz, Z., and Dai, W. (2008). *Dev. Cell* **14**, 331–341.

Waters, J. C., Chen, R. H., Murray, A. W., and Salmon, E. D. (1998). *J. Cell Biol.* **141**, 1181–1191.

Weaver, B. A., and Cleveland, D. W. (2006). *Curr. Opin. Cell Biol.* **18**, 658–667.

Weaver, B. A., Silk, A. D., and Cleveland, D. W. (2008). *Cell Oncol.* **30**, 453.

Weaver, B. A., Silk, A. D., Montagna, C., Verdier-Pinard, P., and Cleveland, D. W. (2007). *Cancer Cell* **11**, 25–36.

Weiss, E., and Winey, M. (1996). *J. Cell Biol.* **132**, 111–123.

Whitfield, W. G., Gonzalez, C., Maldonado-Codina, G., and Glover, D. M. (1990). *J. Eur. Mol. Biol. Org.* **9**, 2563–2572.

Williams, B. R., Prabhu, V. R., Hunter, K. E., Glazier, C. M., Whittaker, C. A., Housman, D. E., and Amon, A. (2008). *Science* **322**, 703–709.

Winey, M., Goetsch, L., Baum, P., and Byers, B. (1991). *J. Cell Biol.* **114**, 745–754.

Winge, O. (1930). *Zeitschrift fuer Zellforschung und Mikroskopische Anatomie* **10**, 683–735.

Wise, D. A., and Brinkley, B. R. (1997). *Cell Motil. Cytoskeleton* **36**, 291–302.

Yamaguchi, K., Okami, K., Hibi, K., Wehage, S. L., Jen, J., and Sidransky, D. (1999). *Cancer Lett.* **139**, 183–187.

Yoon, D. S., Wersto, R. P., Zhou, W., Chrest, F. J., Garrett, E. S., Kwon, T. K., and Gabrielson, E. (2002). *Am. J. Pathol.* **161**, 391–397.

Yu, H. G., and Dawe, R. K. (2000). *J. Cell Biol.* **151**, 131–142.

Yuan, B., Xu, Y., Woo, J. H., Wang, Y., Bae, Y. K., Yoon, D. S., Wersto, R. P., Tully, E., Wilsbach, K., and Gabrielson, E. (2006). *Clin. Cancer Res.* **12**, 405–410.

Yun, C., Cho, H., Kim, S. J., Lee, J. H., Park, S. Y., and Chan, G. K. (2004). *Mol. Cancer Res.* **2**, 159–169.

Zhai, Y., Kronebusch, P. J., and Borisy, G. G. (1995). *J. Cell Biol.* **131**, 721–734.

Zhang, L., Rothman, N., Wang, Y., Hayes, R. B., Li, G., Dosemeci, M., Yin, S., Kolachana, P., Titenko-Holland, N., and Smith, M. T. (1998). *Carcinogenesis* **19**, 1955–1961.

Zhivotovsky, B., and Kroemer, G. (2004). *Nat. Rev. Mol. Cell Biol.* **5**, 752–762.

Zou, H., McGarry, T. J., Bernal, T., and Kirschner, M. W. (1999). *Science* **285**, 418–422.

Nuclear Remodeling as a Mechanism for Genomic Instability in Cancer

Macoura Gadji*,†, Rhea Vallente*, Ludger Klewes*, Christiaan Righolt*,‡, Landon Wark*, Narisorn Kongruttanachok*, Hans Knecht§ and Sabine Mai*

**The University of Manitoba, Manitoba Institute of Cell Biology, Cancer Care Manitoba, The Genomic Centre for Cancer Research and Diagnosis, Winnipeg, Manitoba, Canada*

† The Cheikh Anta Diop University of Dakar, Laboratory of Haematology and Immunology, National Centre of Blood Transfusion of Dakar, Senegal

‡ Department of Imaging Science and Technology, Delft University of Technology, Delft, The Netherlands

§ Division of Haematology/Oncology, CHUS University Hospital, Sherbrooke (Québec), Canada

Advances in CANCER RESEARCH, Volume 112
Copyright 2011, Elsevier Inc. All right reserved.
0065-230X/10 $35.00
DOI: 10.1016/B978-0-12-387688-1.00004-1

This chapter focuses on the three-dimensional organization of the nucleus in normal, early genomically unstable, and tumor cells. A cause–consequence relationship is discussed between nuclear alterations and the resulting genomic rearrangements. Examples are presented from studies on conditional *Myc* deregulation, experimental tumorigenesis in mouse plasmacytoma, nuclear remodeling in Hodgkin's lymphoma, and in adult glioblastoma. A model of nuclear remodeling is proposed for cancer progression in multiple myeloma. Current models of nuclear remodeling are described, including our model of altered nuclear architecture and the onset of genomic instability. © 2011 Elsevier Inc.

I. INTRODUCTION

Cancer is a genetic disease that is directly influenced by the microenvironment. At the start of the twentieth century, Theodor Boveri described concepts and the underlying principles of malignancy in cancer cells (Boveri, 1914, 1929). Landmark studies have since confirmed many aspects of his theories and work. For example, cancer is recognized as a "disease of DNA organization and dynamic cell structure" (Pienta *et al.*, 1989).

Genomic instability is a dynamic process that ultimately generates karyotypic variation in cells (Bayani *et al.*, 2007). Genomic instability is a common feature of most cancer cells, wherein it may sometimes lead to the initiation and promotion of cancer (Fest *et al.*, 2005; Hanahan and Weinberg, 2000, 2011; Holland and Cleveland, 2009; Lazebnik, 2010; Mai, 2010; Rajagopalan and Lengauer, 2004; Weaver and Cleveland, 2009). Interestingly, high levels of instability are inhibitory to cancer development, suggesting that not all genomic changes progress to tumor development and that specific changes are essential to drive carcinogenesis (Fest *et al.*, 2005; Holland and Cleveland, 2009; Weaver and Cleveland, 2009).

Two major types of genomic instability have been described based on the nature of changes in the genomic material of cells. Nonrandom genomic instability, for example, is characteristic for c-*Myc*/*Ig* translocations in mouse plasmacytoma, rat immunocytoma, and Burkitt's lymphoma (Potter and Wiener, 1992) or for *bcr*/*abl* translocations in chronic myeloid leukemia (Rowley, 2001). On the other hand, random genomic instability created via breakage-bridge fusion cycles and centrosome aberrations is common in Reed–Sternberg (RS) cells of Hodgkin's lymphoma (HL) (Guffei *et al.*, 2010). A combination of the two types of genomic instability is also possible, such as recurrent chromosomal rearrangements in chromosomes 6p, 8q, and 17p that are found in osteosarcomas with otherwise complex and dynamic karyotypic instability (Selvarajah *et al.*, 2008).

The nucleus exhibits nonrandom and evolutionarily conserved nuclear chromosome positions (Bolzer *et al.*, 2005; Cremer and Cremer, 2010;

Tanabe *et al.*, 2002). Chromosomes may, under certain conditions, change their nuclear positioning (Guffei *et al.*, 2010; Lacoste *et al.*, 2010; Louis *et al.*, 2005; Mai and Garini, 2005, 2006; Mehta *et al.*, 2010). This has also been documented in specific cell types during cellular differentiation. Examples include adipocytes (Kuroda *et al.*, 2004) and keratinocytes (Marella *et al.*, 2009).

Repositioning of chromosomes was observed after c-Myc deregulation (Louis *et al.*, 2005), after *ex vivo* infection of human B cells with Epstein–Barr virus (EBV) (Lacoste *et al.*, 2010) and in Hodgkin's lymphoma, during the transition of mono-nucleated Hodgkin to multinucleated RS cells (Guffei *et al.*, 2010). Investigations on the 3D organization of telomeres in nuclei of cancer cells indicated that specific 3D nuclear telomeric profiles are associated with cancer and allow for the identification of patient subgroups (Chuang *et al.*, 2004; Gadji *et al.*, 2010; Knecht *et al.*, 2009, 2010a, 2010b; Mai and Garini, 2006).

II. IMAGING TECHNIQUES EMPLOYED IN NUCLEAR REMODELING STUDIES

Nuclear changes have been documented since the late nineteenth century (Hansemann, 1858–1920 and Boveri, 1862–1915). To date, we still focus on nuclear architecture and have developed methods to objectively quantify these cellular components, including other subcellular elements, in order to determine linkages between the changes in the nuclear organization and genomic instability. A number of techniques and tools have been developed to visualize and measure the cellular, subcellular, and molecular organization of biological samples. These technologies and tools will be discussed in this section.

A. Microscopy

The knowledge gained in many areas of cell biology has gone hand in hand with the development of light microscopy since Robert Hooke first coined the term "cell" in the seventeenth century (Hooke, 1665). Antonie van Leeuwenhoek discovered and described a wide range of microbiological concepts in the years to follow, mainly due to his improvement of microscope lenses (Schierbeek, 1959). These early lenses were far from perfect, the image quality was greatly improved over the years by better lens manufacturing techniques. These improvements eventually led to the conclusion that finite lenses have limits and that the image of an infinitesimal point is a finite-sized disk (Airy, 1835).

This so-called blurring fundamentally limits the resolution of any lens (Abbe, 1873; Rayleigh, 1879).

The development of the epifluorescence microscope (Ploem, 1967) and techniques in fluorescent labeling (Chalfie *et al.*, 1994; Herzog and Schutze, 1968; Kapuscinski and Skoczylas, 1978; Shaner *et al.*, 2005) enabled the visualization of a wide variety of subnuclear components of cells. Methods such as fluorescent *in situ* hybridization (Bauman *et al.*, 1980) and spectral karyotyping (SKY; Schröck *et al.*, 1996) widened the range of nuclear components that could be imaged with optical microscopy. During the last decades of the twentieth century, (fluorescence) microscopy evolved into the 3D methods used today. The introduction of confocal microscopy (Minsky, 1961; White *et al.*, 1987) and optical sectioning (Agard and Sedat, 1983) provided higher resolution 3D images.

A new generation of high-resolution microscopes is slowly finding its way to biological laboratories and will become a more common tool in cell biology in the next decade. A comprehensive overview of microscope techniques is beyond the scope of this paper and others have reviewed some of the novel techniques (Garini *et al.*, 2005; Hell, 2007). The latest methods in microscopy are worth mentioning in this chapter. For example, structured illumination microscopy (SIM) uses a structured illumination pattern and utilizes knowledge of this pattern to reconstruct a higher resolution image (Gustafsson, 2000). Another imaging method is stimulated emission depletion, which effectively decreases the fluorescent spot size to achieve a higher resolution image (Hell and Wichmann, 1994). Stochastic optical reconstruction microscopy and photo-activated localization microscopy, on the other hand, utilizes the stochastic switching of fluorophores to achieve a higher-resolution image (Betzig *et al.*, 2006; Rust *et al.*, 2006).

B. Quantitative Image Analysis

The actual imaging of the nuclear part of interest is only the first step in the analysis chain. These recorded data have to be processed, either manually by a human interpreter, computationally or by a hybrid form of the two. The main human role is the identification of objects of interest. This is usually termed segmentation when this process is (semi)-automated (Gonzalez and Woods, 2008). Although the segmentation process is an important part of the analytical methodology, it does not have a high biological relevance. Relevant for biological questions is the image analysis, the actual interpretation of data, including the size, intensity, and morphology of the objects of interest. The emphasis in the remainder of this review section will be on the analysis, the actual segmentation steps will usually not be mentioned and can generally be found in the references for each analysis method (see Smal *et al.*, 2010) for a comparison of different methods to segment small objects. Different excitation and emission spectra for different fluorescent dyes mean that one cell can be stained with multiple dyes to mark different components of the cell whose spatial relation can be examined.

The methodology to quantitatively compare cells falls into two broad categories. First, there are specific, tailor-made measurements that determine a small set of particular properties of cellular components. Second, there is a general method, in which a wide selection of nonspecific features is measured for one cell. The measurements span an *N*-dimensional hyperspace, which is subsequently analyzed by a variety of statistical methods (Young *et al.*, 2008). High throughput scanners (Walter *et al.*, 2010; Zanella *et al.*, 2010) are usually used to obtain a large amount of data to ensure that the conclusions are statistically significant. The actual features and analyses differ from study to study and an overview of the differences is beyond the scope of this review.

C. Imaging of Nuclear and Subnuclear Structures

The fluorescent dyes diaminophenyl-indole (DAPI) and Hoechst are most commonly used to visualize interphase cells and chromosomes on a microscope slide. The volume of the cell nucleus can be measured from the image of DNA-based dyes, although the nuclear morphology (Rohde *et al.*, 2008) or nuclear lamina (Righolt *et al.*, 2011a) could also be used to measure the nuclear volume. The chromatin fibers themselves can be seen with high-resolution microscopes, such as SIM (Schermelleh *et al.*, 2008). High-resolution images also show chromosome fusions. Such dicentric chromosomes are known to initiate breakage-bridge fusion cycles [Fig. 4, in Guffei *et al.* (2010)].

1. CHROMOSOME TERRITORIES

The term chromosome territory refers to the preference of specific chromosomes to be located in a certain position within the nuclear space (Cremer and Cremer, 2010). Whether chromosome territories are distant from each other, separated by an interchromosomal space (interchromosomal domain model) or intermingling with each other inside the nucleus (interchromosomal network model) still remains elusive (Branco and Pombo, 2006; Nunez *et al.*, 2009). To complicate matters, chromosomes can change their positions during cellular differentiation (Kuroda *et al.*, 2004; Marella *et al.*, 2009), as a result of serum stimulation or removal (Mehta *et al.*, 2010), viral infection (Apostolou and Thanos, 2008), and as a direct consequence of c-Myc deregulation (Louis *et al.*, 2005; Mai and Garini, 2005). How chromosomes move is not completely understood, but their movement, at least in the case of serum removal or stimulation, seems to depend on myosin (probably nuclear myosin 1β) and actin in the presence of ATP/GTP (Mehta *et al.*, 2010). Whether chromosomes are in close contact all the time, as proposed in the interchromosomal network model, or establish this contact due to development-, environment-, and agent-specific induction is not yet known for all cells and tissues. In order to promote genomic rearrangements, contacts have to be either present or be temporarily established. The term "chromosome kissing" was established to describe this chromosome contact (Cavalli, 2007; Kioussis, 2005). A direct interaction of maternal and paternal chromosomes 15 was reported by LaSalle and LaLande (1996). This interaction was suggested to be linked to imprinting in the 15q11-13 regions carrying the Prader–Willi syndrome and the Angelman syndrome loci. However, data by Teller *et al.* (2007) do not confirm this finding.

Imaging and analysis will help to assess the above models in further detail. The imaging of chromosome territories involves measuring the size and shape of different fluorescent regions, which represent different chromosomes. One way to determine chromosome position is by simply measuring its center of mass (Kozubek *et al.*, 1999a). An average over the whole territory can, however, yield a better estimate of the average radial position (Cremer *et al.*, 2001; Cremer and Cremer, 2001). The mutual distance between chromosomes can be measured as well and compared between normal and cancer cells (Righolt *et al.*, 2011b). The contacts between chromosomes could not be resolved using the capabilities of the current models of microscopes. The introduction of higher resolution microscopes in the laboratory will lead to a better insight of chromosome organization and how they interact.

2. GENE LOCI

The position of a gene locus is given by the center of mass of the point spread of the locus in the image. The relative radial position is then given by measuring the locus position with respect to the centre of the nucleus (Roix *et al.*, 2003; Fig. 1A). The position of loci also yields the mutual distance between two loci, either as an absolute measure (Osborne *et al.*, 2007), or relative to the diameter of the nucleus (Fig. 1B). This measurement is applicable to both alleles of one gene, or to the four pairs of alleles for two different genes in a cell. This can be used to investigate the occurrence of gene colocalization (Osborne *et al.*, 2007). In the case of comparative studies between normal and tumor cells, the gene positions could serve as a parameter for detecting genomic instability. For example, Osborne and colleagues have followed the interaction between *Myc* and *IgH* loci in mouse B cells and observed their colocalization in transcription factories containing RNA polymerase II. This is worth mentioning since

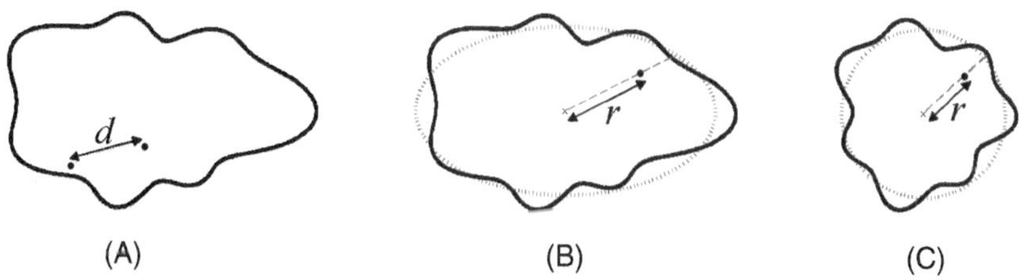

Fig. 1 Techniques in measuring subnuclear features. Measurement of the volume of subnuclear components such as genes and telomeres are visualized in two dimensions, the three-dimensional case is analogous to this. The solid line illustrates the outline of the nucleus in all images, while the dotted line shows the general outline of the nuclear shape. Different scenarios are measured during image analysis. (A) The physical distance between two dots, *d*, can either be absolute or normalized to the cell size. (B) The radial position in relation to the centre of mass is shown and can be normalized by measuring: (1) the maximum distance to the centre of mass, (2) the length of the line from the centre of mass to the edge of the nucleus (dashed line), or (3) a reference radius based on the volume and/or surface area of the nucleus. (C) The ellipsoidal shape of the nucleus is transformed to a circle (or sphere) before calculating the relative radius.

the translocation between *Myc* and *IgH* is the most common translocation (>90%) in mouse plasmacytomas (Potter and Wiener, 1992) and the close vicinity of these genes in the transcription factory could assist in illegitimate recombination events.

3. TELOMERES AND CENTROMERES

The 3D organization of telomeres and centromeres can also be studied with fluorescence microscopy, with the total fluorescent intensity serving as the direct measurement of the length of these heterochromatic regions (Gonzalez-Suarez *et al.*, 2009; Lansdorp *et al.*, 1996; Poon *et al.*, 1999). The positions of the telomeres and centromeres are roughly distributed in an ellipsoid, which is based on the lengths of the largest and shortest principal axes that would pass through the nucleus' centre of mass (Chuang *et al.*, 2004; Mai and Garini, 2006; Sarkar *et al.*, 2007; Vermolen *et al.*, 2005). When the fluorescent intensity of a telomere is significantly higher than normal for the specific tissue in question, then this fluorescent signal is often categorized as a telomeric aggregate (Mai, 2010; Mai and Garini, 2005, 2006; Raz *et al.*, 2006). Telomeric aggregates are clusters of telomeres that could not be further resolved with the employed imaging system, due to the resolution limit of the lens (for review, see Mai, 2010).

Centromeres were shown to alter their nuclear positioning during the transformation of mouse B lymphocytes. These changes were distinct from cell-cycle-dependent positional alterations that do occur in these cells (Sarkar *et al.*, 2007). During cellular transformation, telomere characteristics also change. Four parameters consistently differ between normal and tumor cells and within tumor cell subpopulations, namely the size of the telomeres, the number of telomeres, the overall 3D telomere organization, and frequency of telomeric aggregates.

The 3D telomere organization was measured using a semiautomated program, TeloView (Vermolen *et al.*, 2005). This program was fully automated as TeloScan, a high-throughput system for both imaging and analysis. The system is able to scan 10,000–15,000 cells per hour in 3D, and can detect one tumor cell in 1000 normal cells (Klewes *et al.*, 2011). It can also be used to determine the subclassification of patient groups, as shown in adult glioblastomas (Gadji *et al.*, 2010). This study identified three patient subgroups with intermediate, short-term, and long-term survival, based on their distinct 3D telomeric profiles.

III. MYC-DEPENDENT GENOMIC INSTABILITY AND NUCLEAR REMODELING

Myc deregulation occurs in >70% of all cancers (Nesbit *et al.*, 1999). The conditional deregulation of Myc induces genomic instability (Felsher and Bishop, 1999; Mai, 1994; Mai *et al.*, 1996a, 1996b), resulting in changes in copy numbers of specific genes (Kuschak *et al.*, 1999; Mai *et al.*, 1996a, 1999). Myc-driven genomic instability also leads to dynamic karyotypic rearrangements (Felsher and Bishop, 1999; Mai *et al.*, 1996b). How are these changes brought about? Does Myc deregulation affect the organization of the nucleus? Is this instability a direct cause or a consequence of nuclear changes? In an attempt to address these questions, Louis *et al.* (2005) investigated conditional Myc deregulation, resulting in nuclear changes and genomic instability. The authors found that Myc deregulation led to nuclear remodeling of telomere and chromosome positions. Under all conditions studied, these changes preceded the onset of dynamic chromosomal rearrangements (Louis *et al.*, 2005). Telomeres formed aggregates, which were not observed in non-Myc deregulated cells (ibid). These aggregates represented, in part, dicentric chromosomes that caused the initiation breakage-bridge fusion cycles. The latter induced ongoing and more complex chromosomal instability. This type of telomere dysfunction occurs in many cancers including osteosarcoma (Selvarajah *et al.*, 2006), prostate cancer (Vukovic *et al.*, 2003, 2007), breast cancer (Meeker *et al.*, 2004), and colon cancer (Stewenius *et al.*, 2005) (for reviews see, DePinho and Polyak, 2004; Lansdorp, 2009; Murnane and Sabatier, 2004).

Robertsonian fusions of mouse chromosomes have also been observed after Myc deregulation (Guffei *et al.*, 2007). Telomeric signals were present at all fusion points suggesting that telomere dysfunction had enabled the joining of telocentric mouse chromosomes. While telomeres change their positions, chromosomes perform spatial movements as well. Although not investigated for all chromosomes, Louis *et al.* (2005) showed that chromosomes that are involved in structural rearrangements move toward each other after Myc deregulation over a 96-h period. In contrast, a control chromosome that was never involved in rearrangements did not undergo any positional change despite Myc deregulation. These spatiotemporal changes suggest that Myc has the ability to remodel the nuclear organization of the genome by promoting chromosomal rearrangements (Mai and Garini, 2005).

A follow-up study determined that *Myc box II* of the Myc protein is required for the nuclear remodeling features of Myc (Caporali *et al.*, 2007). A *Myc box* II deletion mutant (Δ106) was unable to induce telomeric aggregates or Robertsonian fusions (Guffei *et al.*, 2007). Moreover, deregulated expression of *Myc box* II deleted Myc protein was also unable to promote *in vivo* tumorigenesis in mice (Fest *et al.*, 2005). Figure 2

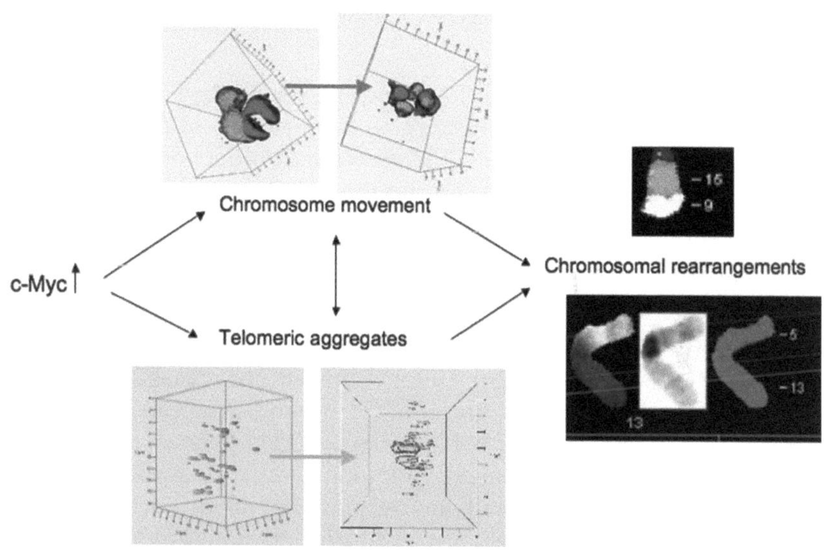

Fig. 2 Myc deregulation-dependent nuclear remodeling. Conditional Myc deregulation alters the 3D nuclear telomere organization (green dots: telomeres) resulting in the formation of aggregates (green arrow: change in 3D telomere organization). In addition, the positions of specific chromosomes (red and green shapes) within the nucleus are altered, allowing chromosomes to come into closer contact, share illegitimate territories and promote chromosomal rearrangements (Louis *et al.*, 2005; Mai and Garini, 2005). The examples illustrated in this scheme depict an unbalanced translocation of mouse chromosomes 9 and 15, T(9;15) and a Robertsonian fusion of mouse chromosomes 5 and 13, Rb5.13, (Guffei *et al.*, 2007; Louis *et al.*, 2005; Mai and Garini, 2005). (See Page 3 in Color Section at the back of the book.)

summarizes our current knowledge on Myc-dependent nuclear remodeling and its consequences. This figure illustrates one cycle of Myc-dependent nuclear remodeling and resulting chromosomal instability with unbalanced translocation and Robertsonian fusion. In reality, one such cycle induces ongoing instability (Mai and Garini, 2005; Louis *et al.*, 2005).

Several questions regarding the mechanisms of Myc-dependent nuclear remodeling are yet to be addressed. Does Myc activate nuclear motor activities that are required for chromosome movements (Mehta *et al.*, 2010)? Does Myc enable the detachment of telomeres from the nuclear matrix (de Lange, 1992)? Future studies will surely address these questions. However, with respect to the question of cause and consequence, we can argue convincingly that Myc-induced nuclear remodeling precedes genomic instability.

IV. NUCLEAR REMODELING AND GENOMIC INSTABILITY IN PRIMARY TUMORS

In this section, we will focus on nuclear remodeling that results in genomic instability. Our examples come from studies of plasmacytoma, Hodgkin's lymphoma, and adult glioblastoma. A model of nuclear remodeling and cancer progression will be presented for multiple myeloma (MM).

A. Mouse Plasmacytoma

A well-known model of Myc-induced tumorigenesis is the mouse plasmacytoma (Potter and Wiener, 1992). Conventional plasmacytomas of BALB/c mice and congenics are induced by an agent causing chronic inflammation, pristane (2,6,10,14-tetra-methyl-pentadecane). This method induces plasmacytomas with long latency (up to 300 days). Over 90% of pristane-induced mouse plasmacytomas carry the *Myc/IgH* translocation that causes the constitutive deregulation of the oncoprotein. In a study of early plasmacytotic foci at days 7–14 after pristane, preneoplastic cell clusters with telomeric aggregates were identified (Mai and Garini, 2006). Myc was found upregulated in peritoneal cavity-derived lymphocytes of such pristane-induced mice (Taylor and Mai, 1998). A locus-specific *dihydrofolate reductase* (*DHFR*) gene copy number increase was also observed.

These data suggest that nuclear remodeling is an early event *in vivo* and precedes the development of a fully developed tumor. Similarly, early premalignant lesions in cervical cancer show nuclear remodeling of telomeres and telomere aggregate formation (Mai and Garini, 2006). Interestingly, these early lesions concomitantly showed Myc deregulation and *DHFR* gene copy number increases (Guijon *et al.*, 2007).

B. Hodgkin's Lymphoma

Hodgkin's lymphoma is characterized by the presence of CD15+, CD30+ mono-nucleated H cells, and bi-, tri-, tetra- to multinucleated RS cells. The CD15+, CD30+ RS cell is the diagnostic cell of Hodgkin's lymphoma. About 50% of Hodgkin's lymphomas are EBV positive and express the latent membrane protein 1 (LMP1) or its deletion variants in the cytoplasm and the cell membrane of the H and RS cells (Knecht *et al.*, 1993; Pallesen *et al.*, 1991). Hodgkin's lymphoma is cytogenetically complex as illustrated in Table I.

Our study of Hodgkin's lymphoma focused on the 3D nuclear organization of telomeres and chromosomes in H and RS cells. The analysis of telomeric profiles of both types of cells has indicated telomere shortening, aggregation, and aberrant telomere numbers, as well as an overall aberrant 3D nuclear distribution of telomeres (Knecht *et al.*, 2009). Upon comparative examination of mono-, bi-, tri-, tetra-, and multinucleated cells, it became apparent that the mono-nucleated H cells are the precursors of the bi- and multinucleated RS cells. Accordingly, H cells display the above parameters of telomere dysfunction and nuclear remodeling, and RS cells also share these same features, but in a much more advanced stage. With each attempted additional cell division, they increasingly lose telomeric sequences and thus exhibit increasingly very short or unmeasurable telomeric sequences ("t-stumps," Xu and Blackburn, 2007) and showed significantly increased numbers of telomeric aggregates and nuclear volumes (Knecht *et al.*, 2009). These findings apply to EBV positive, LMP1 expressing and EBV negative Hodgkin's lymphoma (Knecht *et al.*, 2010a).

Increased telomere dysfunction is also obvious when the composition of the shelterin complex is investigated in H and RS cells. Both mono-nucleated H and multinucleated RS cells display the loss of TRF2 and TRF1 at the telomeric ends. Some of TRF1 and TRF2 proteins are mis-allocated to the cytoplasm. In addition, Mre11, Rad51c, and γH2AX were present in H and RS cells at distinct levels of expression. Significantly higher levels of Mre11, Rad51c, and γH2Ax staining were seen in RS cells, which concurrently display a more extensive level of nuclear remodeling and instability (Knecht *et al.*, 2009, 2010b).

Hodgkin cells do not only suffer from telomere dysfunction, but also show aberrations in centrosome duplication and consecutively, aberrant

Table I Genetic Defects in Hodgkin's Lymphoma

Genetic defect	Specific aberrations	Marks	Reference
Aneuploidy			(Falzetti *et al.*, 1999; Guffei *et al.*, 2010; Jansen *et al.*, 1999; MacLeod *et al.*, 2000; Martin-Subero *et al.*, 2006a)
Chromosomal rearrangements	Translocations, deletions, inversions, and duplications		(Falzetti *et al.*, 1999; Guffei *et al.*, 2010; MacLeod *et al.*, 2000; Martin-Subero *et al.*, 2006a)
Chromosomal breakpoints	1p36, 4q33, 6q15, 6q21, 7q22, 7q31, 7q32, 11q13, 11q23, 12q24, 13p11, 14p10-11, 15p10-11, 17p11, and 20q13		(Falzetti *et al.*, 1999)
	Acrocentric chromosome short arm, 2p13, 3q21, 3q26–27, 4q24, 5p11, 5p15.3, 5q11, 6q24, 8q24, 9p24, 19q12–13		(MacLeod *et al.*, 2000)
	2p16, 3q27, 8q24.1, 14q24.3, 16p13.1, 17q12, 18q21, and 19q13.2	Chromosomal breakpoints affected immunoglobulin loci	(Martin-Subero *et al.*, 2006b; Szymanowska *et al.*, 2008)
Chromosomal gains	2p, 12q, 17p, 9p,16p,17q, and 20q	*REL*(2p15-p16)	(Barth *et al.*, 2003; Joos *et al.*, 2002)
	Most frequently regions at 17q, 2p, 12q, 17p, 22q, 9p, 14q, and 16p, with minimal overlapping regions at 17q21, 2p23–13, 12q24, 17p13, 22q13, 9p24–23, 14q32, 16p13.3, and 16p11.2.	The most frequent abnormality was gain on 17q.	(Chui *et al.*, 2003)
	12q13.3	*STAT*(12q13.3)	(Feys *et al.*, 2007)
	2p12-16, 5q15-23, 6p22, 8q13, 8q24, 9p21-24, 9q34, 12q13-14, 17q12, 19p13, 19q13, and 20q11	*STAT*(12q13), *NOTCH1*(9q34), and *JUNB*(19p13)	(Hartmann *et al.*, 2008)
	2p15-16.1, 9p21.1, 9p24.1-24.3, 17q21.31-32, 20q13.11-13.12, and 20q13.2		(Steidl *et al.*, 2010)

Chromosomal losses	13q		(Joos *et al.*, 2002)
	Most frequently regions at 13q, 6q, 11q, and 4q, with minimal overlapping regions at 13q21, 6q22, 11q22, and 4q32.		(Chui *et al.*, 2003)
	6q, 3q, 12p, and 2p	The high frequency of chromosomal deletions on 6q25	(Re *et al.*, 2003)
	15q26.2 and 16q12.1	*RGMA* and *CHD2* (15q26.2)	(Feys *et al.*, 2007)
	Xp21, 6q23-24, and 13q22		(Hartmann *et al.*, 2008)
	6q23.2, 11q22.3, and 13q14.3-21.1		(Steidl *et al.*, 2010)
Centrosome aberrations			(Knecht *et al.*, 2009; Martin-Subero *et al.*, 2006a)
Subtelomere and telomere dysfunctions	Interstitial subtelomeres	Metaphases	(MacLeod *et al.*, 2000)
	Telomere-free ends	Metaphases	(Guffei *et al.*, 2010)
	Interstitial telomeres	Metaphases	(Guffei *et al.*, 2010)
	Telomere losses	Interphases	(Knecht *et al.*, 2009, 2010a)
	Short telomeres	Interphases	(Knecht *et al.*, 2009, 2010a)
	Telomere aggregates	Interphases	(Knecht *et al.*, 2009, 2010a)

spindle formation (Knecht *et al.*, 2009). These defects predominantly affect RS cells and lead to the mis-segregation of chromosomes. While some RS cells still divide, those with highly multinucleated DNA content fail to perform karyokinesis and in turn, undergo repeated cycles of endomitosis.

The organization of chromosome territories is already altered in H cells when compared to normal B cells, but is increasingly complex and remodeled in RS cells (Guffei *et al.*, 2010). Chromosomes 9 and 22, which are often juxtaposed in normal human B cells, are found to have undergone multiple breakage-bridge fusion cycles with each other and with additional chromosomes in H and increasingly so in RS cells (Guffei *et al.*, 2010). Spectral karyotyping confirmed the increasing complexity of rearrangements exemplified by the study of just two chromosomes. In fact, "zebra-type" chromosomes with multiple breakage-bridge fusion cycles involving a variety of chromosome partners are found in RS cells (Guffei *et al.*, 2010). This is schematically illustrated in Fig. 3. 3D-SIM super resolution imaging enabled further insights into the nuclear remodeling of H and RS cells confirming the complex nature of genomic remodeling and chromosome fusions in Hodgkin's lymphoma cells (Guffei *et al.*, 2010).

In conclusion, nuclear remodeling in Hodgkin's lymphoma is associated with telomere dysfunction coupled with aberrant centrosome duplication and spindle formation. These nuclear changes initiate the dynamic and complex genomic rearrangements seen in this cancer (Fig. 3).

C. Glioblastoma

Human gliomas represent the most common primary brain tumors in adults, accounting for approximately 35% of all cancers of the central nervous system (Friedman *et al.*, 2000; Murat *et al.*, 2008). According to World Health Organization classification, gliomas are divided into astrocytomas, oligodendrogliomas, and oligoastrocytomas, based on the tumor-cell phenotype. The highly malignant and rapidly growing anaplastic astrocytoma (grade III) and glioblastoma multiforme (GBM; grade IV) constitute about 50–60% of primary brain tumors with an incidence ranging from 5 to 8 per 100,000 in the human population (Friedman *et al.*, 2000). GBM constitutes 55% of all primary gliomas (Collins, 1998; Gurney and Kadan-Lottick, 2001).

The pathological classification of gliomas remains controversial due to the lack of specific immunohistochemical biomarkers to distinguish each subtype. O6- MGMT (O6-methylguanine-DNA methyltransferase) promoter methylation has been reported as a potentially predictive factor for response to alkylating agents in a fraction of glioblastomas (Esteller *et al.*, 2000; Hegi *et al.*, 2005; Stupp *et al.*, 2009). Further, recurrent mutations in the active site of IDH1 (Isocitrate dehydrogenase 1) have been noted in

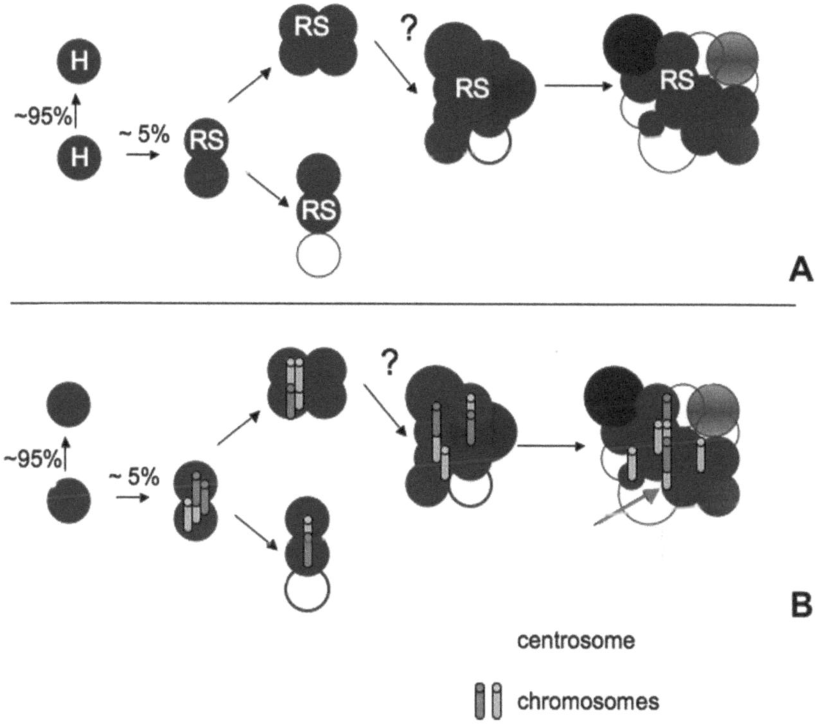

Fig. 3 Cell division in Hodgkin's lymphoma. (A) Schematic diagram illustrating the generation of mono-nucleated Hodgkin (H) and bi, tri-, tetra-, and multinucleated Reed–Sternberg (RS) cells. Each blue circle represents the nucleus of an H cell, as well as nuclear units that remain associated with each other in RS cells. The faint blue or empty blue circles represent DNA-poor RS nuclei, so-called ghost nuclei (Guffei *et al.*, 2010; Knecht *et al.*, 2009, 2010a). Most of the time, H cells produce H cells, yet it is still possible for H cells to generate bi-nucleated RS cells that are precursors of multinucleated RS cells. (B) Schematic diagram representing centrosome duplication defects and chromosomal instability in H and RS cells according to Knecht *et al.* (2009) and Guffei *et al.* (2010). Yellow dots: centrosomes; blue dots: nuclei; pink dots: chromosome pair 1; green dots: chromosome pair 2. RS cells develop centrosome duplication defects, such as the increase in number of centrosomes, forming aberrant mitotic spindles that promote mis-segregation of chromosomes (Knecht *et al.*, 2009). Chromosomes undergo dynamic rearrangements due to telomere dysfunction, causing end-to-end fusions and breakage-bridge-fusion cycles (Guffei *et al.*, 2010). This is illustrated in the model by the increase in the number of translocations and fusions between chromosome pair 1 and 2. A red arrow points to a long breakage-bridge fusion chromosome. (See Page 4 in Color Section at the back of the book.)

12% of patients bearing GBM (Parsons *et al.*, 2008), mostly in secondry GBM Based on clinical and genetic parameters, GBMs are classified into primary and secondary GBMs. Primary GBM originates *de novo* and account for up to 95% of all GBM and is characterized by EGF (epidermal growth factor) receptor amplification and mutations (Gadji *et al.*, 2009), loss of heterozygosity of 10q, deletion of the phosphatase and tensin homologue on chromosome 10 (PTEN 10) and p16 deletion (Murat *et al.*, 2008). Secondary GBM slowly evolves from preexisting low-grade diffuse astrocytoma (grades I and II) and passing through anaplastic changes (grade III) reaches the GBM stage after several years (Dropcho and Soong, 1996; Furnari *et al.*, 2007; Murat *et al.*, 2008; Ohgaki and Kleihues, 2007). During this evolutional process, secondary GBM originates from low-grade astrocytomas (grade I to II) with p53 mutation (>65%), overexpression of PDGF-A and PDGFR alpha (60%), LOH of 19q (50%), and RB alteration. However, TP53 mutations are the earliest detectable alteration.

From the high grade astrocytomas (grade III to IV: GBM), secondary GBMs accumulate LOH of chromosome 10, loss of expression of DCC (50%), and PDGFR alpha amplification (10%) (Gadji *et al.*, 2009, 2010; Kleihues *et al.*, 1999; Louis *et al.*, 2007).

About 30% of glioblastoma, however, show molecular alterations that neither corresponds to primary nor secondary GBM pathways (Collins, 1998; Holland, 2001; Kleihues and Ohgaki, 1999; Martinez *et al.*, 2004; von Deimling *et al.*, 1995; Table II).

Investigations of the 3D nuclear telomeric architecture in GBM showed that three GBM categories can be distinguished for the first time according to their nuclear telomeric patterns, and these profiles corresponded to patient survival and disease progression (Fig. 4; Gadji *et al.*, 2010). GBM was considered fatal, with a median survival of generally only 12 to 15 months (Louis *et al.*, 2002). The variability of patient responses to treatment and outcome implies an underlying biological heterogeneity and the existence of patient subgroups, thus highlighting the clinical application of telomere signatures as a new prognostic, predictive, and potentially pharmacodynamic biomarker in GBM (Fig. 4).

D. Multiple Myeloma

Multiple myeloma (MM) is an incurable disease, with a prevalence rate of 1–4 per 100,000 people per year (Bladé *et al.*, 2010). This haematological disease is more prevalent in men, and is twice as common in blacks as it is in Caucasians (Landgren and Weiss, 2009). Patients survive, on average, for 3–4 years with conventional treatment and this can be extended to 5–7 years with advanced treatment (Raab *et al.*, 2009). Signs and symptoms of this malignancy include anemia, hypercalcemia, hyperviscosity, renal failure, and osteolytic bone lesions (Shapiro-Shelef and Calame, 2005).

MM is mostly preceded by premalignant stages of monoclonal gammopathy of undetermined significance (MGUS) and or smoldering multiple

Table II Genomic Instability in Glioblastoma

Genomic instability	Primary GBM	Secondary GBM	References
Gains	7q; 8q	7q; 8q	(Ueki *et al.*, 1996)
Deletions and loss of heterozygosis	10p and 10q; 19q; p16; 17p	10q; 19q; 17p	(Fujisawa *et al.*, 1999; Louis *et al.*, 2007; Ueki *et al.*, 1996)
Amplifications	EGFR; MDM2; CDK4;	PDGFRα	(Fleming *et al.*, 1992; Fults and Pedone, 1993; Louis *et al.*, 2007; Schmidt *et al.*, 2002; Ueki *et al.*, 1996)
Mutations	PTEN; RB	P53; PTEN; RB	(Ichimura *et al.*, 1996; Louis *et al.*, 2007; Ueki *et al.*, 1996; Zhou *et al.*, 1999)
Overexpression	EGFR; MDM2	PDGFα; PDGFRα	(Louis *et al.*, 2007; Newcomb *et al.*, 1998; Reifenberger *et al.*, 1993; Ueki *et al.*, 1996)
Methylations	p14ARF; IDH1, 2; O6MGMT	IDH1 and 2; O6MGMT	(Esteller *et al.*, 2000; Nakamura *et al.*, 2001; Parsons *et al.*, 2008; Stupp *et al.*, 2009; Ueki *et al.*, 1996; Yamini *et al.*, 2007;)
SNPs	5p15.33 (*TERT*); 20q13.33 (*RTEL1*); 11q23.3 (*PHLDB1*), 8q24.21 (*CCDC26*); 9p21.3 (*CDKN2A-CDKN2B*).		(Shete *et al.*, 2009; Wrensch *et al.*, 2009)

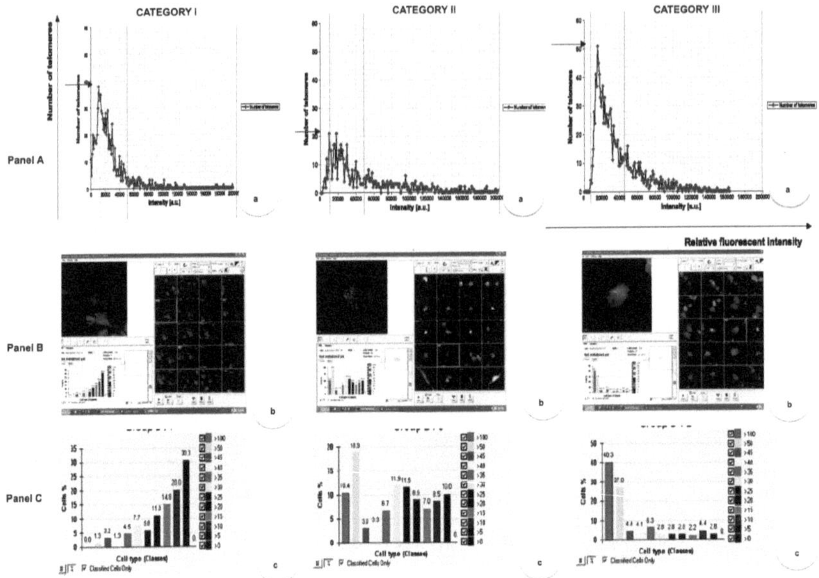

Fig. 4 Three-dimensional nuclear telomeric profiles of glioblastoma (GBM) patients. 3D nuclear telomere analyses revealed three GBM patient categories characterized by distinct and clinically predictive three-dimensional nuclear-telomeric architectures defined by telomere number, size, and frequency of telomeric aggregates. These measurements identified three patient groups (categories I–III), displaying significant differences in telomere numbers/ nucleus, telomere length, and number of telomeric aggregates (a–c in Panels A, B, and C, respectively). Panel A: Number of telomeres versus relative intensities of fluorescent signals analyzed by TeloView (Vermolen *et al.*, 2005); Panels B and C: Screenshots of human GBM nuclei (blue) and telomeres (red) with TeloScan and screenshot of a selected field of the gallery of 20 scanned cells (Panel B). The corresponding histogram showing the distribution of classified defined cells based on the number of telomeres/cell (Panel C). These categories corresponded to patients with long-term, intermediate, and short-term survival, respectively: Category I identifies patients with long-term survival, category II with intermediate and category III with short-term survival, respectively. (Lines and arrows define the three groups of cell populations). Figure 4 is reproduced with permission from Neoplasia (Gadji *et al.*, 2010). (See Page 5 in Color Section at the back of the book.)

myeloma (SMM). MGUS is characterized by the clonal expansion of plasma cells in the bone marrow associated with a monoclonal globulin in the serum or urine (Bence–Jones proteins) (Kyle and Rajkumar, 2004; Palumbo *et al.*, 2009; Zhan *et al.*, 2002). Patients diagnosed with MGUS are at an increased risk of developing MM (Kyle *et al.*, 2010), at a rate of about 1–2% a year (Zhan *et al.*, 2002). It has also been reported that the relative rate of progression from MGUS to MM increases over time (Kyle *et al.*, 2006; Fig. 5).

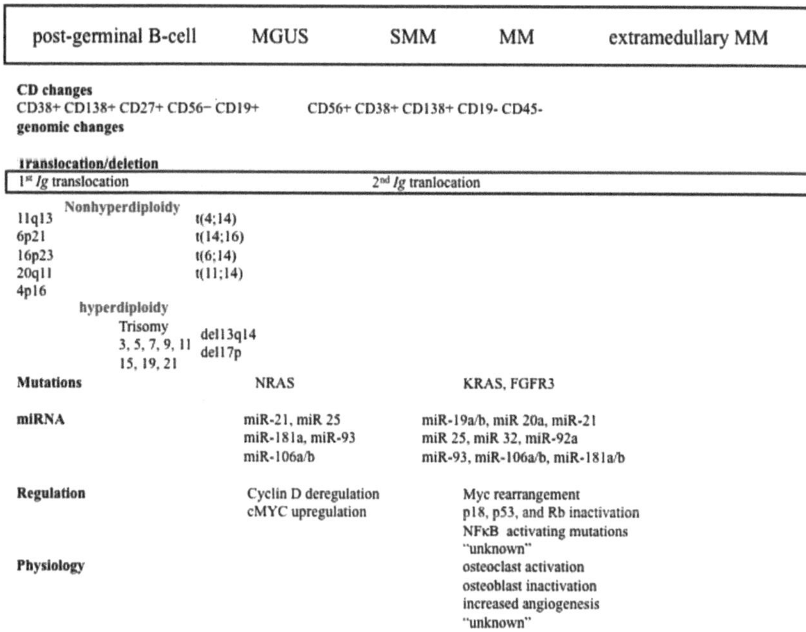

Fig. 5 Genetic events associated with cancer progression in MM. The transition from normal postgerminal B-cell to MM requires several oncogenic events. Early translocations involve 11q13, 6p21, 16p23, 20q11, and 4p16 and multiple chromosome partners. These cells are nonhyperdiploid. *Ig* translocations [t(4;14), t(14;16), t(6;14), and t(11;14)] and hyperploidy are also considered early events. Del(13), which is common to 80–90% of all patients (MGUS and MM), occurs concomitantly with either t(4;14), t(14;16), or t(14;20) translocations. A common step during myeloma progression is the dysregulation of the *Cyclin D* gene and the upregulation of *Myc*. Secondary translocations and deletions can occur at any stage of myelomagenesis, encompassing SMM and MM. *NRAS*-gene mutations have been observed in both MGUS and MM; however, the *KRAS*-gene mutation was only described for MM. *Myc* rearrangements, the inactivation of the *RB1*-gene, the inactivation or deletion of the *TP53*- or the *CDKN2C* genes, and *NFKB1* activating mutations are typical for advanced stages of MM. Furthermore, there are miRNA pattern changes in both MGUS and MM (Horvilleur *et al.*, 2010; Pichiorri *et al.*, 2008; Zhou *et al.*, 2010), with the potential of modulating the expression of proteins critical to myeloma pathogenesis.

The hallmark genetic aberrations linked to the development of MM and MGUS have not been established, although there are a number of genetic descriptors that have been published in seminal reports (for review, see Kyle *et al.*, 2006). Table III provides a list of recent research reports that summarize the findings of genetic analyses of patients diagnosed with MM or MGUS. One of the first reports of the gross chromosomal rearrangements involved in this haematological cancer was released in 2002 and implicated the long arms of chromosomes 11, 13, or 14 (Harrison *et al.*,

Table III Genetic Aberrations Associated with MM and MGUS

Genetic defect	Specific aberration	Genomic regions involved	Reference
Gross chromosomal rearrangement	Translocation	t(11;14)(q13;q32)	(Harrison *et al.*, 2002; Lopez-Corral *et al.*, 2011; Nahi *et al.*, 2011; Robillard *et al.*, 2003)
		t(4:14)	(Chang *et al.*, 2011; Karlin *et al.*, 2011; Martinez-Garcia *et al.*, 2011; Neben *et al.*, 2010)
		t(14;20)	(Ross *et al.*, 2010; Vekemans *et al.*, 2010)
	Deletion	1p21	(Chang *et al.*, 2010, 2011)
		8p	(Sutlu *et al.*, 2009)
		12p	(Tapper *et al.*, 2011)
		13q	(Chang *et al.*, 2011; Facon *et al.*, 2001; Zojer *et al.*, 2000)
		17p; 17p13	(Avet-Loiseau *et al.*, 2007, 2010; Chang *et al.*, 2005, 2011; Gertz *et al.*, 2005; Neben *et al.*, 2010)
	Duplication/amplification	1q21	(Chang *et al.*, 2011; Fonseca *et al.*, 2006; Nemec *et al.*, 2010)
		Chr5	(Tapper *et al.*, 2011)
Copy number variation	Deletion	1p, 6q, 8p, 12p, 13q, 14q, 16q, 17p, 20, and 22 (18%)	(Walker *et al.*, 2010)
		1p31–32	(Chng *et al.*, 2010)
Epigenetic defect	Variable methylation patterns	t(4;14)	(Walker *et al.*, 2011)
		p16	(Park *et al.*, 2011)
Post-translational defect	Loss-of-function	IRF4	(Shaffer *et al.*, 2008)
Single nucleotide polymorphisms	Based on genome-wide association	Locus: 10q23, within gene CYP2C8, rs1934951, rs1934980, rs1341162, rs17110453	(Sarasquete *et al.*, 2008)
	Based on pairwise linkage-disequilibrium-based algorithm	Within CD4 (variant rs11064392) and LAG3 genes:	(Lee *et al.*, 2010)
		PSMA6-8C>G; G allele	(Bachmann *et al.*, 2010)

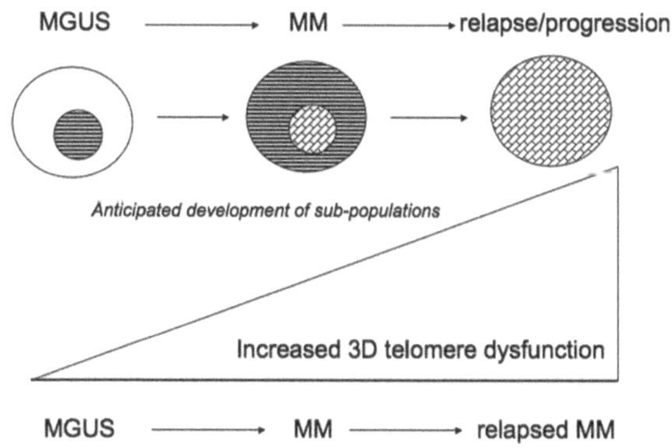

Fig. 6 Proposed model for telomere dysfunction and clonal evolution in MM. The current proposed model for the development of MM is driven by an increase in telomere dysfunction from monoclonal gammopathy of undetermined significance (MGUS) through relapsed MM. 3D nuclear profiles suggest the existence of subpopulations in plasma cell populations for MGUS, MM, and during disease relapse and progression, possibly representing the progression of this cancer through distinct nuclear remodeling steps.

2002). Further genetic analysis of this chromosomal region identified putative genes involves in the aetiology of MM and MGUS, namely the immunoglobulin heavy chain genes *IGHG1* (MIM ID#147100) and *IGH/CCND1* (MIM ID#168461).

MM requires new approaches to diagnosis allowing for optimized treatment upfront. Preliminary studies with plasma cells from MGUS, MM, and relapsed MM have been initiated (Klewes *et al.*, unpublished data). We propose a model of telomere dysfunction in MGUS and MM as shown in Fig. 6. In this model, telomere dysfunction is seen in all stages of MM development but has an increasingly disturbed 3D phenotype as patients transition from MGUS to MM and from MM in remission to relapse. This model also anticipates the presence of subpopulations within MGUS and MM that have different 3D telomeric signatures and are linked to disease progression. If validated, such profiles would enable the detection of individual risk to rapid progression and would allow for personalized treatment decisions.

V. MODELS OF GENOMIC INSTABILITY OCCURRING WITHIN THE NUCLEAR SPACE

A. DNA Damage and Nuclear Organization

Studies employing various types of DNA damage have assessed the result of such insult to interphase nuclei and the genome. There are two types of studies, namely biological studies that assess the damage to the 3D nucleus using both interphase and metaphase chromosomes, and mathematical studies in which models were generated to predict the outcome of genotoxic treatments. Combinations of the two also exist (Holley *et al.*, 2002; Kreth *et al.*, 1998; Vazquez *et al.*, 2002). Several models for nuclear organization emerged based on DNA damage and resulting aberrations (for review, see Kreth *et al.*, 1998). However, the data that best reflect the known nuclear organization of the genome concluded that chromosomes are nonrandomly organized within the nuclear space.

Kreth *et al.* investigated all models of nuclear organization (random and nonrandom, see Fig. 1, in Kreth *et al.*, 1998) using computer simulations and experimental results. Simulation of localized irradiation best fit with our current understanding of nonrandom nuclear organization. In subsequent analyses, Bickmore and Teague (2002) proposed that chromosome size, gene density, and nuclear position affect the frequencies of constitutional translocations in the human population. Larger chromosomes have a higher frequency of interchromosomal rearrangements (Fig. 1, in Bickmore and Teague, 2002). Similar results of higher rearrangement frequencies between large chromosomes have been reported by Cornforth *et al.* (2002). For smaller chromosomes, such as the gene-dense human chromosomes 17, 19, and 22, the nuclear position is more relevant to the frequency of observed translocations than the size. All three chromosomes were found in close proximity to the nuclear centre (Bickmore and Teague, 2002). Another observation was that radial nuclear organization influences the degree of DNA damage observed. According to studies conducted by Gazave *et al.* (2005), human chromosome 19 was preferentially located in the nuclear interior, while human chromosome 18 was observed at the periphery. It should be noted that each of these chromosomes exhibit different rates of DNA damage, with chromosome 19 undergoing genomic changes at higher rates than that of chromosome 18. The underlying factors contributing to this finding remain elusive; however, it is possible that the high gene content of chromosome 19 (as opposed to chromosome 18) and differences in chromatin structure may contribute to this observation (Gazave *et al.*, 2005).

To date, hot spots for radiation-induced double-strand breaks on specific chromosomes have not been identified and thus impact of the territorial proximity of chromosomes strengthens its influence on the outcome of irradiation damage (Folle, 2008; Sachs *et al.*, 2000). Two models have been proposed to understand the occurrence of translocations in the 3D nuclear space, namely the contact-first and the breakage-first models (for review, see Misteli and Soutoglou, 2009). The contact first model appears to be a more plausible scenario of chromosome translocations, mainly due to the relatively higher prevalence of this specific chromosomal rearrangement among neighboring chromosomes (Kozubek *et al.*, 1999b).

B. Nuclear Matrix and Nuclear Organization

The nuclear matrix, which is regarded as the key organizer of a cell's nuclear structure, was first described by Ronald Berezney and Donald Coffey in the 1970s (Berezney and Coffey, 1974, 1977). This nuclear "skeleton" maintains the structural integrity of the normal nucleus, and is commonly associated with cancer-specific changes that involve matrix alterations (Coffey, 2002; Partin *et al.*, 1993). Proteomic analyses have later established that the consistency of the nuclear matrix is cancer type specific (Coffey, 2002). A study conducted by Cannon *et al.* (2007) highlighted the differences in nuclear matrix proteins (NMPs) associated with renal cell carcinomas (RCCs). Five previously identified NMPs were further characterized, as well as the alternative splicing routes for RCCA-1 and albumin alterations in RCC (Cannon *et al.*, 2007).

The nuclear matrix high mobility group protein I(Y) [(HMGI(Y)] is another key protein player that dynamically interacts with and regulates histone H1 and other transcription factors (Catez *et al.*, 2004) as well as promotes anchorage-independent growth of Rat and Rat1a fibroblasts (Hommura *et al.*, 2004; Rothermund *et al.*, 2005). HMGI(Y) is a Myc-dependent NMP (Takaha *et al.*, 2002) that is significantly overexpressed during neoplasia (Leman *et al.*, 2003). The extent of unbalanced translocations correlates with HMGI expression levels (Takaha *et al.*, 2002). Transfection of the *HMGI* gene into the LNCAP cell line, which is known to engage solely in balanced translocations, resulted in the generation of unbalanced translocations (ibid).

The seminal work by de Lange (1992) showed that telomeres are associated with the nuclear matrix. Similarly, work by Ma *et al.* (1999) described the association of chromosome territories with the nuclear matrix. Many questions regarding the nuclear matrix and its impact on the organization of the nucleus are yet to be addressed. Based on current data one can postulate that any change to the matrix and/or the attachments between telomeres, chromosomes, and the matrix will enhance positional changes in nuclei and enable new contacts between these components. This, in turn, may influence the frequency of genetic rearrangements.

C. The Chromatin Scaffold Theory of Genomic Stability

Another theory that has been proposed to explain the mechanism of genomic instability is largely based on the structural properties and spatial conformations of chromosomes (Mateos-Langerak *et al.*, 2009; Wang *et al.*, 2009). The chromatin scaffold can influence the degree and location of gene expression, as well as the repair of stretches of DNA (Fitzgerald *et al.*, 2011; Wang *et al.*, 2011). The higher order of structural organization of chromatin, specifically that of the 30-nm fiber, together with the underlying loops, axes, and connections between other chromosomal structures, have been strongly associated with major cellular activities such as recombination, gene expression, and genomic instability (Chadwick, 2008; Hu *et al.*, 2009).

According to the scaffold model of chromosomal organization initially reported in 2002, the extent of scaffolding and looping of chromatin provides inherent information on the physical properties of the genetic material (Dekker *et al.*, 2002; Fransz *et al.*, 2002). More importantly, the qualitative features of chromatin largely influence the stability of the cell. In the natural state, chromatin is described to be highly flexible throughout the entire stretch of the DNA (Bohn and Heermann, 2011; Mozziconacci *et al.*, 2006). However, each region of the chromatin fiber has a specific capacity to exhibit a specific conformation, such as a contorted ring or a loop, which allows the interaction of nuclear proteins that play roles in replication, repair, and recombination (Dekker, 2008). The scaffold model of genome organization also describes that specific chromosomes interact to a higher degree than others, thus showing an intracellular form of relationship between the chromatin material within the nucleus (Dekker *et al.*, 2002). One frequent interaction is that occurring between two nonhomologous chromosomes, and this can be observed as the close proximity of specific chromosomal regions such as telomeres and centromeres within the nuclear space (Del Rey *et al.*, 2010; Kulkarni *et al.*, 2010; Pobiega and Marcand, 2010). In as much as such aggregation is observed, there are reports that support this theory, such as that of the telomeric bouquet (Chikashige *et al.*, 2009) and more recently, telomere aggregates (Mai, 2010). Another interesting example would include the recurrent chromosomal rearrangements that generate Robertsonian translocations, which usually involve centromeric regions of nonhomologous proteins (Kalitsis *et al.*, 2006).

The scaffold model is also applicable in the case of recombination, wherein homologous chromosomes within the nucleus could find each other within nuclear space and time (Yue *et al.*, 2009; for review, see Koszul and Kleckner, 2009). It has been proposed that the pairing of chromosomes within the nucleus could be due to DNA and protein features of the chromosomes, such as the complementary sequences between the stretches of DNA within two homologous chromosomes (Kleckner *et al.*, 2004; Scialdone and Nicodemi, 2009). In addition, the presence of structural chromosomal proteins such as cohesins, ligases, and separases may also play essential roles in establishing chromosomes pairs and allowing the next chromosomal activities such as cross-linking and recombination to occur (for review, see Mannini *et al.*, 2010; Strunnikov, 2010).

Subsequent studies related to the scaffold model of genomic organization has resulted in reports that chromosomal function is mainly based on the mechanical forces that are generated by the expansion of the chromatin fiber, especially in terms of constraining features (Kleckner *et al.*, 2004). The mechanical basis of the chromatin organization is largely based on the idea wherein homologues interact with each other and that one chromatin strand could suppress the activities of the other strand (Rach *et al.*, 2011). The condition, also known as interference, has been employed as the main mechanism behind the control of cross-linking, the generation of recombination sites and hotspots and the production of Holliday junctions (Bohn and Heermann, 2011; Cook and Marenduzzo, 2009). Furthermore, the presence of an initial event within the chromatin fiber significantly decreases the chance that an additional, yet similar event would occur within at least one centimorgan from the site. This classic requirement has extensively been utilized in linkage studies, and has also served as the assumption for linkage disequilibrium conditions.

The scaffold theory of genomic organization also takes into account that constraining features within the nucleus could also allow or prevent other nuclear activities from taking place (Groehler and Lannigan, 2010). For example, when a chromatin strand is highly compressed, the accessibility of nuclear proteins into specific regions of the genome may be significantly lower, thus suppressing gene expression, repair, and recombination (de Nooijer *et al.*, 2009). On the other hand, less compression stress along the chromatin fiber would increase the likelihood that nuclear proteins could attach to the genetic material, thus facilitating the replication, repair, and recombination of specific genetic regions (Bohn and Heermann, 2011).

The scaffold model of genomic organization has been applied to both mitotic and meiotic cells, with specific applications in the events that are inherent in each type of cell cycle phase (Groehler and Lannigan, 2010). In the mitotic program, including that of cancer cells, the scaffold model could be localized within the S phase, wherein the chromatin loops and arrays are generated (for review, see Crepaldi and Riccio, 2009). Other S-phase-specific events also occur at this time, including nucleosome installation and intersister connections (Rosa and Everaers, 2008). Any perturbation that would occur within a chromatin region with a specific degree of scaffolding could result in the induction of instability within the genome. The specific capacity of chromatin in the nuclear space to exhibit a specific conformation thus allows its interaction with other nuclear structures and entities that are essential in establishing or disrupting genomic stability (Fig. 7).

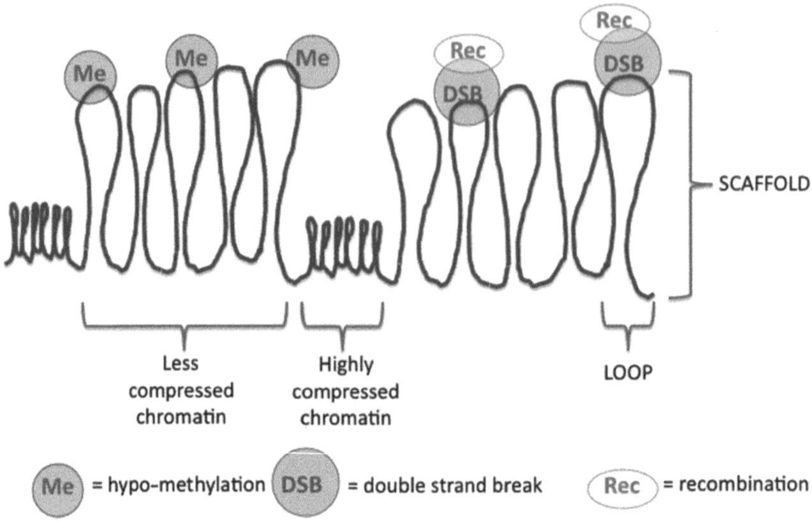

Fig. 7 Chromatin scaffold model of genomic instability. Nuclear chromatin is arranged as scaffolds and loops that provide information on the physical properties of genetic material. Each region of the chromatin fiber has a specific capacity to exhibit a specific conformation, which allows the interaction of nuclear proteins that play roles in replication, repair, and recombination. ***Me***: methylation; ***DSB***: double strand break; ***Rec***: recombination. (See Page 5 in Color Section at the back of the book.)

D. Nuclear Lamins and Genomic Stability

Nuclear lamins refer to the 10- to 20-nm thick protein meshwork associated with the inner nuclear membrane that mainly plays a role in stabilizing the nuclear membrane of eukaryotic cells. It is predominantly composed of lamins A and B, which are members of the intermediate filament family (Stuurman *et al.*, 1998). These also assist in DNA and RNA synthesis, as well as in cell division (Smith *et al.*, 2005; Vlcek and Foisner, 2007; Worman and Courvalin, 2005). Furthermore, lamins are also involved in the organization of chromatin and the anchorage of the nuclear pore complexes located within the nuclear membrane (Gerace and Burke, 1988).

Lamin type A provides a venue for chromatin scaffolding, as well as correcting chromosome positioning (Meaburn *et al.*, 2007). In aging cells, lamin A dysfunction was associated with an increase in the sensitivity to mechanical stress of the nucleus (Smith *et al.*, 2005; Worman and Courvalin, 2005). Mutations in lamin and lamin-associated proteins are the main causative factors of laminopathies (Capell and Collins, 2006; Jacob and Garg, 2006; Mounkes *et al.*, 2003), most notably the Hutchinson–Gilford progeria syndrome (HGPS), a disease characterized by accelerated aging, which has been first described by Hutchinson

(Hutchinson, 1886), and Type 2 (Dunnigan's) familial partial lipodystrophy (FPLD) (Boschmann *et al.*, 2010; Hegele *et al.*, 2000a, b; Shackleton *et al.*, 2000), an autosomal-dominant disease caused by mutations in the Lamin A (LMNA), ataxia telangiectasia, mutated (ATM), ATM- and Rad3-related (ATR) gene (Cao and Hegele, 2000; Gruenbaum *et al.*, 2005; Speckman *et al.*, 2000).

Progerin, a truncated version of lamin A with an internal 50 amino acid deletion, is the result of a point mutation (C1824T) (De Sandre-Giovannoli *et al.*, 2003) that is associated with tumor development in prostate (Tang *et al.*, 2010). The accumulation of progerin, as detected in HGPS patients, causes DNA damage and consequently activates DNA damage response pathways, including the activation of ATM and ATR, and the phosphorylation of Chk1, Chk2, and p53 (Liu and Chang, 2006; Varela *et al.*, 2005). At the same time, progerin accumulation increases the sensitivity to DNA damaging agents.

Mammalian telomeres are distributed similarly throughout the nucleoplasm during G1 and S-phase. In late G2, the telomeres are organized in the centre of the nucleus, forming a telomeric disk (Chuang *et al.*, 2004). An interaction between nuclear matrix and telomeres was proposed (de Lange, 1992; Luderus *et al.*, 1996). Recent experiments demonstrate the binding of telomeres to A-type lamin spanning throughout the nucleus, contributing to the nuclear compartmentalization (Gonzalez-Suarez *et al.*, 2009). This interaction would explain the organization of chromosomes in microterritories (Rouquette *et al.*, 2010; Sullivan *et al.*, 1999). Furthermore, the telomeres are shortened and predominately localized at the nuclear periphery. The overall distribution of telomeres is also altered, highlighted by telomeric migration and increased motility (Gonzalez-Suarez *et al.*, 2009). Other studies show that unprotected telomeres exhibit increased motility, covering larger territories than capped telomeres (Dimitrova *et al.*, 2008; Wang *et al.*, 2008). These data suggest that A-type lamins play an important role in the nuclear organization of telomeres.

E. The Role of Euchromatic and Heterochromatic Regions in Nuclear Remodeling

Chromatin, which is mainly comprised of nuclear DNA and its associated proteins, is technically classified according to the degree of condensation in the cell (Woodcock and Ghosh, 2010). Chromatin that is interchangeably found in condensed and decondensed forms based on the requirements of the cell is regarded as euchromatin, while genomic material that is constantly maintained in a condensed state is considered as heterochromatin (Sharma *et al.*, 2007). Euchromatin contains genes and interspersed repeat sequences, such as the Alus, SINES, and LINES (Cui *et al.*, 2011; Lunyak *et al.*, 2007). There are also deactivated genes within the euchromatic region, also known as pseudogenes (Cotton and Page, 2005).

Heterochromatin can be further classified according to the propensity to condense during the cell cycle (Huberman, 2011). Constitutive heterochromatin, namely the telomeres and the centromeres, are commonly localized to the nuclear periphery (Krull *et al.*, 2010). These regions are comprised of tandem repeat sequences, such as the TTAGGG cassettes in the telomeres and the alpha, beta and classical satellites in the centromeric regions (Gent *et al.*, 2011; Stimpson *et al.*, 2010). On the other hand, facultative heterochromatin is often situated across the rest of the genome, as these regions are required for gene silencing activities (Luo *et al.*, 2009). Despite the major differences in the degree of compaction of chromatin, the conditions and interactions of these regions can influence the stability of the nucleus, as well as of the rest of the cell (Costa-Nunes *et al.*, 2010; Tsang *et al.*, 2010).

Constitutive and facultative heterochromatic regions are strongly associated with transcriptional silencing, although the processes for each type may be different (Iijima *et al.*, 2008; Sun *et al.*, 2010). In constitutive heterochromatin, silencing is generally an indirect and unspecific process that occurs within the nucleus (Yu *et al.*, 2011). On the other hand, constitutive heterochromatic regions mainly serve as genome stabilizers, specifically through the prevention of chromosomal and gene rearrangements between highly homologous sequences (Jansen *et al.*, 2007). Constitutive heterochromatic regions also ensure that chromosomal segregation occurs with utmost precision (Kang *et al.*, 2011). The phenomenon of gene silencing based on the presence of constitutive heterochromatin in adjacent regions of genes is known as position-effect variegation (Lemos *et al.*, 2010).

Heterochromatic regions of the genome generally serve two roles, namely to maintain certain genes in a silent state and to confer stability to the rest of the nuclear material (Dimitriadis *et al.*, 2010). Telomeres are thus considered as the physical caps of chromosomes, preventing these from attaching to other chromosomes of the cell (Rhodin Edsö *et al.*, 2011). Centromeres, in turn, serve as regions for the attachment of kinetochore proteins, which are the key attachment points of microtubules during cell division cycles (Shivaraju *et al.*, 2011). Any disruption in the centromere–kinetochore complex may result in chromosomal nondisjunctions, laggards, and bridges (Corbett *et al.*, 2010; Dumont *et al.*, 2010).

The repetitive nature of heterochromatic regions also allows for the occurrence of misalignments, deletions, and translocations, all of which are hallmarks of cancer (Kanikarla-Marie *et al.*, 2011; Marteau *et al.*, 2010). The extreme shortening of telomeres has also been associated with the development of cancer, as well as in conditions of ageing (for review, see Oberdoerffer and Sinclair, 2007). Extremely short telomeric regions could increase the capacity of chromosomes to fuse with other chromosomes.

The role of euchromatic and heterochromatic regions in nuclear remodeling largely stems from their capacity to confer genomic stability and to assist in DNA repair and in epigenetic modifications (for review, see Lo and Sukumar, 2008; Tanaka *et al.*, 2010). During the process of DNA repair, it is necessary for chromatin to be unpacked, thus allowing repair and histone modification enzymes to interact with the DNA sequence (Halicka *et al.*, 2009; Mitsiades and Anderson, 2009; Thyagarajan *et al.*, 2007). In addition, it is also essential for repair processes to occur within specific stretches of the chromatin, thus correcting any disruptions within the actual sequence of the DNA. Several reports have described that a single break within the stretch of the DNA strand can induce the cell to release proteins that are responsible for repair (for review, see Gerlitz, 2010; Messick and Greenberg, 2009).

The composition of heterochromatic regions, namely that of tandem repeat sequences, may also be transcriptionally activated in cancer cells, thus indicating that the silencing effect of these chromatin regions has been disrupted (Oberdoerffer and Sinclair, 2007). In addition, it is also possible that the number of repeats within the heterochromatic regions is decreased and this could be observed as a change in the nuclear architecture of an aberrant cell. This change in length of the heterochromatic regions could also affect its position within the perinuclear space, which in turn could change the positions of chromosome territories. For example, defective telomeric organization may disrupt the interactions between the telomeres with the nuclear matrix, thus generating an aberrant nuclear architecture (Olins *et al.*, 2010).

Changes in the positions and gene ordering of euchromatic and heterochromatic regions may also alter the subnuclear localization of genes within the nucleus (Cremer and Cremer, 2010; Solovei *et al.*, 2009). It is thus possible that certain essential genes are repressed, while others are activated to produce significant amounts of proteins that are unnecessary for cell survival and maintenance, thus triggering the development of malignancies and other tumorigenic tissues. An example of these specific chromatin regions in the nucleus is the senescence-associated heterochromatin foci, which are considered as repressed chromatin regions that are commonly located within the promoter regions of cell-cycle regulator proteins (Jiang *et al.*, 2011). Another example would involve hypermethylation of CpG islands within promoter regions, contributing to a neoplastic phenotype (Chim *et al.*, 2007).

The redistribution of chromatin regions within the nuclear space may thus induce changes in the transcription, gene expression, repair, and protection of specific genes within the nucleus (Black and Whetstine,

2011; Oberdoerffer and Sinclair, 2007). In combination with other nuclear remodeling events such as epigenetic modifications (e.g., methylation, histone acetylation/deacetylation, RNA interference), as well as changes in the nuclear lamina and alterations in DNA damage and repair response pathways, the entire nuclear architecture could be altered, thus enhancing the cell's susceptibility to tumorigenesis.

F. Our Model of the Onset of Genomic Instability

The nucleus, with its established components and typical architecture, is one of the most important components of the cell. The packaged chromatin experiences multiple modifications and mutations during all stages of its existence. As such, eukaryotic cells have evolved a dynamic nuclear organization that assists in the cell's requirements for survival and homeostasis. In the last few decades, we have witnessed major developments in the fields of microscopy, genetics, and molecular biology, thus paving the way for more detailed investigations on specific components of the cell. Although the linkage between nuclear organization and cancer has been proposed for more than a century, it is still imperative that specific studies be performed to better understand the mechanisms behind each nuclear event.

We propose a model that addresses the chain of events that leads to genomic instability. Fig. 8 presents the linkages between structure and function in normal cells (Figs. 8A and C) and at the onset of genomic instability (Figs. 8B and D). In most normal cells, the heterochromatic telomeres of chromosomes are positioned close to the nuclear lamina through its association with nuclear matrix attachment sites. Thus, the chromosomes are retained within specific territories of the 3D nuclear space (Figs. 8A and C). DNA alterations in the nucleus are generally repaired using specific checkpoint and response mechanisms, conducted as a concerted effort to conserve the nuclear organization. In addition, specific compaction levels are maintained at distinct chromosomal regions, thus allowing epigenetic modifications to occur during certain cellular conditions and stages for homeostasis and survival.

Cancer cells deviate from the concept of genomic stability as a consequence of multiple crucial events that alter the nuclear architecture and function. Mutations in lamin genes often promote the detachment of chromosomes from the inner nuclear periphery, thus promoting illegitimate migration of chromosomes and redefining chromosomal territories within the 3D nuclear space (Figs. 8B and D). Furthermore, these relocated chromosomes may be situated close to other chromosomes, and if experiencing damage, increasing the chances of engaging in unbalanced translocations, fusions, and nonhomologous recombination.

It is important to emphasize that the impact of mutations on the nuclear matrix and lamina genes is amplified by physical features of the chromatin

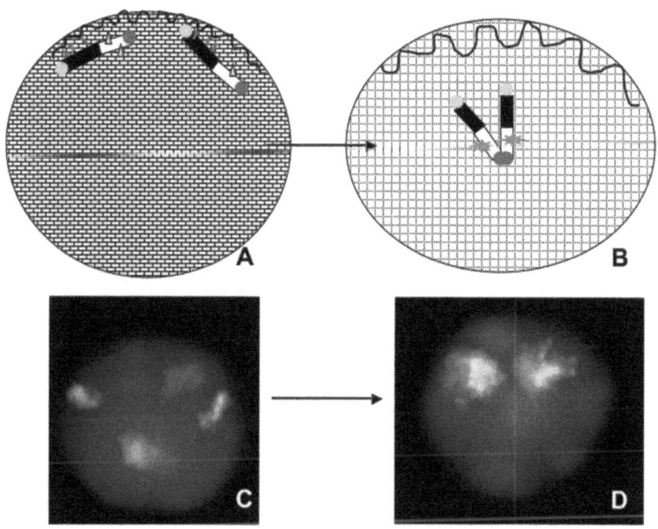

Fig. 8 Model of nuclear remodeling in genomic instability. Initial nuclear remodeling events confer genomic instability and reprogramming of gene expression. The chromosomal organization of heterochromatic regions (black boxes) and euchromatin (white boxes) is shown for normal and early nuclear changes prior to the onset of chromosomal rearrangements (A and B, respectively). While heterochromatin is anchored in the normal cell (A), it is often released in the premalignant cell (B). The nuclear matrix (dark mesh in A; gray mash in B), generally undergoes alterations from (A) to (B). In addition, the epigenetic marks are changed (blue triangles in A and orange stars in B), and chromosomes have moved within the nuclear space, and telomeres (red) have formed aggregates. Centromeres (green) are also repositioned. C and D show examples of chromosome painting in cells where chromosomes 15 (red) and 16 (green) are not positioned close to each other (C, normal cell), while they are in very close proximity in tumor cells (D, cancer cell). This scenario may lead to the formation of dicentric chromosomes, unbalanced translocations and ongoing breakage-bridge fusion cycles. The occurrence of DNA damage in two neighboring chromosomes may result in nonhomologous end-joining and balanced translocation events. Matrix changes, epigenetic changes, and dissociation from the matrix/lamin/scaffold may also affect gene regulation. For details, see the text. (See Page 6 in Color Section at the back of the book.)

scaffold and the degree of condensation of the chromosomes. The extent of looping and tightness of the chromatin cores may facilitate epigenetic modifications within the chromatin, while in certain regions may prevent alterations in specific stretches of DNA.

The roles of each nuclear event to nuclear remodelling may contribute to the establishment of genomic instability, yet it is equally important to focus on the causes and consequences of alternations in the nuclear organization through time. It is highly likely that the capacity of a cell to initiate DNA repair activities to correct any mutations and DNA strand breaks serves as the forefront for normal nuclear organization. This cellular response could subsequently induce modifications in the chromatin structure, including that of the architecture of the scaffold, degree of condensation, which in turn could influence the levels of transcription and gene expression to prevent further nuclear damage. The accumulation of nuclear insults over time, in combination with substandard or decreased levels of repair, transcription, and chromatin compaction, may thus be conducive to genomic instability, ultimately resulting in a cell's transformation into a malignant cell.

VI. CONCLUSIONS

Major developments in the fields of microscopy, genetics, and cellular biology have set the path toward defining the linkages between nuclear structure and function. Normal cells exhibit a defined, ordered nuclear organization and genome stability. Changes to this nuclear organization lead to the onset of genomic instability. According to current research reports that have employed high-resolution fluorescence imaging techniques coupled with quantitative optical analyses, the nucleus harbours chromosomes that are nonrandomly positioned within the 3D nuclear space. If cells experience DNA damage, and/or any insult affecting the regular order of the nucleus, the maintenance of the nuclear matrix and lamina, chromatin scaffolding, de/condensation and epigenetic modifications, cells may either die or develop genomic instability.

Nuclear remodeling studies have permitted biomedical researchers to examine the spatiotemporal dynamics of chromosomes and genes according to various *in vitro* conditions and based on defined disease entities. Data presented in this review indicate that nuclear changes precede the onset of genomic instability, whether they are induced through the oncoprotein *Myc*, viral infection, or genotoxic stress. The nuclei of primary tumors discussed in this review clearly displayed a nuclear organization that was different from that of normal cells. This happened during the first week of plasmacytoma development. Alterations in the nuclear order were accompanied by complex genomic rearrangements as mononucleated Hodgkin cells transitioned to multinucleated RS cells. The 3D nuclear telomeric organization in adult glioblastomas allowed for the subgrouping of patients into short-term, intermediate-, and long-term survivors.

Our studies conclude that the 3D nuclear space is at the origin of stability/instability of the genome. Any change that alters the nuclear organization potentially induces changes in its organization that precede the onset of genomic instability. It is anticipated that changes in the 3D nuclear organization may, in the future, assist in the diagnosis, treatment and monitoring of specific types of cancers.

ACKNOWLEDGMENTS

The authors thank the Canadian Institutes of Health Research (CIHR) (SM), Fonds Bourgouin du CHUS 2010-2011 (HK), and the Département de médecine interne du CHUS 2011-2012 (HK) for research funding. CR would like to thank the research programs Cyttron I and Cyttron II and the Delft Health Research Initiative for their support, MG and NK would like to thank the Manitoba Health Research Council for support.

REFERENCES

Abbe, E. K. (1873). Beiträge zur Theorie des Mikroskops und der mikroskopischen Wahrnehmung. *Archiv für Mikroskopische Anatomie* **9**, 413–468.

Agard, D. A., and Sedat, J. W. (1983). Three-dimensional architecture of a polytene nucleus. *Nature* **302**, 676–681.

Airy, G. B. (1835). On the diffraction of an object-glass with circular aperture. *Trans. Camb. Phil. Soc.* **5**, 283–292.

Apostolou, E., and Thanos, D. (2008). Virus infection induces NF-kappa-B-dependent interchromosomal associations mediating monoallelic IFN-beta gene expression. *Cell* **134**, 85–96.

Avet-Loiseau, H., Attal, M., Moreau, P., Charbonnel, C., Garban, F., Hulin, C., Leyvraz, S., Michallet, M., Yakoub-Agha, I., Garderet, L., Marit, G., Michaux, L., Voillat, L., Renaud, M., Grosbois, B., Guillerm, G., Benboubker, L., Monconduit, M., Thieblemont, C., Casassus, P., Caillot, D., Stoppa, A. M., Sotto, J. J., Wetterwald, M., Dumontet, C., Fuzibet, J. G., Azais, I., Dorvaux, V., Zandecki, M., Bataille, R., Minvielle, S., Harousseau, J. L., Facon, T., and Mathiot, C. (2007). Genetic abnormalities and survival in multiple myeloma: the experience of the Intergroupe Francophone du Myelome. *Blood* **109**, 3489–3495.

Avet-Loiseau, H., Leleu, X., Roussel, M., Moreau, P., Guerin-Charbonnel, C., Caillot, D., Marit, G., Benboubker, L., Voillat, L., Mathiot, C., Kolb, B., Macro, M., Campion, L., Wetterwald, M., Stoppa, A. M., Hulin, C., Facon, T., Attal, M., Minvielle, S., and Harousseau, J. L. (2010). Bortezomib plus dexamethasone induction improves outcome of patients with t(4;14) myeloma but not outcome of patients with del(17p). *J. Clin. Oncol.* **28**, 4630–4634.

Bachmann, H. S., Novotny, J., Sixt, S., Liebisch, P., Frey, U. H., Duhrsen, U., Siffert, W., and Nuckel, H. (2010). The G-Allele of the PSMA6-8C>G polymorphism is associated with poor outcome in multiple myeloma independently of circulating proteasome serum levels. *Eur. J. Haematol.* **85**, 108–113.

Barth, T. F., Martin-Subero, J. I., Joos, S., Menz, C. K., Hasel, C., Mechtersheimer, G., Parwaresch, R. M., Lichter, P., Siebert, R., and Mooller, P. (2003). Gains of 2p involving the REL locus correlate with nuclear c-Rel protein accumulation in neoplastic cells of classical Hodgkin lymphoma. *Blood* **101**, 3681–3686.

Bauman, J. G., Wiegant, J., Borst, P., and van Duijn, P. (1980). A new method for fluorescence microscopical localization of specific DNA sequences by *in situ* hybridization of fluorochromelabelled RNA. *Exp. Cell Res.* **128**, 485–490.

Bayani, J., Selvarajah, S., Maire, G., Vukovic, B., Al-Romaih, K., Zielenska, M., and Squire, J. A. (2007). Genomic mechanisms and measurement of structural and numerical instability in cancer cells. *Semin. Cancer Biol.* **17**, 5–18.

Berezney, R., and Coffey, D. S. (1974). Identification of a nuclear protein matrix. *Biochem Biophys. Res. Commun.* **60**, 1410–1417.

Berezney, R., and Coffey, D. S. (1977). Nuclear matrix. Isolation and characterization of a framework structure from rat liver nuclei. *J. Cell Biol.* **73**, 616–637.

Betzig, E., Patterson, G. H., Sougrat, R., Lindwasser, O. W., Olenych, S., Bonifacino, J. S., Davidson, M. W., Lippincott-Schwartz, J., and Hess, H. F. (2006). Imaging intracellular fluorescent proteins at nanometer resolution. *Science* **313**, 1642–1645.

Bickmore, W. A., and Teague, P. (2002). Influences of chromosome size, gene density and nuclear position on the frequency of constitutional translocations in the human population. *Chromosome Res.* **10**, 707–715.

Black, J. C., and Whetstine, J. R. (2011). Chromatin landscape: methylation beyond transcription. *Epigenetics* **6**, 9–15.

Bladé, J., Cibeira, M. T., Fernández de Larrea, C., and Rosiñol, L. (2010). Multiple myeloma. *Ann. Oncol.* **21**(Suppl 7), vii313–vii319.

Bohn, M., and Heermann, D. W. (2011). Repulsive forces between looping chromosomes induce entropy-driven segregation. *PLoS One* **6**, e14428.

Bolzer, A., Kreth, G., Solovei, I., Koehler, D., Saracoglu, K., Fauth, C., Muller, S., Eils, R., Cremer, C., Speicher, M. R., and Cremer, T. (2005). Three-dimensional maps of all chromosomes in human male fibroblast nuclei and prometaphase rosettes. *PLoS Biol.* **3**, e157.

Boschmann, M., Engeli, S., Moro, C., Luedtke, A., Adams, F., Gorzelniak, K., Rahn, G., Mahler, A., Dobberstein, K., Kruger, A., Schmidt, S., Spuler, S., Luft, F. C., Smith, S. R., Schmidt, H. H., and Jordan, J. (2010). LMNA mutations, skeletal muscle lipid metabolism, and insulin resistance. *J. Clin. Endocrinol. Metab.* **95**, 1634–1643.

Boveri, T. (1914). Zur Frage der Entstehung maligner Tumoren. Fischer, Jena.

Boveri, T. (1929). The origin of malignant tumors. Translated by Marcella Boveri. Baltimore: The Williams & Wilkens Company.

Branco, M. R., and Pombo, A. (2006). Intermingling of chromosome territories in interphase suggests role in translocations and transcription-dependent associations. *PLoS Biol.* **4**, e138.

Cannon Jr, G. M., Balasudramani, M., and Getzenberg, R. H. (2007). Characterization of nuclear matrix protein alterations associated with renal cell carcinoma. *Urology* **69**, 1227–1230.

Cao, H., and Hegele, R. A. (2000). Nuclear lamin A/C R482Q mutation in canadian kindreds with Dunnigan-type familial partial lipodystrophy. *Hum. Mol. Genet.* **9**(1), 109–112.

Capell, B. C., and Collins, F. S. (2006). Human laminopathies: nuclei gone genetically awry. *Nat. Rev. Genet.* **7**, 940–952.

Caporali, A., Wark, L., Vermolen, B. J., Garini, Y., and Mai, S. (2007). Telomeric aggregates and end-to-end chromosomal fusions require Myc box II. *Oncogene* **26**, 1398–1406.

Catez, F., Yang, H., Tracey, K. J., Reeves, R., Misteli, T., and Bustin, M. (2004). Network of dynamic interactions between histone H1 and high-mobility-group proteins in chromatin. *Mol. Cell Biol.* **24**, 4321–4328.

Cavalli, G. (2007). Chromosome kissing. *Curr. Opin. Genet. Dev.* **17**, 443–450.

Chadwick, B. P. (2008). DXZ4 chromatin adopts an opposing conformation to that of the surrounding chromosome and acquires a novel inactive X-specific role involving CTCF and antisense transcripts. *Genome Res.* **18**, 1259–1269.

Chalfie, M., Tu, Y., Euskirchen, G., Ward, W. W., and Prasher, D. C. (1994). Green fluorescent protein as a marker for gene expression. *Science* **263**, 802–805.

Chang, H., Qi, X., Jiang, A., Xu, W., Young, T., and Reece, D. (2010). 1p21 deletions are strongly associated with 1q21 gains and are an independent adverse prognostic factor for the outcome of high-dose chemotherapy in patients with multiple myeloma. *Bone Marrow Transplant* **45**, 117–121.

Chang, H., Qi, C., Yi, Q. L., Reece, D., and Stewart, A. K. (2005). p53 gene deletion detected by fluorescence *in situ* hybridization is an adverse prognostic factor for patients with multiple myeloma following autologous stem cell transplantation. *Blood* **105**, 358–360.

Chang, H., Trieu, Y., Qi, X., Jiang, N. N., Xu, W., and Reece, D. (2011). Impact of cytogenetics in patients with relapsed or refractory multiple myeloma treated with bortezomib: adverse effect of 1q21 gains. *Leuk. Res.* **35**, 95–98.

Chikashige, Y., Yamane, M., Okamasa, K., Tsutsumi, C., Kojidani, T., Sato, M., Haraguchi, T., and Hiraoka, Y. (2009). Membrane proteins Bqt3 and -4 anchor telomeres to the nuclear envelope to ensure chromosomal bouquet formation. *J. Cell Biol.* **187**, 413–427.

Chim, C. S., Wong, K. Y., Loong, F., Lam, W. W., and Srivastava, G. (2007). Frequent epigenetic inactivation of Rb1 in addition to p15 and p16 in mantle cell and follicular lymphoma. *Hum. Pathol.* **38**, 1849–1857.

Chng, W. J., Gertz, M. A., Chung, T. H., Van Wier, S., Keats, J. J., Baker, A., Bergsagel, P. L., Carpten, J., and Fonseca, R. (2010). Correlation between array-comparative genomic hybridization-defined genomic gains and losses and survival: identification of 1p31-32 deletion as a prognostic factor in myeloma. *Leukemia* **24**, 833–842.

Chuang, T. C., Moshir, S., Garini, Y., Chuang, A. Y., Young, I. T., Vermolen, B., van den Doel, R., Mougey, V., Perrin, M., Braun, M., Kerr, P. D., Fest, T., Boukamp, P., and Mai, S. (2004). The three-dimensional organization of telomeres in the nucleus of mammalian cells. *BioMed Central Biol.* **2**, 12.

Chui, D. T., Hammond, D., Baird, M., Shield, L., Jackson, R., and Jarrett, R. F. (2003). Classical Hodgkin lymphoma is associated with frequent gains of 17q. *Genes Chromosomes Cancer* **38**, 126–136.

Coffey, D. S. (2002). Nuclear matrix proteins as proteomic markers of preneoplastic and cancer lesions: commentary re: G. Brunagel *et al.*, nuclear matrix protein alterations associated with colon cancer metastasis to the liver. *Clin. Cancer Res.* **8**, 3031–3033.

Collins, V. P. (1998). Gliomas. *Cancer Surv.* **32**, 37–51.

Cornforth, M. N., Greulich-Bode, K. M., Loucas, B. D., Arsuaga, J., Vazquez, M., Sachs, R. K., Bruckner, M., Molls, M., Hahnfeldt, P., Hlatky, L., and Brenner, D. J. (2002). Chromosomes are predominantly located randomly with respect to each other in interphase human cells. *J. Cell Biol.* **159**, 237–244.

Cook, P. R., and Marenduzzo, D. (2009). Entropic organization of interphase chromosomes. *J. Cell Biol.* **186**, 825–834.

Corbett, K. D., Yip, C. K., Ee, L. S., Walz, T., Amon, A., and Harrison, S. C. (2010). The monopolin complex crosslinks kinetochore components to regulate chromosome-microtubule attachments. *Cell* **142**, 556–567.

Costa-Nunes, P., Pontes, O., Preuss, S. B., and Pikaard, C. S. (2010). Extra views on RNA-dependent DNA methylation and MBD6-dependent heterochromatin formation in nucleolar dominance. *Nucleus* **1**(3), 254–259.

Cotton, J. A., and Page, R. D. (2005). Rates and patterns of gene duplication and loss in the human genome. *Proc. Biol. Sci.* **272**, 277–283.

Cremer, M., von Hase, J., Volm, T., Brero, A., Kreth, G., Walter, J., Fischer, C., Solovei, I., Cremer, C., and Cremer, T. (2001). Non-random radial higher-order chromatin arrangements in nuclei of diploid human cells. *Chromosome Res.* **9**, 541–567.

Cremer, T., and Cremer, C. (2001). Chromosome territories, nuclear architecture and gene regulation in mammalian cells. *Nat. Rev. Genet.* **2**, 292–301.

Cremer, T., and Cremer, M. (2010). Chromosome territories. *Cold Spring Harb. Perspect. Biol.* **2**, a003889.

Crepaldi, L., and Riccio, A. (2009). Chromatin learns to behave. *Epigenetics* **4**, 23–26.

Cui, F., Sirotin, M. V., and Zhurkin, V. B. (2011). Impact of Alu repeats on the evolution of human p53 binding sites. *Biol. Direct.* **6**, 2.

de Lange, T. (1992). Human telomeres are attached to the nuclear matrix. *J. Eur. Mol. Biol. Org.* **11**, 717–724.

de Nooijer, S., Wellink, J., Mulder, B., and Bisseling, T. (2009). Non-specific interactions are sufficient to explain the position of heterochromatic chromocenters and nucleoli in interphase nuclei. *Nucleic Acids Res.* **37**, 3558–3568.

Dekker, J. (2008). Mapping in vivo chromatin interactions in yeast suggests an extended chromatin fiber with regional variation in compaction. *J. Biol. Chem.* **283**, 34532-34540.

Dekker, J., Rippe, K., Dekker, M., and Kleckner, N. (2002). Capturing chromosome conformation. *Science* **295**, 1306–1311.

Del Rey, J., Prat, E., Ponsa, I., Lloreta, J., Gelabert, A., Algaba, F., Camps, J., and Miro, R. (2010). Centrosome clustering and cyclin D1 gene amplification in double minutes are common events in chromosomal unstable bladder tumors. *BioMed Central Cancer* **10**, 280.

DePinho, R. A., and Polyak, K. (2004). Cancer chromosomes in crisis. *Nat. Genet.* **36**, 932–934.

De Sandre-Giovannoli, A., Bernard, R., Cau, P., Navarro, C., Amiel, J., Boccaccio, I., Lyonnet, S., Stewart, C. L., Munnich, A., Le Merrer, M., and Levy, N. (2003). Lamin a truncation in Hutchinson–Gilford progeria. *Science* **300**, 2055.

Dimitriadis, E. K., Weber, C., Gill, R. K., Diekmann, S., and Dalal, Y. (2010). Tetrameric organization of vertebrate centromeric nucleosomes. *Proc. Natl. Acad. Sci. U.S.A.* **107**, 20317–20322.

Dimitrova, N., Chen, Y. C., Spector, D. L., and de Lange, T. (2008). 53BP1 promotes nonhomologous end joining of telomeres by increasing chromatin mobility. *Nature* **456**, 524–528.

Dropcho, E. J., and Soong, S. J. (1996). The prognostic impact of prior low grade histology in patients with anaplastic gliomas: a case-control study. *Neurology* **47**, 684–690.

Dumont, J., Oegema, K., and Desai, A. (2010). A kinetochore-independent mechanism drives anaphase chromosome separation during acentrosomal meiosis. *Nat. Cell Biol.* **12**, 894–901.

Esteller, M., Garcia-Foncillas, J., Andion, E., Goodman, S. N., Hidalgo, O. F., Vanaclocha, V., Baylin, S. B., and Herman, J. G. (2000). Inactivation of the DNA-repair gene MGMT and the clinical response of gliomas to alkylating agents. *N. Engl. J. Med.* **343**, 1350–1354.

Facon, T., Avet-Loiseau, H., Guillerm, G., Moreau, P., Genevieve, F., Zandecki, M., Lai, J. L., Leleu, X., Jouet, J. P., Bauters, F., Harousseau, J. L., Bataille, R., and Mary, J. Y. (2001). Chromosome 13 abnormalities identified by FISH analysis and serum beta2-microglobulin produce a powerful myeloma staging system for patients receiving high-dose therapy. *Blood* **97**, 1566–1571.

Falzetti, D., Crescenzi, B., Matteuci, C., Falini, B., Martelli, M. F., Van Den Berghe, H., and Mecucci, C. (1999). Genomic instability and recurrent breakpoints are main cytogenetic findings in Hodgkin's disease. *Haematologica* **84**, 298–305.

Felsher, D. W., and Bishop, J. M. (1999). Transient excess of Myc activity can elicit genomic instability and tumorigenesis. *Proc. Natl. Acad. Sci. U.S.A.* **96**, 3940–3944.

Fest, T., Guffei, A., Williams, G., Silva, S., and Mai, S. (2005). Uncoupling of genomic instability and tumorigenesis in a mouse model of Burkitt's lymphoma expressing a conditional box II-deleted Myc protein. *Oncogene* **24**, 2944–2953.

Feys, T., Poppe, B., De Preter, K., Van Roy, N., Verhasselt, B., De Paepe, P., De Paepe, A., and Speleman, F. (2007). A detailed inventory of DNA copy number alterations in four commonly used Hodgkin's lymphoma cell lines. *Haematologica* **92**, 913–920.

Fitzgerald, J., Moureau, S., Drogaris, P., O'Connell, E., Abshiru, N., Verreault, A., Thibault, P., Grenon, M., and Lowndes, N. F. (2011). Regulation of the DNA damage response and gene expression by the Dot1L histone methyltransferase and the 53Bp1 tumour suppressor. *PLoS One* **6**, e14714.

Fleming, T. P., Saxena, A., Clark, W. C., Robertson, J. T., Oldfield, E. H., Aaronson, S. A., and Ali, I. U. (1992). Amplification and/or overexpression of platelet-derived growth

factor receptors and epidermal growth factor receptor in human glial tumors. *Cancer Res.* **52**, 4550–4553.
Folle, G. A. (2008). Nuclear architecture, chromosome domains and genetic damage. *Mutat. Res.* **658**, 172–183.
Fonseca, R., Van Wier, S. A., Chng, W. J., Ketterling, R., Lacy, M. Q., Dispenzieri, A., Bergsagel, P. L., Rajkumar, S. V., Greipp, P. R., Litzow, M. R., Price-Troska, T., Henderson, K. J., Ahmann, G. J., and Gertz, M. A. (2006). Prognostic value of chromosome 1q21 gain by fluorescent *in situ* hybridization and increase CKS1B expression in myeloma. *Leukemia* **20**, 2034–2040.
Fransz, P., De Jong, J. H., Lysak, M., Castiglione, M. R., and Schubert, I. (2002). Interphase chromosomes in Arabidopsis are organized as well defined chromocenters from which euchromatin loops emanate. *Proc. Natl. Acad. Sci. U.S.A.* **99**, 14584–14589.
Friedman, H. S., Kerby, T., and Calvert, H. (2000). Temozolomide and treatment of malignant glioma. *Clin. Cancer Res.* **6**, 2585–2597.
Fujisawa, H., Kurrer, M., Reis, R. M., Yonekawa, Y., Kleihues, P., and Ohgaki, H. (1999). Acquisition of the glioblastoma phenotype during astrocytoma progression is associated with loss of heterozygosity on 10q25-qter. *Am. J. Pathol.* **155**, 387–394.
Fults, D., and Pedone, C. (1993). Deletion mapping of the long arm of chromosome 10 in glioblastoma multiforme. *Genes Chromosomes Cancer* **7**, 173–177.
Furnari, F. B., Fenton, T., Bachoo, R. M., Mukasa, A., Stommel, J. M., Stegh, A., Hahn, W. C., Ligon, K. L., Louis, D. N., Brennan, C., Chin, L., DePinho, R. A., and Cavenee, W. K. (2007). Malignant astrocytic glioma: genetics, biology, and paths to treatment. *Genes Dev.* **21**, 2683–2710.
Gadji, M., Crous, A. M., Fortin, D., Krcek, J., Torchia, M., Mai, S., Drouin, R., and Klonisch, T. (2009). EGF receptor inhibitors in the treatment of glioblastoma multiform: old clinical allies and newly emerging therapeutic concepts. *Eur. J. Pharmacol.* **625**, 23–30.
Gadji, M., Fortin, D., Tsanaclis, A. M., Garini, Y., Katzir, N., Wienburg, Y., Yan, J., Klewes, L., Klonisch, T., Drouin, R., and Mai, S. (2010). Three-dimensional nuclear telomere architecture is associated with differential time to progression and overall survival in glioblastoma patients. *Neoplasia* **12**, 183–191.
Garini, Y., Vermolen, B. J., and Young, I. T. (2005). From micro to nano: recent advances in high-resolution microscopy. *Curr. Opin. Biotechnol.* **16**, 3–12.
Gazave, E., Gautier, P., Gilchrist, S., and Bickmore, W. A. (2005). Does radial nuclear organisation influence DNA damage? *Chromosome Res.* **13**, 377–388.
Gent, J. I., Schneider, K. L., Topp, C. N., Rodriguez, C., Presting, G. G., and Dawe, R. K. (2011). Distinct influences of tandem repeats and retrotransposons on CENH3 nucleosome positioning. *Epigenetics Chromatin* **4**, 3.
Gerace, L., and Burke, B. (1988). Functional organization of the nuclear envelope. *Annu. Rev. Cell Biol.* **4**, 335–374.
Gerlitz, G. (2010). HMGNs, DNA repair and cancer. *Biochem. Biophys. Acta* **1799**, 80–85.
Gertz, M. A., Lacy, M. Q., Dispenzieri, A., Greipp, P. R., Litzow, M. R., Henderson, K. J., Van Wier, S. A., Ahmann, G. J., and Fonseca, R. (2005). Clinical implications of t(11;14)(q13;q32), t(4;14)(p16.3;q32), and -17p13 in myeloma patients treated with high-dose therapy. *Blood* **106**, 2837–2840.
Gonzalez, R. C., and Woods, R. E. (2008). Digital Image Processing. Pearson, New York.
Gonzalez-Suarez, I., Redwood, A. B., Perkins, S. M., Vermolen, B., Lichtensztejin, D., Grotsky, D. A., Morgado-Palacin, L., Gapud, E. J., Sleckman, B. P., Sullivan, T., Sage, J., Stewart, C. L., Mai, S., and Gonzalo, S. (2009). Novel roles for A-type lamins in telomere biology and the DNA damage response pathway. *J. Eur. Mol. Biol. Org.* **28**, 2414–2427.

Groehler, A. L., and Lannigan, D. A. (2010). A chromatin-bound kinase, ERK8, protects genomic integrity by inhibiting HDM2-mediated degradation of the DNA clamp PCNA. *J. Cell Biol.* **190**, 575–586.

Gruenbaum, Y., Margalit, A., Goldman, R. D., Shumaker, D. K., and Wilson, K. L. (2005). The nuclear lamina comes of age. *Nat. Rev. Mol. Cell Biol.* **6**, 21–31.

Guffei, A., Lichtensztejn, Z., Goncalves Dos Santos Silva, A., Louis, S. F., Caporali, A., and Mai, S. (2007). c-Myc-dependent formation of Robertsonian translocation chromosomes in mouse cells. *Neoplasia* **9**, 578–588.

Guffei, A., Sarkar, R., Klewes, L., Righolt, C., Knecht, H., and Mai, S. (2010). Dynamic chromosomal rearrangements in Hodgkin's lymphoma are due to ongoing three-dimensional nuclear remodeling and breakage-bridge fusion cycles. *Haematologica* **95**, 2038–2046.

Guijon, F. B., Greulich-Bode, K., Paraskevas, M., Baker, P., and Mai, S. (2007). Premalignant cervical lesions are characterized by dihydrofolate reductase gene amplification and c-Myc overexpression: possible biomarkers. *J. Low Genit. Tract Dis.* **11**, 265–272.

Gurney, J. G., and Kadan-Lottick, N. (2001). Brain and other central nervous system tumors: rates, trends, and epidemiology. *Curr. Opin. Oncol.* **13**, 160–166.

Gustafsson, M. G. (2000). Surpassing the lateral resolution limit by a factor of two using structured illumination microscopy. *J. Microsc.* **198**, 82–87.

Halicka, H. D., Zhao, H., Podhorecka, M., Traganos, F., and Darzynkiewicz, Z. (2009). Cytometric detection of chromatin relaxation, an early reporter of DNA damage response. *Cell Cycle* **8**, 2233–2237.

Hanahan, D., and Weinberg, R. A. (2000). The hallmarks of cancer. *Cell* **100**, 57–70.

Hanahan, D., and Weinberg, R. A. (2011). Hallmarks of cancer: the next generation. *Cell* **144**, 646–674.

Harrison, C. J., Mazzullo, H., Ross, F. M., Cheung, K. L., Gerrard, G., Harewood, L., Mehta, A., Lachmann, H. J., Hawkins, P. N., and Orchard, K. H. (2002). Translocations of 14q32 and deletions of 13q14 are common chromosomal abnormalities in systemic amyloidosis. *Br. J. Haematol.* **117**, 427–435.

Hartmann, S., Martin-Subero, J. I., Gesk, S., Husken, J., Giefing, M., Nagel, I., Riemke, J., Chott, A., Klapper, W., Parrens, M., Merlio, J. P., Kuppers, R., Brauninger, A., Siebert, R., and Hansmann, M. L. (2008). Detection of genomic imbalances in microdissected Hodgkin and Reed–Sternberg cells of classical Hodgkin's lymphoma by array-based comparative genomic hybridization. *Haematologica* **93**, 1318–1326.

Hegele, R. A., Cao, H., Anderson, C. M., and Hramiak, I. M. (2000a). Heterogeneity of nuclear lamin A mutations in Dunnigan-type familial partial lipodystrophy. *J. Clin. Endocrinol. Metab.* **85**, 3431–3435.

Hegele, R. A., Cao, H., Huff, M. W., and Anderson, C. M. (2000b). LMNA R482Q mutation in partial lipodystrophy associated with reduced plasma leptin concentration. *J. Clin. Endocrinol. Metab.* **85**, 3089–3093.

Hegi, M. E., Diserens, A. C., Gorlia, T., Hamou, M. F., de Tribolet, N., Weller, M., Kros, J. M., Hainfellner, J. A., Mason, W., Mariani, L., Bromberg, J. E., Hau, P., Mirimanoff, R. O., Cairncross, J. G., Janzer, R. C., and Stupp, R. (2005). MGMT gene silencing and benefit from temozolomide in glioblastoma. *N. Engl. J. Med.* **352**, 997–1003.

Hell, S. W. (2007). Far-field optical nanoscopy. *Science* **316**, 1153–1158.

Hell, S. W., and Wichmann, J. (1994). Breaking the diffraction resolution limit by stimulated emission: stimulated-emission-depletion fluorescence microscopy. *Opt. Lett.* **19**, 780–782.

Herzog, A., and Schutze, H. R. (1968). [Fluorescent staining of mammalian chromosomes with a basically substituted bis-benzimidazole derivative (preparation 33258 "Hoechst")]. *Dtsch Tierarztl Wochenschr* **75**, 476–479.

Holland, A. J., and Cleveland, D. W. (2009). Boveri revisited: chromosomal instability, aneuploidy and tumorigenesis. *Nat. Rev. Mol. Cell Biol.* **10**, 478–487.

Holland, E. C. (2001). Gliomagenesis: genetic alterations and mouse models. *Nat. Rev. Genet.* **2**, 120–129.

Holley, W. R., Mian, I. S., Park, S. J., Rydberg, B., and Chatterjee, A. (2002). A model for interphase chromosomes and evaluation of radiation-induced aberrations. *Radiat. Res.* **158**, 568–580.

Hommura, F., Katabami, M., Leaner, V. D., Donninger, H., Sumter, T. F., Resar, L. M., and Birrer, M. J. (2004). HMG-I/Y is a c-Jun/activator protein-1 target gene and is necessary for c-Jun-induced anchorage-independent growth in Rat1a cells. *Mol. Cancer Res.* **2**, 305–314.

Horvilleur, E., Wilson, L. A., and Willis, A. E. (2010). Translation deregulation in B-cell lymphomas. *Biochem. Soc. Trans.* **38**, 1593–1597.

Hooke, R. (1665). Micrographia: or Some physiological descriptions of minute bodies made by magnifying glasses. J. Martyn and J. Allestry, London.

Hu, Y., Kireev, I., Plutz, M., Ashourian, N., and Belmont, A. S. (2009). Large-scale chromatin structure of inducible genes: transcription on a condensed, linear template. *J. Cell Biol.* **185**, 87–100.

Huberman, J. A. (2011). Subtle interactions between heterochromatin and DNA replication timing. *Cell Cycle.* **10**, 873–878.

Hutchinson, J. (1886). Case of congenital absence of hair, with atrophic condition of the skin and its apendages. *Lancet* **1**, 923.

Ichimura, K., Schmidt, E. E., Goike, H. M., and Collins, V. P. (1996). Human glioblastomas with no alterations of the CDKN2A (p16INK4A, MTS1) and CDK4 genes have frequent mutations of the retinoblastoma gene. *Oncogene* **13**, 1065–1072.

Iijima, K., Ohara, M., Seki, R., and Tauchi, H. (2008). Dancing on damaged chromatin: functions of ATM and the RAD50/MRE11/NBS1 complex in cellular responses to DNA damage. *J. Radiat. Res. (Tokyo)* **49**, 451–464.

Jacob, K. N., and Garg, A. (2006). Laminopathies: multisystem dystrophy syndromes. *Mol. Genet. Metab.* **87**, 289–302.

Jansen, L. E., Black, B. E., Foltz, D. R., and Cleveland, D. W. (2007). Propagation of centromeric chromatin requires exit from mitosis. *J. Cell Biol.* **176**, 795–805.

Jansen, M. P., Hopman, A. H., Bot, F. J., Haesevoets, A., Stevens-Kroef, M. J., Arends, J. W., Jox, A., Wolf, J., Ramaekers, F. C., and Schouten, H. C. (1999). Morphologically normal, CD30-negative B-lymphocytes with chromosome aberrations in classical Hodgkin's disease: the progenitor cell of the malignant clone? *J. Pathol.* **189**, 527–532.

Jiang, W. Q., Nguyen, A., Cao, Y., Chang, A. C., and Reddel, R. R. (2011). HP1-mediated formation of alternative lengthening of telomeres-associated PML bodies requires HIRA but not ASF1a. *PLoS One* **6**, e17036.

Joos, S., Menz, C. K., Wrobel, G., Siebert, R., Gesk, S., Ohl, S., Mechtersheimer, G., Trumper, L., Moller, P., Lichter, P., and Barth, T. F. (2002). Classical Hodgkin lymphoma is characterized by recurrent copy number gains of the short arm of chromosome 2. *Blood* **99**, 1381–1387.

Kalitsis, P., Griffiths, B., and Choo, K. H. (2006). Mouse telocentric sequences reveal a high rate of homogenization and possible role in Robertsonian translocation. *Proc. Natl. Acad. Sci. U.S.A.* **103**, 8786–8791.

Kang, J., Chaudhary, J., Dong, H., Kim, S., Brautigam, C. A., and Yu, H. (2011). Mitotic centromeric targeting of HP1 and its binding to Sgo1 are dispensable for sister–chromatid cohesion in human cells. *Mol. Biol. Cell* **8**, 1181–1190.

Kanikarla-Marie, P., Ronald, S., and De Benedetti, A. (2011). Nucleosome resection at a double-strand break during non-homologous ends joining in mammalian cells—implications from repressive chromatin organization and the role of ARTEMIS. *BioMed. Central Res. Notes* **4**, 13.

Kapuscinski, J., and Skoczylas, B. (1978). Fluorescent complexes of DNA with DAPI 4',6-diamidine-2-phenyl indole.2HCl or DCI 4',6-dicarboxyamide-2-phenyl indole. *Nucleic Acids Res.* **5**(10), 3775–3799.

Karlin, L., Soulier, J., Chandesris, O., Choquet, S., Belhadj, K., Macro, M., Bouscary, D., Porcher, R., Ghez, D., Malphettes, M., Asli, B., Brouet, J. C., Bories, J. C., Hermine, O., Fermand, J. P., and Arnulf, B. (2011). Clinical and biological features of t(4;14) multiple myeloma: a prospective study. *Leuk. Lymphoma* **52**, 238–246.

Kioussis, D. (2005). Gene regulation: kissing chromosomes. *Nature* **435**, 579–580.

Kleckner, N., Zickler, D., Jones, G. H., Dekker, J., Padmore, R., Henle, J., and Hutchinson, J. (2004). A mechanical basis for chromosome function. *Proc. Natl. Acad. Sci. U S A* **101** (34), 12592–12597.

Kleihues, P., and Ohgaki, H. (1999). Primary and secondary glioblastomas: from concept to clinical diagnosis. *Neuro. Oncol.* **1**(1), 44–51.

Klewes, L., Hobsch, C., Katzir, N., Rourke, D., Garini, Y., and Mai, S. (2011). Novel automated three-dimensional genome scanning based on the nuclear architecture of telomeres. *Cytometry A* **79**, 159–166.

Knecht, H., Bruderlein, S., Mai, S., Moller, P., and Sawan, B. (2010a). 3D structural and functional characterization of the transition from Hodgkin to Reed–Sternberg cells. *Ann. Anat.* **192**, 302–308.

Knecht, H., Sawan, B., Lichtensztejn, Z., Lichtensztejn, D., and Mai, S. (2010b). 3D Telomere FISH defines LMP1-expressing Reed–Sternberg cells as end-stage cells with telomere-poor 'ghost' nuclei and very short telomeres. *Lab. Invest.* **90**, 611–619.

Knecht, H., Sawan, B., Lichtensztejn, D., Lemieux, B., Wellinger, R. J., and Mai, S. (2009). The 3D nuclear organization of telomeres marks the transition from Hodgkin to Reed–Sternberg cells. *Leukemia* **23**, 565–573.

Knecht, H., Bachmann, E., Brousset, P., Sandvej, K., Nadal, D., Bachmann, F., Odermatt, B. F., Delsol, G., and Pallesen, G. (1993). Deletions within the LMP1 oncogene of Epstein–Barr virus are clustered in Hodgkin's disease and identical to those observed in nasopharyngeal carcinoma. *Blood* **82**, 2937–2942.

Koszul, R., and Kleckner, N. (2009). Dynamic chromosome movements during meiosis: a way to eliminate unwanted connections? *Trends Cell Biol.* **19**, 716–724.

Kozubek, M., Kozubek, S., Lukasova, E., Mareckova, A., Bartova, E., Skalnikova, M., and Jergova, A. (1999a). High-resolution cytometry of FISH dots in interphase cell nuclei. *Cytometry* **36**, 279–293.

Kozubek, S., Lukasova, E., Mareckova, A., Skalnikova, M., Kozubek, M., Bartova, E., Kroha, V., Krahulcova, E., and Slotova, J. (1999b). The topological organization of chromosomes 9 and 22 in cell nuclei has a determinative role in the induction of t(9,22) translocations and in the pathogenesis of t(9,22) leukemias. *Chromosoma* **108**, 426–435.

Kreth, G., Munkel, C., Langowski, J., Cremer, T., and Cremer, C. (1998). Chromatin structure and chromosome aberrations: modeling of damage induced by isotropic and localized irradiation. *Mutat. Res.* **404**, 77–88.

Krull, S., Dörries, J., Boysen, B., Reidenbach, S., Magnius, L., Norder, H., Thyberg, J., and Cordes, V. C. (2010). Protein Tpr is required for establishing nuclear pore-associated zones of heterochromatin exclusion. *J. Eur. Mol. Biol. Org.* **29**, 1659–1673.

Kulkarni, A., Zschenker, O., Reynolds, G., Miller, D., and Murnane, J. P. (2010). Effect of telomere proximity on telomere position effect, chromosome healing, and sensitivity to DNA double-strand breaks in a human tumor cell line. *Mol. Cell. Biol.* **30**, 578–589.

Kuroda, M., Tanabe, H., Yoshida, K., Oikawa, K., Saito, A., Kiyuna, T., Mizusawa, H., and Mukai, K. (2004). Alteration of chromosome positioning during adipocyte differentiation. *J. Cell Sci.* **117**, 5897–5903.

Kuschak, T. I., Taylor, C., McMillan-Ward, E., Israels, S., Henderson, D. W., Mushinski, J. F., Wright, J. A., and Mai, S. (1999). The ribonucleotide reductase R2 gene is a non-transcribed target of c-Myc-induced genomic instability. *Gene* **238**, 351–365.
Kyle, R. A., Durie, B. G., Rajkumar, S. V., Landgren, O., Blade, J., Merlini, G., Kroger, N., Einsele, H., Vesole, D. H., Dimopoulos, M., San Miguel, J., Avet-Loiseau, H., Hajek, R., Chen, W. M., Anderson, K. C., Ludwig, H., Sonneveld, P., Pavlovsky, S., Palumbo, A., Richardson, P. G., Barlogie, B., Greipp, P., Vescio, R., Turesson, I., Westin, J., and Boccadoro, M. (2010). Monoclonal gammopathy of undetermined significance (MGUS) and smoldering (asymptomatic) multiple myeloma: IMWG consensus perspectives risk factors for progression and guidelines for monitoring and management. *Leukemia* **24**, 1121–1127.
Kyle, R. A., and Rajkumar, S. V. (2004). Multiple myeloma. *N. Engl. J. Med.* **351**, 1860–1873.
Kyle, R. A., Therneau, T. M., Rajkumar, S. V., Larson, D. R., Plevak, M. F., Offord, J. R., Dispenzieri, A., Katzmann, J. A., and Melton 3rd, L. J. (2006). Prevalence of monoclonal gammopathy of undetermined significance. *N Engl. J. Med.* **354**, 1362–1369.
Lacoste, S., Wiechec, E., Dos Santos Silva, A. G., Guffei, A., Williams, G., Lowbeer, M., Benedek, K., Henriksson, M., Klein, G., and Mai, S. (2010). Chromosomal rearrangements after ex vivo Epstein–Barr virus (EBV) infection of human B cells. *Oncogene* **29**, 503–515.
Landgren, O., and Weiss, B. M. (2009). Patterns of monoclonal gammopathy of undetermined significance and multiple myeloma in various ethnic/racial groups: support for genetic factors in pathogenesis. *Leukemia* **23**, 1691–1697.
Lansdorp, P. M. (2009). Telomeres and disease. *J. Eur. Mol. Biol. Org.* **28**, 2532–2540.
Lansdorp, P. M., Verwoerd, N. P., van de Rijke, F. M., Dragowska, V., Little, M. T., Dirks, R. W., Raap, A. K., and Tanke, H. J. (1996). Heterogeneity in telomere length of human chromosomes. *Hum. Mol. Genet.* **5**, 685–691.
LaSalle, J. M., and Lalande, M. (1996). Homologous association of oppositely imprinted chromosomal domains. *Science* **272**, 725–728.
Lazebnik, Y. (2010). What are the hallmarks of cancer? *Nat. Rev. Cancer* **10**, 232–233.
Lee, K. M., Baris, D., Zhang, Y., Hosgood 3rd, H. D., Menashe, I., Yeager, M., Zahm, S. H., Wang, S. S., Purdue, M. P., Chanock, S., Zheng, T., Rothman, N., and Lan, Q. (2010). Common single nucleotide polymorphisms in immunoregulatory genes and multiple myeloma risk among women in Connecticut. *Am. J. Hematol.* **85**, 560–563.
Leman, E. S., Madigan, M. C., Brunagel, G., Takaha, N., Coffey, D. S., and Getzenberg, R. H. (2003). Nuclear matrix localization of high mobility group protein I(Y) in a transgenic mouse model for prostate cancer. *J. Cell Biochem.* **88**, 599–608.
Lemos, B., Branco, A. T., and Hartl, D. L. (2010). Epigenetic effects of polymorphic Y chromosomes modulate chromatin components, immune response, and sexual conflict. *Proc. Natl. Acad. Sci. U.S.A.* **107**, 15826–15831.
Liu, Y., and Chang, A. (2006). DNA damage responses in progeroid syndromes arise from defective maturation of prelamin A. *J. Cell Sci.* **119**, 360–369.
Lo, P. K., and Sukumar, S. (2008). Epigenomics and breast cancer. *Pharmacogenomics* **12**, 1879–1902.
Lopez-Corral, L., Gutierrez, N. C., Vidriales, M. B., Mateos, M. V., Rasillo, A., Garcia-Sanz, R., Pavia, B., and San Miguel, J. (2011). The progression from MGUS to smoldering myeloma and eventually to multiple myeloma involves a clonal expansion of genetically abnormal plasma cells. *Clin. Cancer Res.* **17**, 1692–1700.
Louis, D. N., Ohgaki, H., Wiesler, O. D., and Cavenee, W. K. (2007). WHO classifiation of tumours of the central nervous system. WHO Press, Geneva.
Louis, D. N., Pomeroy, S. L., and Cairncross, J. G. (2002). Focus on central nervous system neoplasia. *Cancer Cell* **1**, 125–128.

Louis, S. F., Vermolen, B. J., Garini, Y., Young, I. T., Guffei, A., Lichtensztejn, Z., Kuttler, F., Chuang, T. C., Moshir, S., Mougey, V., Chuang, A. Y., Kerr, P. D., Fest, T., Boukamp, P., and Mai, S. (2005). c-Myc induces chromosomal rearrangements through telomere and chromosome remodeling in the interphase nucleus. *Proc. Natl. Acad. Sci. U.S.A.* **102**, 9613–9618.

Luderus, M. E., van Steensel, B., Chong, L., Sibon, O. C., Cremers, F. F., and de Lange, T. (1996). Structure, subnuclear distribution and nuclear matrix association of the mammalian telomeric complex. *J. Cell Biol.* **135**, 867–881.

Lunyak, V. V., Prefontaine, G. G., Núñez, E., Cramer, T., Ju, B. G., Ohgi, K. A., Hutt, K., Roy, R., García-Díaz, A., Zhu, X., Yung, Y., Montoliu, L., Glass, C. K., and Rosenfeld, M. G. (2007). Developmentally regulated activation of a SINE B2 repeat as a domain boundary in organogenesis. *Science* **317**, 248–251.

Luo, L., Gassman, K. L., Petell, L. M., Wilson, C. L., Bewersdorf, J., and Shopland, L. S. (2009). The nuclear periphery of embryonic stem cells is a transcriptionally permissive and repressive compartment. *J. Cell Sci.* **122**(Pt 20), 3729–3737.

Ma, H., Siegel, A. J., and Berezney, R. (1999). Association of chromosome territories with the nuclear matrix. Disruption of human chromosome territories correlates with the release of a subset of nuclear matrix proteins. *J. Cell Biol.* **146**, 531–542.

MacLeod, R. A., Spitzer, D., Bar-Am, I., Sylvester, J. E., Kaufmann, M., Wernich, A., and Drexler, H. G. (2000). Karyotypic dissection of Hodgkin's disease cell lines reveals ectopic subtelomeres and ribosomal DNA at sites of multiple jumping translocations and genomic amplification. *Leukemia* **14**, 1803–1814.

Mai, S. (1994). Overexpression of c-Myc precedes amplification of the gene encoding dihydrofolate reductase. *Gene* **148**, 253–260.

Mai, S. (2010). Initiation of telomere-mediated chromosomal rearrangements in cancer. *J. Cell Biochem.* **109**, 1095–1102.

Mai, S., Fluri, M., Siwarski, D., and Huppi, K. (1996a). Genomic instability in MycER-activated Rat1A-MycER cells. *Chromosome Res.* **4**, 365–371.

Mai, S., and Garini, Y. (2005). Oncogenic remodeling of the three-dimensional organization of the interphase nucleus: c-Myc induces telomeric aggregates whose formation precedes chromosomal rearrangements. *Cell Cycle* **4**, 1327–1331.

Mai, S., and Garini, Y. (2006). The significance of telomeric aggregates in the interphase nuclei of tumor cells. *J. Cell Biochem.* **97**, 904–915.

Mai, S., Hanley-Hyde, J., and Fluri, M. (1996b). c-Myc overexpression associated DHFR gene amplification in hamster, rat, mouse and human cell lines. *Oncogene* **12**, 277–288.

Mai, S., Hanley-Hyde, J., Rainey, G. J., Kuschak, T. I., Paul, J. T., Littlewood, T. D., Mischak, H., Stevens, L. M., Henderson, D. W., and Mushinski, J. F. (1999). Chromosomal and extrachromosomal instability of the cyclin D2 gene is induced by Myc overexpression. *Neoplasia* **1**, 241–252.

Mannini, L., Menga, S., and Musio, A. (2010). The expanding universe of cohesin functions: a new genome stability caretaker involved in human disease and cancer. *Hum. Mutat.* **31**, 623–630.

Marella, N. V., Seifert, B., Nagarajan, P., Sinha, S., and Berezney, R. (2009). Chromosomal rearrangements during human epidermal keratinocyte differentiation. *J. Cell Physiol.* **221**, 139–146.

Marteau, J. B., Rigaud, O., Brugat, T., Gault, N., Vallat, L., Kruhoffer, M., Orntoft, T. F., Nguyen-Khac, F., Chevillard, S., Merle-Beral, H., and Delic, J. (2010). Concomitant heterochromatinisation and down-regulation of gene expression unveils epigenetic silencing of RELB in an aggressive subset of chronic lymphocytic leukemia in males. *BioMed Central Med. Genomics* **3**, 53.

Martin-Subero, J. I., Klapper, W., Sotnikova, A., Callet-Bauchu, E., Harder, L., Bastard, C., Schmitz, R., Grohmann, S., Hoppner, J., Riemke, J., Barth, T. F., Berger, F., Bernd, H. W., Claviez, A., Gesk, S., Frank, G. A., Kaplanskaya, I. B., Moller, P., Parwaresch, R. M., Rudiger, T., Stein, H., Kuppers, R., Hansmann, M. L., and Siebert, R. (2006a). Chromosomal breakpoints affecting immunoglobulin loci are recurrent in Hodgkin and Reed–Sternberg cells of classical Hodgkin lymphoma. *Cancer Res.* **66**, 10332–10338.

Martin-Subero, J. I., Wlodarska, I., Bastard, C., Picquenot, J. M., Hoppner, J., Giefing, M., Klapper, W., and Siebert, R. (2006b). Chromosomal rearrangements involving the BCL3 locus are recurrent in classical Hodgkin and peripheral T-cell lymphoma. *Blood* **108**, 401–402.

Martinez, R., Schackert, H. K., Plaschke, J., Baretton, G., Appelt, H., and Schackert, G. (2004). Molecular mechanisms associated with chromosomal and mcirosatellite instability in sporadic gliobalstoma multiforme. *Oncology* **66**, 395–403.

Martinez-Garcia, E., Popovic, R., Min, D. J., Sweet, S. M., Thomas, P. M., Zamdborg, L., Heffner, A., Will, C., Lamy, L., Staudt, L. M., Levens, D. L., Kelleher, N. L., and Licht, J. D. (2011). The MMSET histone methyl transferase switches global histone methylation and alters gene expression in t(4;14) multiple myeloma cells. *Blood* **117**, 211–220.

Mateos-Langerak, J., Bohn, M., de Leeuw, W., Giromus, O., Manders, E. M., Verschure, P. J., Indemans, M. H., Gierman, H. J., Heermann, D. W., van Driel, R., and Goetze, S. (2009). Spatially confined folding of chromatin in the interphase nucleus. *Proc. Natl. Acad. Sci. U.S.A.* **106**, 3812–3817.

Meaburn, K. J., Cabuy, E., Bonne, G., Levy, N., Morris, G. E., Novelli, G., Kill, I. R., and Bridger, J. M. (2007). Primary laminopathy fibroblasts display altered genome organization and apoptosis. *Aging Cell* **6**, 139–153.

Meeker, A. K., Hicks, J. L., Gabrielson, E., Strauss, W. M., De Marzo, A. M., and Argani, P. (2004). Telomere shortening occurs in subsets of normal breast epithelium as well as *in situ* and invasive carcinoma. *Am. J. Pathol.* **164**, 925–935.

Mehta, I. S., Amira, M., Harvey, A. J., and Bridger, J. M. (2010). Rapid chromosome territory relocation by nuclear motor activity in response to serum removal in primary human fibroblasts. *Genome Biol.* **11**, R5.

Messick, T. E., and Greenberg, R. A. (2009). The ubiquitin landscape at DNA double-strand breaks. *J. Cell Biol.* **187**, 319–326.

Minsky M. (1961). Microscopy apparatus. U.S. Patent 3, 467, editor.

Misteli, T., and Soutoglou, E. (2009). The emerging role of nuclear architecture in DNA repair and genome maintenance. *Nat. Rev. Mol. Cell Biol.* **10**, 243–254.

Mitsiades, C. S., and Anderson, K. C. (2009). Epigenetic modulation in hematologic malignancies: challenges and progress. *J. Natl. Compr. Canc. Netw.* **7**(Suppl. 8), S1–12.

Mounkes, L., Kozlov, S., Burke, B., and Stewart, C. L. (2003). The laminopathies: nuclear structure meets disease. *Curr. Opin. Genet. Dev.* **13**, 223–230.

Mozziconacci, J., Lavelle, C., Barbi, M., Lesne, A., and Victor, J. M. (2006). A physical model for the condensation and decondensation of eukaryotic chromosomes. *Fedration Eur. Biochem. Soc. Lett.* **580**, 368–372.

Murat, A., Migliavacca, E., Gorlia, T., Lambiv, W. L., Shay, T., Hamou, M. F., de Tribolet, N., Regli, L., Wick, W., Kouwenhoven, M. C., Hainfellner, J. A., Heppner, F. L., Dietrich, P. Y., Zimmer, Y., Cairncross, J. G., Janzer, R. C., Domany, E., Delorenzi, M., Stupp, R., and Hegi, M. E. (2008). Stem cell-related "self-renewal" signature and high epidermal growth factor receptor expression associated with resistance to concomitant chemoradiotherapy in glioblastoma. *J. Clin. Oncol.* **26**, 3015–3024.

Murnane, J. P., and Sabatier, L. (2004). Chromosome rearrangements resulting from telomere dysfunction and their role in cancer. *Bioessays* **26**, 1164–1174.

Nahi, H., Sutlu, T., Jansson, M., Alici, E., and Gahrton, G. (2011). Clinical impact of chromosomal aberrations in multiple myeloma. *J. Intern. Med.* **269**, 137–147.

Nakamura, M., Watanabe, T., Klangby, U., Asker, C., Wiman, K., Yonekawa, Y., Kleihues, P., and Ohgaki, H. (2001). p14ARF deletion and methylation in genetic pathways to glioblastomas. *Brain Pathol.* **11**, 159–168.

Neben, K., Jauch, A., Bertsch, U., Heiss, C., Hielscher, T., Seckinger, A., Mors, T., Muller, N. Z., Hillengass, J., Raab, M. S., Ho, A. D., Hose, D., and Goldschmidt, H. (2010). Combining information regarding chromosomal aberrations t(4;14) and del(17p13) with the International Staging System classification allows stratification of myeloma patients undergoing autologous stem cell transplantation. *Haematologica* **95**, 1150–1157.

Nemec, P., Zemanova, Z., Greslikova, H., Michalova, K., Filkova, H., Tajtlova, J., Kralova, D., Kupska, R., Smetana, J., Krejci, M., Pour, L., Zahradova, L., Sandecka, V., Adam, Z., Buchler, T., Spicka, I., Gregora, E., Kuglik, P., and Hajek, R. (2010). Gain of 1q21 is an unfavorable genetic prognostic factor for multiple myeloma patients treated with high-dose chemotherapy. *Biol. Blood Marrow Transplant* **16**, 548–554.

Nesbit, C. E., Tersak, J. M., and Prochownik, E. V. (1999). Myc oncogenes and human neoplastic disease. *Oncogene* **18**, 3004–3016.

Newcomb, E. W., Cohen, H., Lee, S. R., Bhalla, S. K., Bloom, J., Hayes, R. L., and Miller, D. C. (1998). Survival of patients with glioblastoma multiforme is not influenced by altered expression of p16, p53, EGFR. *MDM2 or Bcl-2 genes. Brain Pathol.* **8**, 655–667.

Nunez, E., Fu, X. D., and Rosenfeld, M. G. (2009). Nuclear organization in the 3D space of the nucleus - cause or consequence? *Curr. Opin. Genet. Dev.* **19**, 424–436.

Oberdoerffer, P., and Sinclair, D. A. (2007). The role of nuclear architecture in genomic stability and ageing. *Nat. Rev. Mol. Cell Biol.* **8**, 692–702.

Ohgaki, H., and Kleihues, P. (2007). Genetic pathways to primary and secondary glioblastoma. *Am. J. Pathol.* **170**, 1445–1453.

Olins, A. L., Rhodes, G., Welch, D. B., Zwerger, M., and Olins, D. E. (2010). Lamin B receptor: Multi-tasking at the nuclear envelope. *Nucleus* **1**, 53–70.

Osborne, C. S., Chakalova, L., Mitchell, J. A., Horton, A., Wood, A. L., Bolland, D. J., Corcoran, A. E., and Fraser, P. (2007). Myc dynamically and preferentially relocates to a transcription factory occupied by Igh. *PLoS Biol.* **5**, e192.

Pallesen, G., Hamilton-Dutoit, S. J., Rowe, M., and Young, L. S. (1991). Expression of Epstein-Barr virus latent gene products in tumour cells of Hodgkin's disease. *Lancet* **337**, 320–322.

Palumbo, A., Sezer, O., Kyle, R., Miguel, J. S., Orlowski, R. Z., Moreau, P., Niesvizky, R., Morgan, G., Comenzo, R., Sonneveld, P., Kumar, S., Hajek, R., Giralt, S., Bringhen, S., Anderson, K. C., Richardson, P. G., Cavo, M., Davies, F., Bladé, J., Einsele, H., Dimopoulos, M. A., Spencer, A., Dispenzieri, A., Reiman, T., Shimizu, K., Lee, J. H., Attal, M., Boccadoro, M., Mateos, M., Chen, W., Ludwig, H., Joshua, D., Chim, J., Hungria, V., Turesson, I., Durie, B. G., and Lonial, S. (2009). International Myeloma Working Group guidelines for the management of multiple myeloma patients ineligible for standard high-dose chemotherapy with autologous stem cell transplantation. *Leukemia* **10**, 1716–1730.

Park, G., Kang, S. H., Lee, J. H., Suh, C., Kim, M., Park, S. M., Kim, T. Y., Oh, B., Min, H. J., Yoon, S. S., Yang, I. C., Cho, H. I., and Lee, D. S. (2011). Concurrent p16 methylation pattern as an adverse prognostic factor in multiple myeloma: a methylation-specific polymerase chain reaction study using two different primer sets. *Ann. Hematol.* **90**, 73–79.

Parsons, D. W., Jones, S., Zhang, X., Lin, J. C., Leary, R. J., Angenendt, P., Mankoo, P., Carter, H., Siu, I. M., Gallia, G. L., Olivi, A., McLendon, R., Rasheed, B. A., Keir, S., Nikolskaya, T., Nikolsky, Y., Busam, D. A., Tekleab, H., Diaz Jr, L. A., Hartigan, J., Smith, D. R., Strausberg, R. L., Marie, S. K., Shinjo, S. M., Yan, H., Riggins, G. J., Bigner, D. D., Karchin, R., Papadopoulos, N., Parmigiani, G., Vogelstein, B., Velculescu, V. E., and Kinzler, K. W. (2008). An integrated genomic analysis of human glioblastoma multiforme. *Science* **321**, 1807–1812.

Partin, A. W., Getzenberg, R. H., CarMichael, M. J., Vindivich, D., Yoo, J., Epstein, J. I., and Coffey, D. S. (1993). Nuclear matrix protein patterns in human benign prostatic hyperplasia and prostate cancer. *Cancer Res.* **53**, 744–746.

Pichiorri, F., Suh, S. S., Ladetto, M., Kuehl, M., Palumbo, T., Drandi, D., Taccioli, C., Zanesi, N., Alder, H., and Hagan, J. P., et al., (2008). MicroRNAs regulate critical genes associated with multiple myeloma pathogenesis. *Proc. Natl. Acad. Sci. U.S.A.* **105**(35), 12885–12890.

Pienta, K. J., Partin, A. W., and Coffey, D. S. (1989). Cancer as a disease of DNA organization and dynamic cell structure. *Cancer Res.* **49**, 2525–2532.

Ploem, J. S. (1967). The use of a vertical illuminator with interchangeable dichroic mirrors for fluorescence microscopy with incidental light. *Z. Wiss. Mikrosk.* **68**(3), 129–142.

Pobiega, S., and Marcand, S. (2010). Dicentric breakage at telomere fusions. *Genes Dev.* **24**, 720–733.

Poon, S. S., Martens, U. M., Ward, R. K., and Lansdorp, P. M. (1999). Telomere length measurements using digital fluorescence microscopy. *Cytometry* **36**, 267–278.

Potter, M., and Wiener, F. (1992). Plasmacytomagenesis in mice: model of neoplastic development dependent upon chromosomal translocations. *Carcinogenesis* **13**, 1681–1697.

Raab, M. S., Podar, K., Breitkreutz, I., Richardson, P. G., and Anderson, K. C. (2009). Multiple myeloma. *Lancet* **374**, 324–339.

Rach, E. A., Winter, D. R., Benjamin, A. M., Corcoran, D. L., Ni, T., Zhu, J., and Ohler, U. (2011). Transcription initiation patterns indicate divergent strategies for gene regulation at the chromatin level. *PLoS Genet.* **7**, e1001274.

Rajagopalan, H., and Lengauer, C. (2004). Aneuploidy and cancer. *Nature* **432**, 338–341.

Rayleigh, L. (1879). Investigations in Optics, with special reference to the Spectroscope. Edinburgh, and Dublin, London.

Raz, V., Carlotti, F., Vermolen, B. J., van der Poel, E., Sloos, W. C., Knaan-Shanzer, S., de Vries, A. A., Hoeben, R. C., Young, I. T., Tanke, H. J., Garini, Y., and Dirks, R. W. (2006). Changes in lamina structure are followed by spatial reorganization of heterochromatic regions in caspase-8-activated human mesenchymal stem cells. *J. Cell Sci.* **119**, 4247–4256.

Re, D., Starostik, P., Massoudi, N., Staratschek-Jox, A., Dries, V., Thomas, R. K., Diehl, V., and Wolf, J. (2003). Allelic losses on chromosome 6q25 in Hodgkin and Reed–Sternberg cells. *Cancer Res.* **63**, 2606–2609.

Reifenberger, G., Liu, L., Ichimura, K., Schmidt, E. E., and Collins, V. P. (1993). Amplification and overexpression of the MDM2 gene in a subset of human malignant gliomas without p53 mutations. *Cancer Res.* **53**, 2736–2739.

Rhodin Edsö, J., Gustafsson, C., and Cohn, M. (2011). Single- and double-stranded DNA binding proteins act in concert to conserve a telomeric DNA core sequence. *Genome. Integr.* **2**, 2.

Righolt, C. H., Raz, V., Vermolen, B. J., Dirks, R. W., Tanke, H. J., and Young, I. T. (2011a). Molecular Image Analysis: Quantitative Description and Classification of the Nuclear Lamina in Human Mesenchymal Stem Cells. *International Journal of Molecular Imaging* **2011**, 11. doi: 10.1155/2011/723283 Article ID 723283.

Righolt, C. H., Wiener, F., Taylor-Kashton, C., Harizanova, J., Vermolen, B. J., Garini, Y., Young, I. T., and Mai, S. (2011b). Translocation frequencies and chromosomal proximities for selected mouse chromosomes in primary B lymphocytes. *Cytometry A.* **4**, 276–278.

Robillard, N., Avet-Loiseau, H., Garand, R., Moreau, P., Pineau, D., Rapp, M. J., Harousseau, J. L., and Bataille, R. (2003). CD20 is associated with a small mature plasma cell morphology and t(11;14) in multiple myeloma. *Blood* **102**, 1070–1071.

Rohde, G. K., Ribeiro, A. J., Dahl, K. N., and Murphy, R. F. (2008). Deformation-based nuclear morphometry: capturing nuclear shape variation in HeLa cells. *Cytometry A* **73** (4), 341–350.

Roix, J. J., McQueen, P. G., Munson, P. J., Parada, L. A., and Misteli, T. (2003). Spatial proximity of translocation-prone gene loci in human lymphomas. *Nat. Genet.* **34**, 287–291.

Rosa, A., and Everaers, R. (2008). Structure and dynamics of interphase chromosomes. *PLoS Comput. Biol.* **4**, e1000153.

Ross, F. M., Chiecchio, L., Dagrada, G., Protheroe, R. K., Stockley, D. M., Harrison, C. J., Cross, N. C., Szubert, A. J., Drayson, M. T., and Morgan, G. J. (2010). The t(14;20) is a poor prognostic factor in myeloma but is associated with long-term stable disease in monoclonal gammopathies of undetermined significance. *Haematologica* **95**, 1221–1225.

Rothermund, K., Rogulski, K., Fernandes, E., Whiting, A., Sedivy, J., Pu, L., and Prochownik, E. V. (2005). C-Myc-independent restoration of multiple phenotypes by two C-Myc target genes with overlapping functions. *Cancer Res.* **65**, 2097–2107.

Rouquette, J., Cremer, C., Cremer, T., and Fakan, S. (2010). Functional nuclear architecture studied by microscopy: present and future. *Int. Rev. Cell Mol. Biol.* **282**, 1–90.

Rowley, J. D. (2001). Chromosome translocations: dangerous liaisons revisited. *Nat. Rev. Cancer* **1**, 245–250.

Rust, M. J., Bates, M., and Zhuang, X. (2006). Sub-diffraction-limit imaging by stochastic optical reconstruction microscopy (STORM). *Nat. Methods* **3**, 793–795.

Sachs, R. K., Hlatky, L. R., and Trask, B. J. (2000). Radiation-produced chromosome aberrations: colourful clues. *Trends Genet.* **16**, 143–146.

Sarasquete, M. E., Garcia-Sanz, R., Marin, L., Alcoceba, M., Chillon, M. C., Balanzategui, A., Santamaria, C., Rosinol, L., de la Rubia, J., Hernandez, M. T., Garcia-Navarro, I., Lahuerta, J. J., Gonzalez, M., and San Miguel, J. F. (2008). Bisphosphonate-related osteonecrosis of the jaw is associated with polymorphisms of the cytochrome P450 CYP2C8 in multiple myeloma: a genome-wide single nucleotide polymorphism analysis. *Blood* **112**, 2709–2712.

Sarkar, R., Guffei, A., Vermolen, B. J., Garini, Y., and Mai, S. (2007). Alterations of centromere positions in nuclei of immortalized and malignant mouse lymphocytes. *Cytometry A* **71**, 386–392.

Schermelleh, L., Carlton, P. M., Haase, S., Shao, L., Winoto, L., Kner, P., Burke, B., Cardoso, M. C., Agard, D. A., Gustafsson, M. G., Leonhardt, H., and Sedat, J. W. (2008). Subdiffraction multicolor imaging of the nuclear periphery with 3D structured illumination microscopy. *Science* **320**, 1332–1336.

Schierbeek, A. (1959). Measuring the Invisible World: The Life and Works of Antoni van Leeuwenhoek. Abelard-Schuman, London 226 pp.

Schmidt, M. C., Antweiler, S., Urban, N., Mueller, W., Kuklik, A., Meyer-Puttlitz, B., Wiestler, O. D., Louis, D. N., Fimmers, R., and von Deimling, A. (2002). Impact of genotype and morphology on the prognosis of glioblastoma. *J. Neuropathol. Exp. Neurol.* **61**, 321–328.

Schröck, E., du Manoir, S., Veldman, T., Schoell, B., Wienberg, J., Ferguson-Smith, M. A., Ning, Y., Ledbetter, D. H., Bar-Am, I., Soenksen, D., Garini, Y., and Ried, T. (1996). Multicolor spectral karyotyping of human chromosomes. *Science* **273**, 494–497.

Scialdone, A., and Nicodemi, M. (2009). DNA loci cross-talk through thermodynamics. *J. Biomed. Biotechnol.* **2009**, 516723.

Selvarajah, S., Yoshimoto, M., Ludkovski, O., Park, P. C., Bayani, J., Thorner, P., Maire, G., Squire, J. A., and Zielenska, M. (2008). Genomic signatures of chromosomal instability and osteosarcoma progression detected by high resolution array CGH and interphase FISH. *Cytogenet. Genome Res.* **122**, 5–15.

Selvarajah, S., Yoshimoto, M., Park, P. C., Maire, G., Paderova, J., Bayani, J., Lim, G., Al-Romaih, K., Squire, J. A., and Zielenska, M. (2006). The breakage-fusion-bridge (BFB) cycle as a mechanism for generating genetic heterogeneity in osteosarcoma. *Chromosoma* **115**, 459–467.

Shackleton, S., Lloyd, D. J., Jackson, S. N., Evans, R., Niermeijer, M. F., Singh, B. M., Schmidt, H., Brabant, G., Kumar, S., Durrington, P. N., Gregory, S., O'Rahilly, S., and Trembath, R. C. (2000). LMNA, encoding lamin A/C, is mutated in partial lipodystrophy. *Nat. Genet.* **24**, 153–156.

Shaffer, A. L., Emre, N. C., Lamy, L., Ngo, V. N., Wright, G., Xiao, W., Powell, J., Dave, S., Yu, X., Zhao, H., Zeng, Y., Chen, B., Epstein, J., and Staudt, L. M. (2008). IRF4 addiction in multiple myeloma. *Nature* **454**, 226–231.

Shaner, N. C., Steinbach, P. A., and Tsien, R. Y. (2005). A guide to choosing fluorescent proteins. *Nat. Methods* **2**(12), 905–909.

Sharma, S., Ding, F., and Dokholyan, N. V. (2007). Multiscale modeling of nucleosome dynamics. *Biophys. J.* **92**, 1457–1470.

Shapiro-Shelef, M., and Calame, K. (2005). Regulation of plasma-cell development. *Nat. Rev. Immunol.* **5**, 230–242.

Shete, S., Hosking, F. J., Robertson, L. B., Dobbins, S. E., Sanson, M., Malmer, B., Simon, M., Marie, Y., Boisselier, B., and Delattre, J. Y., et al., (2009). Genome-wide association study identifies five susceptibility loci for glioma. *Nat. Genet.* **41**(8), 899–904.

Shivaraju, M., Camahort, R., Mattingly, M., and Gerton, J. L. (2011). Scm3 is a centromeric nucleosome assembly factor. *J. Biol. Chem.* **286**, 12016–12023.

Smal, I., Loog, M., Niessen, W., and Meijering, E. (2010). Quantitative comparison of spot detection methods in fluorescence microscopy. *IEEE Trans. Med. Imaging* **29**, 282–301.

Smith, E. D., Kudlow, B. A., Frock, R. L., and Kennedy, B. K. (2005). A-type nuclear lamins, progerias and other degenerative disorders. *Mech. Ageing Dev.* **126**, 447–460.

Solovei, I., Kreysing, M., Lanctot, C., Kosem, S., Peichl, L., Cremer, T., Guck, J., and Joffe, B. (2009). Nuclear architecture of rod photoreceptor cells adapts to vision in mammalian evolution. *Cell* **137**, 356–368.

Stewenius, Y., Gorunova, L., Jonson, T., Larsson, N., Hoglund, M., Mandahl, N., Mertens, F., Mitelman, F., and Gisselsson, D. (2005). Structural and numerical chromosome changes in colon cancer develop through telomere-mediated anaphase bridges, not through mitotic multipolarity. *Proc. Natl. Acad. Sci. U.S.A.* **102**, 5541–5546.

Speckman, R. A., Garg, A., Du, F., Bennett, L., Veile, R., Arioglu, E., Taylor, S. I., Lovett, M., and Bowcock, A. M. (2000). Mutational and haplotype analyses of families with familial partial lipodystrophy (Dunnigan variety) reveal recurrent missense mutations in the globular C-terminal domain of lamin A/C. *Am. J. Hum. Genet.* **66**, 1192–1198.

Steidl, C., Telenius, A., Shah, S. P., Farinha, P., Barclay, L., Boyle, M., Connors, J. M., Horsman, D. E., and Gascoyne, R. D. (2010). Genome-wide copy number analysis of Hodgkin Reed-Sternberg cells identifies recurrent imbalances with correlations to treatment outcome. *Blood* **116**, 418–427.

Stupp, R., Hegi, M. E., Mason, W. P., van den Bent, M. J., Taphoorn, M. J., Janzer, R. C., Ludwin, S. K., Allgeier, A., Fisher, B., Belanger, K., Hau, P., Brandes, A. A., Gijtenbeek, J., Marosi, C., Vecht, C. J., Mokhtari, K., Wesseling, P., Villa, S., Eisenhauer, E., Gorlia, T., Weller, M., Lacombe, D., Cairncross, J. G., and Mirimanoff, R. O. (2009). Effects of radiotherapy with concomitant and adjuvant temozolomide versus radiotherapy alone on survival in glioblastoma in a randomised phase III study: 5-year analysis of the EORTC-NCIC trial. *Lancet Oncol.* **10**, 459–466.

Strunnikov, A. V. (2010). One-hit wonders of genomic instability. *Cell Div.* **5**, 15.

Stimpson, K. M., Song, I. Y., Jauch, A., Holtgreve-Grez, H., Hayden, K. E., Bridger, J. M., and Sullivan, B. A. (2010). Telomere disruption results in non-random formation of de novo dicentric chromosomes involving acrocentric human chromosomes. *PLoS Genet.* **6**, pii: e1001061.

Stuurman, N., Heins, S., and Aebi, U. (1998). Nuclear lamins: their structure, assembly, and interactions. *J. Struct. Biol.* **122**, 42–66.

Sullivan, T., Escalante-Alcalde, D., Bhatt, H., Anver, M., Bhat, N., Nagashima, K., Stewart, C. L., and Burke, B. (1999). Loss of A-type lamin expression compromises nuclear envelope integrity leading to muscular dystrophy. *J. Cell Biol.* **147**, 913–920.

Sun, Y., Jiang, X., and Price, B. D. (2010). Tip60: Connecting chromatin to DNA damage signaling. *Cell Cycle* **9**, 930–936.

Sutlu, T., Alici, E., Jansson, M., Wallblom, A., Dilber, M. S., Gahrton, G., and Nahi, H. (2009). The prognostic significance of 8p21 deletion in multiple myeloma. *Br. J. Haematol.* **144**, 266–268.

Szymanowska, N., Klapper, W., Gesk, S., Kuppers, R., Martin-Subero, J. I., and Siebert, R. (2008). BCL2 and BCL3 are recurrent translocation partners of the IGH locus. *Cancer Genet. Cytogenet.* **186**, 110–114.

Takaha, N., Hawkins, A. L., Griffin, C. A., Isaacs, W. B., and Coffey, D. S. (2002). High mobility group protein I(Y): a candidate architectural protein for chromosomal rearrangements in prostate cancer cells. *Cancer Res.* **62**, 647–651.

Tanabe, H., Muller, S., Neusser, M., von Hase, J., Calcagno, E., Cremer, M., Solovei, I., Cremer, C., and Cremer, T. (2002). Evolutionary conservation of chromosome territory arrangements in cell nuclei from higher primates. *Proc. Natl. Acad. Sci. U.S.A.* **99**, 4424–4429.

Tanaka, Y., Yoshimura, I., and Nakai, K. (2010). Positional variations among heterogeneous nucleosome maps give dynamical information on chromatin. *Chromosoma* **119**, 391–404.

Tang, Y., Chen, Y., Jiang, H., and Nie, D. (2010). Promotion of tumor development in prostate cancer by progerin. *Cancer Cell Int.* **10**, 47.

Tapper, W., Chiecchio, L., Dagrada, G. P., Konn, Z. J., Stockley, D. M., Szubert, A. J., Gregory, W. M., Bell, S. E., Jackson, G. H., Child, J. A., Morgan, G. J., and Ross, F. M. (2011). Heterogeneity in the prognostic significance of 12p deletion and chromosome 5 amplification in multiple myeloma. *J. Clin. Oncol.* **29**, e37–e39.

Taylor, C., and Mai, S. (1998). c-Myc-associated genomic instability of the dihydrofolate reductase locus in vivo. *Cancer Detect. Prev.* **22**, 350–356.

Teller, K., Solovei, I., Buiting, K., Horsthemke, B., and Cremer, T. (2007). Maintenance of imprinting and nuclear architecture in cycling cells. *Proc. Natl. Acad. Sci. U.S.A.* **104**, 14970–14975.

Thyagarajan, B., Anderson, K. E., Lessard, C. J., Veltri, G., Jacobs, D. R., Folsom, A. R., and Gross, M. D. (2007). Alkaline unwinding flow cytometry assay to measure nucleotide excision repair. *Mutagenesis* **22**, 147–153.

Tsang, L. W., Hu, N., and Underhill, D. A. (2010). Comparative analyses of SUV420H1 isoforms and SUV420H2 reveal differences in their cellular localization and effects on myogenic differentiation. *PLoS One* **5**(12), e14447.

Ueki, K., Ono, Y., Henson, J. W., Efird, J. T., von Deimling, A., and Louis, D. N. (1996). CDKN2/p16 or RB alterations occur in the majority of glioblastomas and are inversely correlated. *Cancer Res.* **56**, 150–153.

Varela, I., Cadinanos, J., Pendas, A. M., Gutierrez-Fernandez, A., Folgueras, A. R., Sanchez, L. M., Zhou, Z., Rodriguez, F. J., Stewart, C. L., Vega, J. A., Tryggvason, K., Freije, J. M., and Lopez-Otin, C. (2005). Accelerated ageing in mice deficient in Zmpste24 protease is linked to p53 signalling activation. *Nature* **437**, 564–568.

Vazquez, M., Greulich-Bode, K. M., Arsuaga, J., Cornforth, M. N., Bruckner, M., Sachs, R. K., Hlatky, L., Molls, M., and Hahnfeldt, P. (2002). Computer analysis of mFISH chromosome aberration data uncovers an excess of very complicated metaphases. *Int. J. Radiat. Biol.* **78**, 1103–11015.

Vekemans, M. C., Lemmens, H., Delforge, M., Doyen, C., Pierre, P., Demuynck, H., Bries, G., Lemmens, J., Meeus, P., Straetmans, N., Bauwens, D., Vidrequin, S., Rack, K., Vandenberghe, P., Wlodarska, I., and Michaux, L. (2010). The t(14;20)(q32;q12): a rare cytogenetic change in multiple myeloma associated with poor outcome. *Br. J. Haematol.* **149**, 901–904.

Vermolen, B. J., Garini, Y., Mai, S., Mougey, V., Fest, T., Chuang, T. C., Chuang, A. Y., Wark, L., and Young, I. T. (2005). Characterizing the three-dimensional organization of telomeres. *Cytometry A* **67**, 144–150.

Vlcek, S., and Foisner, R. (2007). Lamins and lamin-associated proteins in aging and disease. *Curr. Opin. Cell Biol.* **19**, 298–304.

von Deimling, A., Louis, D. N., and Wiestler, O. D. (1995). Molecular pathways in the formation of gliomas. *Glia* **15**, 328–338.

Vukovic, B., Beheshti, B., Park, P., Lim, G., Bayani, J., Zielenska, M., and Squire, J. A. (2007). Correlating breakage-fusion-bridge events with the overall chromosomal instability and *in vitro* karyotype evolution in prostate cancer. *Cytogenet. Genome Res.* **116**, 1–11.

Vukovic, B., Park, P. C., Al-Maghrabi, J., Beheshti, B., Sweet, J., Evans, A., Trachtenberg, J., and Squire, J. (2003). Evidence of multifocality of telomere erosion in high-grade prostatic intraepithelial neoplasia (HPIN) and concurrent carcinoma. *Oncogene* **22**, 1978–1987.

Walker, B. A., Leone, P. E., Chiecchio, L., Dickens, N. J., Jenner, M. W., Boyd, K. D., Johnson, D. C., Gonzalez, D., Dagrada, G. P., Protheroe, R. K., Konn, Z. J., Stockley, D. M., Gregory, W. M., Davies, F. E., Ross, F. M., and Morgan, G. J. (2010). A compendium of myeloma-associated chromosomal copy number abnormalities and their prognostic value. *Blood* **116**, e56–e65.

Walker, B. A., Wardell, C. P., Chiecchio, L., Smith, E. M., Boyd, K. D., Neri, A., Davies, F. E., Ross, F. M., and Morgan, G. J. (2011). Aberrant global methylation patterns affect the molecular pathogenesis and prognosis of multiple myeloma. *Blood* **117**, 553–562.

Walter, T., Shattuck, D. W., Baldock, R., Bastin, M. E., Carpenter, A. E., Duce, S., Ellenberg, J., Fraser, A., Hamilton, N., and Pieper, S., et al., (2010). Visualization of image data from cells to organisms. *Nat. Methods* 7(3 Suppl), S26–41.

Wang, G., Zhao, J., and Vasquez, K. M. (2009). Methods to determine DNA structural alterations and genetic instability. *Methods* **48**, 54–62.

Wang, Q. E., Milum, K., Han, C., Huang, Y. W., Wani, G., Thomale, J., and Wani, A. A. (2011). Differential contributory roles of nucleotide excision and homologous recombination repair for enhancing cisplatin sensitivity in human ovarian cancer cells. *Mol. Cancer* **10**, 24.

Wang, X., Kam, Z., Carlton, P. M., Xu, L., Sedat, J. W., and Blackburn, E. H. (2008). Rapid telomere motions in live human cells analyzed by highly time-resolved microscopy. *Epigenetics Chromatin* **1**, 4.

Weaver, B. A., and Cleveland, D. W. (2009). The role of aneuploidy in promoting and suppressing tumors. *J. Cell Biol.* **185**, 935–937.

White, J. G., Amos, W. B., and Fordham, M. (1987). An evaluation of confocal versus conventional imaging of biological structures by fluorescence light microscopy. *J. Cell. Biol.* **105**(1), 41–48.

Woodcock, C. L., and Ghosh, R. P. (2010). Chromatin higher-order structure and dynamics. *Cold Spring Harb. Perspect. Biol.* **2**, a000596.

Worman, H. J., and Courvalin, J. C. (2005). Nuclear envelope, nuclear lamina, and inherited disease. *Int. Rev. Cytol.* **246**, 231–279.

Wrensch, M., Jenkins, R. B., Chang, J. S., Yeh, R. F., Xiao, Y., Decker, P. A., Ballman, K. V., Berger, M., Buckner, J. C., and Chang, S., et al., (2009). Variants in the CDKN2B and RTEL1 regions are associated with high-grade glioma susceptibility. *Nat. Genet.* **41**(8), 905–908.

Xu, L., and Blackburn, E. H. (2007). Human cancer cells harbor T-stumps, a distinct class of extremely short telomeres. *Mol. Cell* **28**, 315–327.

Yamini, B., Yu, X., Dolan, M. E., Wu, M. H., Darga, T. E., Kufe, D. W., and Weichselbaum, R. R. (2007). Inhibition of nuclear factor-kappa B activity by temozolomide involves O6-methylguanine induced inhibition of p65 DNA binding. *Cancer Res.* **67**, 6889–6898.

Young, D. W., Bender, A., Hoyt, J., McWhinnie, E., Chirn, G. W., Tao, C. Y., Tallarico, J. A., Labow, M., Jenkins, J. L., and Mitchison, T. J., et al., (2008). Integrating high-content screening and ligand-target prediction to identify mechanism of action. *Nat. Chem. Biol.* **4** (1), 59–68.

Yu, Q., Zhang, X., and Bi, X. (2011). Roles of chromatin remodeling factors in the formation and maintenance of heterochromatin structure. *J. Biol. Chem.* **16**, 14959-14669.

Yue, J., Wang, Q., Lu, H., Brenneman, M., Fan, F., and Shen, Z. (2009). The cytoskeleton protein filamin-A is required for an efficient recombinational DNA double strand break repair. *Cancer Res.* **69**, 7978–7985.

Zanella, F., Lorens, J. B., and Link, W. (2010). High content screening: seeing is believing. *Trends Biotechnol.* **28**, 237–245.

Zhan, F., Hardin, J., Kordsmeier, B., Bumm, K., Zheng, M., Tian, E., Sanderson, R., Yang, Y., Wilson, C., Zangari, M., Anaissie, E., Morris, C., Muwalla, F., van Rhee, F., Fassas, A., Crowley, J., Tricot, G., Barlogie, B., and Shaughnessy Jr, J. (2002). Global gene expression profiling of multiple myeloma, monoclonal gammopathy of undetermined significance, and normal bone marrow plasma cells. *Blood* **99**, 1745–1757.

Zhou, Y., Chen, L., Barlogie, B., Stephens, O., Wu, X., Williams, D. R., Cartron, M. A., van Rhee, F., Nair, B., Waheed, S., Pineda-Roman, M., Alsayed, Y., Anaissie, E., and Shaughnessy Jr, J. D. (2010). High-risk myeloma is associated with global elevation of miRNAs and overexpression of EIF2C2/AGO2. *Proc. Natl. Acad. Sci. U.S.A.* **107**, 7904–7909.

Zhou, X. P., Li, Y. J., Hoang-Xuan, K., Laurent-Puig, P., Mokhtari, K., Longy, M., Sanson, M., Delattre, J. Y., Thomas, G., and Hamelin, R. (1999). Mutational analysis of the PTEN gene in gliomas: molecular and pathological correlations. *Int. J. Cancer* **84**, 150–154.

Zojer, N., Konigsberg, R., Ackermann, J., Fritz, E., Dallinger, S., Kromer, E., Kaufmann, H., Riedl, L., Gisslinger, H., Schreiber, S., Heinz, R., Ludwig, H., Huber, H., and Drach, J. (2000). Deletion of 13q14 remains an independent adverse prognostic variable in multiple myeloma despite its frequent detection by interphase fluorescence *in situ* hybridization. *Blood* **95**, 1925–1930.

Cytogenetic Analysis of Tumor Clonality

Manuel R. Teixeira[*,†,‡] **and Sverre Heim**[‡,§,¶]

[*] *Department of Genetics, Portuguese Oncology Institute, Porto, Portugal*
[†] *Biomedical Sciences Institute (ICBAS), University of Porto, Porto, Portugal*
[‡] *Centre for Cancer Biomedicine, University of Oslo, Oslo, Norway*
[§] *Section for Cancer Cytogenetics, Institute for Medical Informatics, The Norwegian Radium Hospital, Oslo University Hospital, Oslo, Norway*
[¶] *Medical Faculty, University of Oslo, Oslo, Norway*

All or almost all neoplasias subjected to systematic cytogenetic scrutiny have been found to harbor acquired chromosomal aberrations. The paradigm stemming from the study of hematopoietic malignancies and sarcomas is that cancers are of monoclonal origin (i.e., they have developed from a single transformed somatic progenitor) because all the neoplastic parenchyma cells share at least one primary chromosomal abnormality, with subsequent clonal evolution along the lines of Darwinian selection occurring among the various subclones carrying secondary aberrations. When carcinomas began to be studied more extensively by cytogenetic methods, however, sometimes many cytogenetically unrelated clones were found, in seeming contradiction to the monoclonal hypothesis. Also studies of multiple samples from the same patient led to a rethinking of what the cytogenetic evidence really revealed about tumor clonality, both in its early stages and during disease development. The observed cytogenetic heterogeneity in, for example, tumors of the breast and pancreas vastly surpasses that of leukemias, lymphomas, connective tissue tumors, or even most epithelial, including uroepithelial, tumors. Theoretical reasoning as well as the available experimental data we here review show that the clonal evolution of neoplastic cell populations follows either of four principal pathways: (1) initial monoclonality is retained throughout the entire course of the disease with no additional, secondary aberrations accrued as judged by karyotypic appearance; (2) tumorigenesis is monoclonal but additional aberrations develop with time leading to secondary clonal heterogeneity (clonal divergence); (3) polyclonal tumorigenesis exists from the beginning but is followed by an overall reduction in genomic complexity with time (clonal convergence) due to selection

Advances in CANCER RESEARCH, Volume 112
Copyright 2011, Elsevier Inc. All right reserved.
0065-230X/10 $35.00
DOI: 10.1016/B978-0-12-387688-1.00005-3

among cytogenetically unrelated clones during tumor progression, resulting in secondary oligo- or monoclonality; or (4) polyclonal tumorigenesis with early clonal convergence is followed by later clonal divergence due to the acquisition of additional cytogenetic changes by the clone(s) that survived during the middle phases of tumor progression. Further studies of individual tumor cells are necessary to elicit precise information about the cell-to-cell variability that exists in many, especially epithelial, neoplasms and which holds the key to a more profound understanding of the complex issue of tumor clonality during all stages of cancer development. © 2011 Elsevier Inc.

I. INTRODUCTION

According to the now nearly 100-year-old somatic mutation theory of cancer, it is the acquisition of chromosomal mutations by suitably primed target cells that renders them neoplastic; they begin to divide too frequently or die too slowly so that their number increases in an uncontrolled manner. The first clear-cut confirmation of Boveri's theory came half a century ago when Nowell and Hungerford described the specific occurrence of a small marker chromosome, called the Philadelphia (Ph) chromosome in honor of the city in which it was discovered, in bone marrow cells from patients suffering from chronic myeloid leukemia. It was later shown that the Ph-chromosome arose through a reciprocal translocation between chromosomes 9 and 22, t(9;22)(q34;q11), and that it recombined the genes *BCR* (from 22q11) and *ABL1* (from 9q34) to a novel *BCR/ABL1* fusion gene that produced a qualitatively new oncoprotein with abnormal tyrosine kinase properties (Fioretos and Johansson, 2009).

Since the discovery of the Ph-chromosome, similar successes have been repeated over and over again and now literally hundreds of more or less cancer-specific acquired chromosomal abnormalities have been detected in various neoplastic entities (Mitelman *et al.*, 2011). Indeed, all or almost all neoplasias subjected to systematic cytogenetic scrutiny have been found to harbor acquired chromosomal aberrations, and sometimes the pathologic–cytogenetic associations have turned out to be no less specific than the one first seen for chronic myeloid leukemia (Heim and Mitelman, 2009). These discoveries at the chromosomal level have been followed up by studies unraveling the molecular details of the changes taking place at the breakpoints of chromosomal fusions, providing examples of both altered gene regulation and the generation of qualitatively novel oncogenes (fusion genes) as a result of cancer-specific translocations. Thus much new knowledge about the pathogenetic pathways of various neoplasms has been generated—knowledge that is increasingly being brought to bear also therapeutically (Sherbenou and Druker, 2007).

Cancer cytogenetics has during the last few decades grown in two principal directions. First, it has become an invaluable diagnostic tool. The finding of one or more acquired clonal chromosome aberrations in sampled cells proves that one is dealing with a neoplastic disease, and the pattern of aberrations detected sometimes reveals with certainty or near-certainty which kind of neoplasia is present. Furthermore, not rarely is the diagnostic karyotype also prognostically informative, inasmuch as it tells us more about the innate aggressiveness of the disease or about how it is likely to respond to standard treatment than what can be gathered by a scrutiny of the disease's various phenotypic characteristics alone. At the same time, cytogenetics has continued to serve as a potent research tool providing a steady flow of information about new, cancer-specific aberrations that can be examined in more detail by molecular means, thus broadening as well as deepening our understanding of the molecular mechanisms of tumorigenesis. Cytogenetic studies have also contributed research-wise in another manner, namely by allowing reliable assessment of the clonal composition of tumor cell populations. In practice, the existence of a clone of aberrant cells is inferred whenever at least two metaphase cells are found with the same structural chromosomal rearrangement or gain of a whole chromosome, or when at least three cells are seen to have the same chromosome loss (ISCN, 2009). Cells carrying the same chromosome abnormality are presumed to be clonally related since the alternative explanation, that identical aberrations are acquired repeatedly and independently by several cells, is so extremely unlikely. The other main approach to clonality analysis is based on X chromosome inactivation in women (Fialkow, 1976; Vogelstein *et al.*, 1985), where either the paternal or the maternal X chromosome is inactivated in all clonal descendents of a given somatic mother cell. Whereas only two informative states are obtainable with the latter techniques (corresponding to either maternal or paternal X inactivation), the number of acquired chromosomal aberrations a tumor cell may carry is in principle infinite, corresponding to an equally innumerable number of informative states as regards clonality status.

Because all tumor cells from individual cases of leukemia, lymphoma, and sarcoma—the first malignancies to be studied extensively by cytogenetic means—almost invariably had at least one chromosomal abnormality in common (the primary abnormality; not rarely secondary aberrations were also found and these might not always be the same from cell to cell since Darwinian selection took place among the various subclones; see below and Fig. 1), the cytogenetic findings strongly favored a single-cell origin of the neoplastic process, monoclonal tumorigenesis

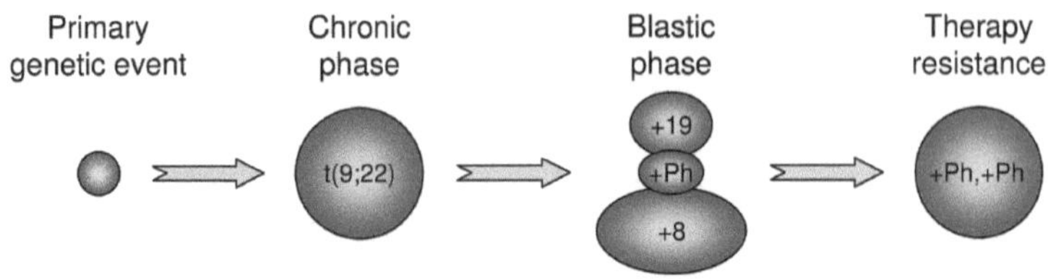

Fig. 1 Chronic myeloid leukemia as the paradigm of a monoclonal neoplasia resulting from a single genetic event, namely the translocation t(9;22)(q34;q11) giving rise to the *BCR/ ABL1* fusion gene, with subsequent disease progression being associated with Darwinian selection for subclones characterized by different secondary chromosome changes (here exemplified by trisomy 8, trisomy 19, or the occurrence of an additional Philadelphia chromosome). The nowadays standard therapy targeting the chimeric protein *BCR/ABL1* (imatinib) is highly effective in more than 95% of patients with chronic phase disease, but some patients eventually develop therapy resistance due to selection for initially minor subclones with amplification of the fusion gene or point mutations in its tyrosine kinase domain.

(Nowell, 1976). When carcinomas began to be studied more systematically by cytogenetic methods during the 1980s, however, the picture became less clear: sometimes many cytogenetically unrelated clones were found, something that clearly seemed to run counter to the monoclonal hypothesis, and also studies of multiple samples from the same patient (either taken from different parts of the same tumor, from different synchronous lesions, or from the primary tumor and later metastases at various time intervals) led to a rethinking of what the cytogenetic evidence really revealed about tumor clonality, both in its early stages and during disease development. Also evidence from studies relying on other principles and techniques, not least modern X chromosome inactivation studies, contributed to showing that the early conclusions of ubiquitous tumor monoclonality had to be reconsidered (Going, 2003; Novelli *et al.*, 2003). We shall look again at some of this evidence from three organ systems of major importance in oncology, both with regard to how the primary tumor arises and with an eye to how the collected data fit with a modern Darwinian view of dynamic clonal tumor development.

II. CYTOGENETIC ASSESSMENT OF CLONAL COMPOSITION OF PANCREATIC AND LARGE BOWEL TUMORS

Some 130 adenocarcinomas of the pancreas with abnormal karyotypes have been reported (Mitelman *et al.*, 2011). Although these tumors generally show considerable karyotypic complexity, a clearly nonrandom overall aberration pattern is discernible with losses of chromosomes 18 (the most common change; 60%), 17, 21, 6, Y, 4, X, 12, 22, and 9, gains of chromosomes 7 and 20, and chromosome bands 13q10, 19q13, 1q10, 8q10, 14q10, 17p11, and 17p10 as frequent (8–13%) sites of structural rearrangements (Bardi and Heim, 2009). Whereas sometimes at least one chromosomal change is common to all karyotypically abnormal cells, indicative of monoclonal tumorigenesis, karyotypic complexity coupled with a relatively high percentage of cytogenetically unrelated clones and extensive intratumor cytogenetic heterogeneity were the most typical features of the carcinomas described by Gorunova *et al.* (1998). Nineteen of the 25 pancreatic carcinomas subjected to in-depth karyotypic analysis in that study exhibited from 2 to 58 clones. Karyotypically documented clonal evolution in the form of related clones was detected in 19 tumors, whereas seemingly unrelated clones were seen in 16 tumors. Although such extensive clonal heterogeneity is otherwise unheard of unless the target tissue has been exposed to genotoxic exposure in the form of, for example, irradiation, the feeling lingers that part of the reason why clonal heterogeneity is often not detected is that it was not diligently looked for, or that it was looked for using methods not up to the task. One cannot but accept the above findings as evidence of polyclonal tumorigenesis in a substantial proportion of pancreatic carcinomas, although this conclusion is by no means unavoidable (see below). But even should the clonal heterogeneity turn out to be secondary rather than primary, it certainly exceeds what has been documented in other tumor types. The diversity it reveals has not been satisfactorily explained.

Both benign and malignant large bowel tumors are common and the former have a tendency to develop into the latter in the so-called adenoma–carcinoma sequence (Bedenne *et al.*, 1992). These adenomas and carcinomas are among the most extensively cytogenetically examined epithelial tumors and more than 500 cases with an abnormal karyotype have been reported (Mitelman *et al.*, 2011). The most common changes in adenomas are the numerical aberrations +7, +13, +20, and −18 as well as deletions from the short arm of chromosome 1, del(1p). The carcinomas typically are more complex with the isochromosomes i(8q), i(17q), and i (13q) and deletions of 17p and 18q as the most common structural rearrangements and gain of chromosome 7 and loss of chromosome 8 as typical numerical aberrations (Bardi and Heim, 2009). In addition, however, and of more interest in the present context, several cytogenetic studies have also examined the clonal relationships within single tumors, between synchronous tumors coexisting within the bowel, and between primary tumors and the metastases they gave rise to. Bomme *et al.* (1996) karyotyped 24 macroscopically distinct colorectal adenomas from 11 patients aiming to find out whether they were clonally related or not. In six patients, similar clones (i.e., they shared at least one chromosome abnormality) were found in separate polyps within the same patient. These polyps were always located within the same part of the large bowel. In the remaining two patients, both with one rectal adenoma and one adenoma of the colon, no karyotypic similarity between the two lesions was seen. The findings indicate that when macroscopically distinct adenomas are growing synchronously in the same part of the large bowel, they are karyotypically similar, in contrast to when they grow in different parts (proximal colon, sigmoid, rectum). In the former situation, two explanations are possible. Either the same etiologic agent has elicited identical pathogenetic responses from several cells within the same anatomical area (i.e., they have acquired the same chromosome aberration), or the morphologically distinct but close adenomas belong to the same clone of neoplastically transformed cells. Evidence that the first-mentioned scenario is feasible is almost nonexistent, although we do have some situations where the etiologic agent leads to one particular genetic rearrangement instead of hitting the genome in a nonspecific manner; for example, cytostatics that inhibit topoisomerase II may induce secondary leukemia by specifically targeting the *MLL* gene in 11q23 (Blanco *et al.*, 2004; Bloomfield *et al.*, 2002). If, on the other hand, the separate adenomas are macroscopic manifestations of parallel but pathogenetically independent neoplastic processes, this could also explain why by and large common loss of heterozygosity (LOH) features were detected in 7–13 adenomas from each of seven patients with sporadic colon cancer examined in the genome-wide allelotyping study conducted by Mao *et al.* (2006).

Another intriguing situation in which the question of possible developmental relatedness comes to the fore concerns synchronous polyps and carcinoma(s) coexisting in the same large bowel. In the light of the aforementioned pathogenetic adenoma–carcinoma sequence and sharing of cytogenetic aberrations among physically separate adenomas, to what extent do such tumors have clonal similarities? To approach this question, Bardi *et al.* (1997a) analyzed cytogenetically 30 tumor lesions of the large bowel—including carcinomas, adenomas, and nonadenomatous polyps—from seven patients with colorectal cancer. Clonal chromosome abnormalities were found in all adenomas and carcinomas but only in 37.5% of the nonadenomatous polyps. Although the majority of the hyperplastic polyps displayed a normal chromosome complement, some showed clonal aberrations that in general seemed to be simpler than those of the dysplastic polyps. It is possible that a subset of hyperplastic polyps with cytogenetic aberrations may have had small dysplastic areas that went undetected by conventional histological examination, or the chromosomal aberrations they carry, which were indistinguishable from those of small tubular adenomas, are not dysplasia-specific but rather related to the hyperproliferation in the intestinal mucosa. Finally, the very occurrence of clonal chromosome aberrations in a proportion of hyperplastic polyps, which constitutes strong evidence that these lesions are neoplastic, could be viewed as the genetic corollary of a hyperplastic polyp–carcinoma sequence even in the absence of corresponding histopathological or clinical indications to this effect. Leggett *et al.* (2001) confirmed our earlier conclusion that some hyperplastic polyps do have acquired genetic changes by finding microsatellite instability in three and *KRAS* mutations in 8 of 47 hyperplastic polyps examined.

Although some chromosome aberrations have been found in large bowel carcinomas but not in adenomas, for example der(8;17)(q10; q10), indicating that they may be specifically associated with malignant transformation of large bowel mucosa, adenomas and carcinomas occurring simultaneously in the same patient by and large share most of their chromosomal features (Bardi *et al.*, 1997a). This karyotypic similarity between malignant and benign tumors in the same patient, and also sometimes among nonmalignant polyps in the same individual, can be interpreted to indicate that the macroscopically distinct lesions arose as part of a single neoplastic clonal expansion, in spite of the fact that the distance between them was always more than 3 cm. The only alternative explanation would be that the same oncogenetic environmental factor(s) induced identical chromosomal rearrangements in more than one cell.

Which are the similarities and differences, in terms of clonality and otherwise, between large bowel carcinomas and their metastases? Bardi *et al.* (1997b) addressed this question by examining cytogenetically 18 tumors from 11 patients with metastatic colorectal cancer. In all cases with matched samples from the primary tumor and lymph node metastases, cytogenetic similarities were found between the primary and the secondary tumor lesions, indicating that many of the chromosomal aberrations were acquired before disease spreading took place. The observed genomic similarities, combined with later data showing close transcriptional resemblance between primary tumors and metastases (D'Arrigo *et al.*, 2005), support the notion that large cell clumps rather than rare sporadic cells from the primary tumor in most instances give rise to the secondaries. The assumed existence of a migrating cancer stem cell (Brabletz *et al.*, 2005) is nevertheless compatible with the genomic and transcriptional similarities between primaries and their matching metastases since cancer stem cell(s) detaching from the primary tumor can form a metastatic lesion with similar genomic and transcription patterns. Compared with the primaries, the metastases appeared to exhibit decreased clonal heterogeneity, probably reflecting clonal selection within the primary tumor from which the metastasis was derived, but, concurrently, an increase in the karyotypic complexity of individual clones

(Bardi *et al.*, 1997b). In addition to the aberrations del(1)(p34), i(17)(q10), −18, −21, +7, and +20, which were found recurrently in both primary and metastatic lesions, the del(10)(q22) found in metastases has not so far been found in primary colorectal carcinomas. The finding of loss of 10q24-qter in two more cases in another series of colorectal metastases to the liver (Parada *et al.*, 1999) provides further support for an association between loss of 10q and tumor progression. In a later LOH study, Fawole *et al.* (2002) used a panel of nine highly polymorphic microsatellite markers spanning the long arm of chromosome 10 to examine 114 colorectal adenocarcinomas. They found the highest LOH frequency at 10q21.1 and suggested that 10q loss is a late event in colorectal carcinogenesis. Using comparative genomic hybridization (CGH), Paredes-Zaglul *et al.* (1998) compared the genetic composition of primary colorectal tumors without distant metastases (TNM stages I-III) with liver metastases (TNM stage IV) and found a distinct predominance of genetic losses in the stage IV lesions. Although not all aberrations were the same as those found by Bardi and coworkers, they too identified two patterns of alterations: (i) changes (+8q, +13q, −4p, −8p, −15q, −17p, −18q, −21q, and −22q) that were more often found in liver metastases than in primary tumors, and (ii) changes (−9q, −11q, and −17q) that were unique to the metastatic lesions.

In a comparison between multiple metastatic lesions in the same patient but at different sites (Bardi *et al.*, 1997b), common karyotypic aberrations in the lesions were demonstrated, but not in a single case was the karyotype of one metastatic lesion exactly identical to that of another, even when both metastatic samples were from lymph nodes. Such genomic differences between primary and secondary tumors and among separate metastatic lesions have also been reported in several CGH studies of paired primary colorectal carcinomas and metastases to either lymph nodes or distant organs such as the liver (Al-Mulla *et al.*, 1999; Alcock *et al.*, 2003; Grade *et al.*, 2006; Knosel *et al.*, 2004) and lung (Jiang *et al.*, 2005; Knosel *et al.*, 2005). However, the data are not sufficient to distinguish specific genomic patterns, either chromosomal or at the gene level, that could predict or explain the metastatic spreading of colorectal tumor cells to particular organs or tissues.

III. CYTOGENETIC ASSESSMENT OF BREAST TUMOR CLONALITY

More than 800 breast carcinomas with clonal chromosome changes identified by chromosome banding analysis have been reported (Mitelman *et al.*, 2011). The pattern of acquired chromosomal aberrations they exhibit is clearly nonrandom, with the structural rearrangements i(1)(q10), der(1;16)(q10;p10), del(3)(p12–13p14–21), del(6)(q21–23), del(1)(p13), del(1)(q11–12), del(1)(p22), i(8)(q10), and del(1)(q42), as well as several numerical cytogenetic changes occurring recurrently (Teixeira *et al.*, 2009). Of relevance to the present discussion is the detection of cytogenetically unrelated clones in a significant proportion of breast carcinomas analyzed after short-term culturing (Pandis *et al.*, 1993, 1995a; Teixeira *et al.*, 2002, 2009). A detailed karyotyping study of multiple samples (one from each tumor quadrant) from 10 breast carcinomas revealed clonal chromosome abnormalities in 32 of 36 (89%) of the tumor samples, with two to nine cytogenetically unrelated clones detected in seven of the 10 carcinomas (Teixeira *et al.*, 1995, 1996). The various clones were always unevenly distributed within the tumor mass in the seven polyclonal breast carcinomas (Teixeira *et al.*, 1995, 1996). Such a zonal distribution of cytogenetically distinct cell populations was also demonstrated by CGH (Torres *et al.*, 2007), a technique that is not subject to culture bias and detects only those imbalances that are present in a significant proportion of the test sample. The fact that CGH also detects the changes seen by karyotyping in cytogenetically unrelated clones (Persson *et al.*, 1999; Teixeira *et al.*, 2001) provides further evidence for polyclonal carcinogenesis in a subset of breast cancers.

Apparently conflicting evidence stems from X chromosome inactivation studies concluding that breast carcinomas are of monoclonal origin (Noguchi *et al.*, 1992, 1994). However, this method can only recognize monoclonal cells against a polyclonal background when the former make up at least 50% of the sample, so smaller clones will inevitably go unnoticed. The central premise of clonality studies based on X-chromosome inactivation was also called into question when it was shown that normal breast epithelium is organized in developmentally discrete regions within which all cells have the same inactive X chromosome (Diallo *et al.*, 2001; Tsai *et al.*, 1996). This notwithstanding, the detailed analysis performed by Going *et al.* (2001) found different X chromosomes inactivated in multiple, morphologically homogeneous tumor samples in one-third of breast carcinomas, supporting the conclusion that a significant proportion of breast carcinomas are polyclonal (Going *et al.*, 2003; Heim *et al.*, 1997).

Because the majority of malignant breast tumors are ductal carcinomas of no special type, histological examination in most instances cannot discriminate between multifocal carcinomas (intramammary spreading from a single primary tumor) and multicentric carcinomas (more than one primary tumor existing within the breast). As even the most common breast cancer-associated chromosome abnormalities occur in no more than 10% of the tumors in the largest study relying on the same technique (Pandis *et al.*, 1995a), cytogenetics can be used to determine whether the tumor cell populations of two macroscopically distinct lesions within the same breast are clonally related. The evolutionary relationship among multiple, ipsilateral breast carcinomas was studied by cytogenetic analysis of 37 tumorous lesions from 17 patients (Pandis *et al.*, 1995b; Teixeira *et al.*, 1994, 1997). Clonal chromosome abnormalities were found in 27 lesions and the overall karyotypic pattern did not differ from that of solitary sporadic breast carcinomas inasmuch as cytogenetic polyclonality and the recurrent structural changes del(3p), der(1;16), and i(1q) were found. Nine of 12 patients with at least two karyotypically abnormal, macroscopically distinct carcinoma foci had an evolutionarily related, cytogenetically abnormal clone in the different tumor lesions from the same breast, showing that the dominant mechanism for the origin of multiple ipsilateral breast tumors is intramammary spreading from a single breast cancer.

The clonal relationship between synchronous or metachronous bilateral breast carcinomas was analyzed by karyotyping 20 tumors from 10 patients (Adeyinka *et al.*, 2000a; Pandis *et al.*, 1995b; Teixeira *et al.*, 1996). The cytogenetic findings in five of the eight cases which were informative indicated that both carcinomas arose independently as a genuinely primary tumor, because the chromosome aberrations they contained were completely different. In the remaining three cases, on the other hand, the cytogenetic data indicated that spreading of the disease from one breast to the other had taken place, since the same rare chromosome abnormalities were shared by the bilateral lesions. The conclusion that most patients with bilateral breast carcinomas have two different neoplastic diseases whereas most ipsilateral breast carcinomas arise through intramammary spreading of a single breast cancer was also reached by unsupervised hierarchical clustering analysis of CGH findings in 26 tumors from 12 breast cancer patients (Teixeira *et al.*, 2004).

The longitudinal clonal evolution in breast cancer can be examined by comparing the karyotypic changes of primary carcinomas with those of their metastases. Of the 17 carcinomas and their corresponding metastatic lesions that have been analyzed by chromosome-banding analysis (Adeyinka *et al.*, 2000a, 2000b; Pandis *et al.*, 1994, 1998; Teixeira *et al.*, 1996), nine primary tumors and eight metastatic lesions had only simple chromosomal abnormalities, including the very changes that are found most frequently as sole aberrations in sporadic carcinomas, whereas at least one clone with complex aberrations was detected in all the remaining cases. Although metastatic lesions tended to have more complex karyotypes than do unselected primary breast carcinomas, 7 out of 17 cases displayed fewer cytogenetically aberrant clones in the metastasis than in the respective primary carcinoma, indicating that the selection pressure facing the neoplastic cells during disease spreading reduced the number of clones surviving from an initially genotypically heterogeneous tumor cell population. The scarce existent data, therefore, indicate that metastatic lesions tend to be karyotypically more complex than primary breast cancers in the sense that the average clone has more extensive changes, but less complex in the sense that cytogenetic polyclonality is rarer. CGH analyses of paired primary-metastatic samples have also revealed extensive clonal divergence, indicating that metastasis may occur relatively early during breast carcinogenesis and is followed by independent clonal evolution from that of the primary tumor (Kuukasjarvi *et al.*, 1997a; Nishizaki *et al.*, 1997; Torres *et al.*, 2007). Although all cases of paired primary carcinoma/lymph node metastasis analyzed by CGH shared some genomic aberrations and no significant difference in the number of genomic imbalances was detected in the primary and secondary lesions, there were often differences in clonal composition (Torres *et al.*, 2007). Our finding of extensive intratumor genetic heterogeneity and divergent clonal evolution in both primary tumors and lymph node metastases is difficult to reconcile with the linear progression model of clonal dominance, which predicts that the spectrum of aberrations in metastatic lesions should be similar to that of the primary tumors from which they originated. The data at hand rather support a scenario in which metastases occur relatively early during breast carcinogenesis, allowing for extensive independent clonal progression of both the primary carcinomas and their metastases (Torres *et al.*, 2007). At the other end of the spectrum of tumor progression, combined tissue microdissection and CGH analyses of lobular or ductal carcinomas *in situ* and adjacent invasive carcinoma areas have shown in most cases an almost identical genetic pattern in the two lesions (Aubele *et al.*, 2000a, 2000b, 2000c; Buerger *et al.*, 1999, 2000; Hwang *et al.*, 2004; Kuukasjärvi *et al.*, 1997b; Nyante *et al.*, 2004), strongly underscoring the *in situ* lesions' precursor role in the development of invasive breast cancer.

IV. CYTOGENETIC ASSESSMENT OF URINARY BLADDER TUMOR CLONALITY

From a theoretical point of view, the epithelial coating of the renal pelvis, ureters, and urinary bladder, the urothelium, comes across as a highly likely tissue in which field cancerization might occur. Carcinogens excreted through the renal filters bathe the cells of the urothelium around the clock making it so likely as to seem almost inevitable that more or less simultaneous transformation of many target cells must occur, one would think. How tumorous lesions occur in the urinary bladder at first glance seems to underpin the reasoning above; around 70% of bladder carcinomas are superficial at diagnosis (Thompson *et al.*, 1993), and of these, 30% are found as multiple tumors and as many as 60% of the patients develop recurrence(s) after transutrethral resection of the primary tumor (s) (Kiemeney *et al.*, 1993). There are three main mechanisms that can be envisaged for this polychronotopicity of urothelial tumors: (1) implantation of cancer cells shed from the primary tumor may occur; (2) growth of new carcinoma(s) may take place from independently neoplastically transformed cells at remote sites; and (3) regrowth of an incompletely resected primary tumor may give rise to later tumor manifestations. Cytogenetic analyses of tumors of the urothelium have provided a wealth of information that has helped assess the relative likelihood of these three hypotheses.

First and foremost, karyotyping of individual transitional cell carcinomas (by far the most common) as well as squamous cell carcinomas of the bladder and ureters has shown that these tumors are monoclonal in composition inasmuch as all karyotypically abnormal cells from each tumor have at least one chromosomal abnormality in common (Fadl-Elmula *et al.*, 1998b, 1999a, 2000). Previous molecular studies based on X chromosome inactivation had also concluded that these tumors are of monoclonal origin (Sidransky *et al.*, 1992). The various numerical aberrations detected by karyotyping typically result in gain of a copy of chromosomes 7, 19, and 20 but loss of chromosomes 9, 10, 16, 18, X, and Y. Structural rearrangements leading to losses of the entire arm or parts of 1p, 6q, 8p, 9p, 11p, 13p, 15p, and 17p and gains of 1q, 3q, 5p, 8q, and 13q are also common. It is in the present context of further interest that the only radiation-induced bladder carcinoma cytogenetically examined and reported had 10 karyotypically unrelated clones (Fadl-Elmula *et al.*, 1998a), indicative of a carcinogenic field effect due to the previous radiation. Postradiation bladder cancers thus seem to be distinct also cytogenetically in addition to their known etiological and clinical distinctive features.

Fadl-Elmula *et al.* (1999b) examined the clonal relationships in multifocal urothelial carcinomas by karyotyping short-term cultured cells from 21 such tumors (mostly of the bladder) from a total of six patients. Karyotypically related, often identical, cytogenetically complex clones were found in all informative tumors from each case, including the recurrent tumors. Rearrangement of chromosome 9 leading to loss of material from the short and/or the long arm was seen in all cases, indicating that this is an early, pathogenetically important event in transitional cell carcinogenesis. The presence of related clones with great karyotypic similarity in anatomically distinct tumors from the same bladder indicates that multifocal urothelial tumors are of monoclonal origin and arise via intraluminal seeding of viable cancer cells shed from the original tumor. Later lesions may develop also from cells shed from the so-called second primary tumors. The relatively complex karyotypes seen in all lesions from most cases argue that the seeding of tumor cells is a late event that succeeds the acquisition by them of multiple secondary genetic abnormalities.

V. MONOCLONAL VERSUS POLYCLONAL TUMORIGENESIS: IS THE JURY STILL OUT?

The reported conclusions of clonality studies in solid tumors are not always in concordance, neither when different technologies were utilized nor even when one and the same technique was used. It may therefore be worth the while to take a closer view at the principal advantages and disadvantages of the two main investigative strategies—karyotyping and X chromosome inactivation analysis—to see if an explanation for the discrepancies can be found there. The main advantage of the cytogenetic approach is that one looks at actual individual cells and that the number of informative states (i.e., the number of clonality markers or different chromosomal aberrations that may be acquired by target cells) is infinitely large. The cytological nature of cytogenetics means that one does not have to rely on advanced statistics to assess clonality; the finding of only two cells with the same gained chromosome or structural rearrangement is sufficient to infer clonality (ISCN, 2009). Admittedly, a proviso in this context is that one does not face a situation with the cells being exposed to a carcinogen with the ability to induce one particular chromosomal change (if one did, the finding of that particular change in many cells would not be a sign of monoclonality). No examples are known, however, where such deviations from the main hypothesis that chromosome changes occur randomly with subsequent selection for cellular fitness deciding which will expand clonally and then become recognizable (in the situation referred to above with etoposide-induced *MLL*-rearrangements in secondary leukemia, the gene may be rearranged through a large number of translocations). In practice, therefore, this caveat can be disregarded, at least when the clonal chromosomal marker is a structural rearrangement of some complexity. The fact that karyotypic analyses of tumor cells are mostly performed on short-term cultured cells also does not represent any real problem for clonality assessments. Large and chromosomally complicated clones cannot evolve *in vitro* in the course of a few days culturing, and sometimes it has also been possible to demonstrate directly that the chromosomal aberrations detected in cultured cells really were present *in vivo* (Teixeira *et al.*, 2001; Torres *et al.*, 2007). A significant problem with the chromosomal approach, however, is that aberrations below the resolution level of cytogenetics (approximately 10 Mb) cannot be seen. It is therefore possible that a unifying, submicroscopic clonal marker (anything from a point mutation to a 10-Mb large chromosomal change) might exist even in tumors with very heterogeneous karyotypic aberration patterns, and consequently that they are monoclonal in spite of their cytogenetically polyclonal appearance. Although this would certainly rescue the consensus view that tumorigenesis is always monoclonal, it would do nothing to explain the observed heterogeneity (now demoted to secondary heterogeneity rather than primary) in, for example, tumors of the breast and pancreas (see above), which vastly surpasses that of leukemias, lymphomas, connective tissue tumors, or even most epithelial, including uroepithelial, tumors. One should also not disregard the nonprincipal but certainly substantial tendency there might be for karyotypic studies to underestimate the cytogenetic heterogeneity of solid tumors by the simple means of not looking carefully and at sufficiently many metaphase cells to be able to detect it. At best, perhaps 10–30 cells may be looked at in all but exceptional cases, which is much fewer than examined in the most meticulous studies (Gorunova *et al.*, 1998), and this certainly may lead to an underreporting of heterogeneity and other "difficult" findings. Here as elsewhere in science, absence of evidence is not necessarily evidence of absence.

Clonality analyses relying on the X chromosome inactivation principle operate with only two informative states; the inactivated chromosome is either maternal or paternal. A crucial hypothesis is furthermore that inactivation of the X is random: theoretically, therefore, in a 1000 cells large random cell population, 500 should have the maternal X and 500 should have the paternal X chromosome inactivated. But because inactivation is stable through cell generations, all descendents of a given somatic mother cell (i.e., the situation corresponding to monoclonal tumorigenesis) will have the same X inactivated. In practice, and because it is not easy to extract DNA from only cells belonging to the tumor parenchyma, investigators of tumor clonality using the X inactivation approach infer that a skewness in X inactivation pattern over a certain level signifies tumor monoclonality and that the deviation from the theoretical 100:0 distribution is attributable to admixture of DNA from stromal elements. However, within this reasoning it becomes impossible to recognize polyclonality with one dominating clone for what it is; the very statistical nature of the approach makes assessment of the true heterogeneity at the cellular level impossible—a limitation that is rarely made explicit in the relevant literature, unfortunately. Another premise behind the X inactivation approach also seems less certain than it once did, namely that the two X chromosomes are evenly inactivated in anatomical areas of more than microscopic size. As it turns out, this is not necessarily so, and either the paternal or the maternal X can be preferentially inactivated in fairly large patches of normal epithelia (Diallo *et al.*, 2001; Novelli *et al.*, 2003; Schmidt and Mead, 1990; Tsai *et al.*, 1996). Many biologically significant clones can develop from such a patch in, for example, breast or colorectal tumorigenesis, and even if they are not part of the same clone they will be misinterpreted as such in X chromosome inactivation analyses (Going, 2003).

We are facing a situation in which we, beyond any doubt, have secondary clonal heterogeneity, as exemplified by the sometime occurrence of secondary chromosome aberrations in leukemias, lymphomas and all sorts of solid tumors, and in some organ systems (perhaps most spectacularly in the breast and pancreas) also what appears to be widespread primary clonal heterogeneity (the clones are not cytogenetically related). Whenever cell-to-cell variation exists within a neoplastic parenchyma, Darwinian selection must also occur so that cells with some aberration patterns have an evolutionary edge and are selected for whereas others may demonstrate lower fitness (although the very fact that we observe them and recognize them at all means that the chromosomal mutations cannot have been wholly deleterious). How does this presumably dynamic clonal interplay within a neoplastic cell population evolve with time?

Theoretical reasoning shows that one may have either of four principal scenarios (Heim *et al.*, 1988; Nowell, 1976, Fig. 2). (1) In some cancers, initial monoclonality is retained throughout the entire course of the disease (Fig. 2A) with no additional, secondary aberrations accrued as judged by karyotypic appearance (only a single chromosome abnormality is seen in the cells of all analyzed samples). This is not an unusual situation in, for example, leukemias and some sarcomas that have been followed longitudinally by cytogenetic analysis. The existence of this clonal subset (one could also envisage clonal stability during progression of an initially polyclonal neoplasm as well as other mixed scenarios, but our focus here is on the most distinctive alternatives only) furthermore underscores that genomic instability with the development of complex karyotypic abnormalities is not ubiquitous even for highly malignant, solid tumors. (2) The perhaps most common situation, however, in both leukemias, lymphomas, sarcomas, and carcinomas is illustrated in Fig. 2B: tumorigenesis is monoclonal but additional aberrations develop with time leading to secondary clonal heterogeneity. (3) If there is initial clonal heterogeneity (polyclonal tumorigenesis), later developments may lead to a reduction in the early complexity resulting in oligo- or monoclonality (Fig. 2C) as evolution works its magic. There are in particular two major steps in a malignant epithelial

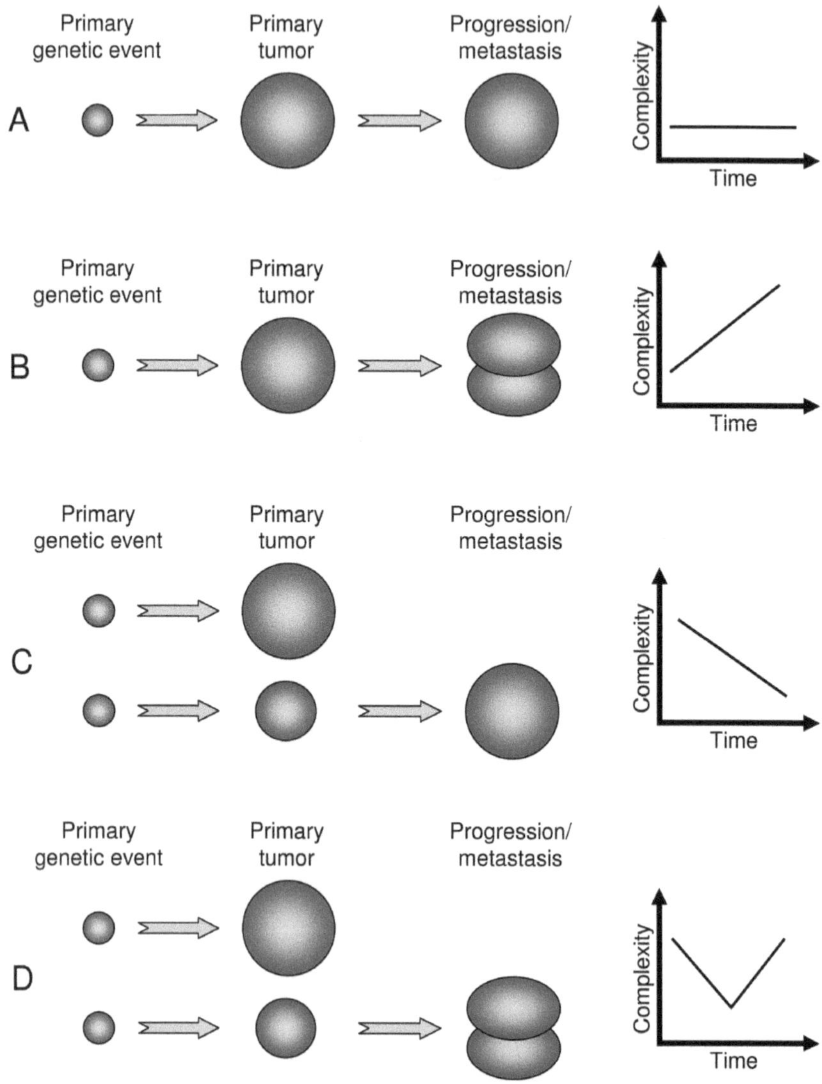

Fig. 2 Genetic stability, divergence, and convergence during tumor progression. (A) Monoclonal tumorigenesis with genomic stability throughout tumor progression; no additional chromosome aberrations develop. (B) Monoclonal tumorigenesis displaying genetic divergence with time as the result of the acquisition and clonal selection for secondary cytogenetic changes (overlapping ellipses represent cytogenetically related clones). (C) Polyclonal tumorigenesis with an overall reduction in genomic complexity with time (clonal convergence) due to selection among cytogenetically unrelated clones during tumor progression and the development of metastases (the areas of nonoverlapping circles represent the relative size of cytogenetically unrelated clones). (D) Polyclonal tumorigenesis with initial clonal convergence followed by later clonal divergence due to the acquisition of additional cytogenetic changes by the clone that was selected during the development of metastases.

tumor's life that are likely to involve a reduction in clonal heterogeneity if any existed to begin with: when the *in situ* lesion penetrates the basal lamina to become an infiltrating carcinoma, and when distant metastases are set up. In addition, iatrogenic intervention may have a major impact, too, not least when effective cytostatics are administered. Leaving stochastic elements aside, although they are certainly likely to be of more than trivial importance in individual cases, both mentioned "natural" events and every introduction of a new anticancer therapy involves a major alteration in the selection pressure the neoplasm faces. After infiltration below the basal lamina, the tumor cells become exposed to the full force of the organism's humoral as well as cellular immune defenses, and it also goes without saying that a change in neoplastic parenchyma cells' immediate milieu from the primary tumor to, for example, a new life in a lymph node or liver or lung tissue involves major environmental changes. Likewise, clones or subclones with different genetic characteristics are likely to manifest different susceptibilities to anticancer drugs. New clones with better fitness in the new locales or new pharmacological environment will inevitably be selected for—something that may help formerly less predominant clones now gain the upper hand. The finding of genomically less complex clones in metastases than in primary tumors in both breast and colon cancer illustrates that this is not only a theoretical necessity; it can also be observed in practice (Adeyinka *et al.*, 2000a, 2000b; Bardi *et al.*, 1997b; Pandis *et al.*, 1994, 1998; Teixeira *et al.*, 1996). Finally, after the initial reduction in clonal complexity outlined above has taken place, a new and secondary complexity may occur (Fig. 2D) in which admittedly fewer cytogenetically unrelated clones are seen (the neoplastic cell population in the metastasis may even have been reduced to pseudomonoclonality), but the occurrence of new secondary abnormalities in the clones that managed to adapt to the new environment leads to a new complexity, whenever the tumor genome is karyotypically unstable. Again, this is something we have been able to observe in cytogenetic studies of primary tumors and their metastases, both in breast cancer and large bowel cancer (Adeyinka *et al.*, 2000a, 2000b; Bardi *et al.*, 1997b; Pandis *et al.*, 1994, 1998; Teixeira *et al.*, 1996).

Can we be certain, then, that the observed cytogenetic polyclonality really in some epithelial tumors corresponds to the actual biological polyclonality inherent in the field cancerization concept? No, we cannot, so in this sense we must admit that the jury is still out. What remains unquestionable, however, is that some carcinomas of, for example, the breast and pancreas exhibit clonal heterogeneity, be it primary or secondary to some unseen and unifying primary change, that sets them radically apart from unquestionably monoclonal tumors. In our opinion, this heterogeneity should be viewed as a clonal corollary of what may pathogenetically be termed clonal collaboration. Differently transformed cells, which lie close to one another in an epithelial sheath, may via paracrine mechanisms contribute in complementary ways to the total phenotype of the neoplasm. In hematological as well as connective tissues, the cells are much farther apart and so comparable paracrine mechanisms cannot come into effect, at least not with the same ease. Although this hypothesis is attractive, especially since it would provide a functional correlate for the logic of field cancerization of epithelia that are exposed to chemical or physical carcinogenic chronic stimuli, it suffers from a near-total lack of understanding of what might constitute the said paracrine effects in molecular terms. Nor do we know why this mechanism seems to work better in exocrine glands such as the breast and pancreas than in the ureters and urinary bladder. One would hope that more studies of individual cells isolated from dysplastic or otherwise partially transformed epithelia might provide the wanted insights into this aspect of carcinogenesis. Whatever the feasibility or outcome of such studies, we feel certain that only examination of individual cells can resolve the age-old questions discussed in this review. Methods that work with samples obtained from a theoretical "mean tumor cell," even if it is a "mean neoplastic parenchyma cell," are intrinsically incapable (Heim, 1992) of revealing information about the cell-to-cell variability that lies at the heart of this fundamental issue.

REFERENCES

Adeyinka, A., Mertens, F., Bondeson, L., Garne, J. P., Borg, A., Baldetorp, B., and Pandis, N. (2000a). Cytogenetic heterogeneity and clonal evolution in synchronous bilateral breast carcinomas and their lymph node metastases from a male patient without any detectable BRCA2 germline mutation. *Cancer Genet. Cytogenet.* **118**, 42–47.

Adeyinka, A., Kytola, S., Mertens, F., Pandis, N., and Larsson, C. (2000 b). Spectral karyotyping and chromosome banding studies of primary breast carcinomas and their lymph node metastases. *Int. J. Mol. Med.* **5**, 235–240.

Alcock, H. E., Stephenson, T. J., Royds, J. A., and Hammond, D. W. (2003). Analysis of colorectal tumor progression by microdissection and comparative genomic hybridization. *Genes Chromosomes Cancer* **37**, 369–380.

Al-Mulla, F., Keith, W. N., Pickford, I. R., Going, J. J., and Birnie, G. D. (1999). Comparative genomic hybridisation analysis of primary colorectal carcinomas and their synchronous metastases. *Genes Chromosomes Cancer* **24**, 306–314.

Aubele, M., Cummings, M., Walsch, A., Zitzelsberger, H., Nährig, J., Höfler, H., and Werner, M. (2000a). Heterogeneous chromosomal aberrations in intraductal breast lesions adjacent to invasive carcinoma. *Anal. Cell. Pathol.* **20**, 17–24.

Aubele, M. M., Cummings, M. C., Mattis, A. E., Zitzelsberger, H. F., Walch, A. K., Kremer, M., Höfler, H., and Werner, M. (2000 b). Accumulation of chromosomal imbalances from intraductal proliferative lesions to adjacent *in situ* and invasive ductal breast cancer. *Diagn. Mol. Pathol.* **9**, 14–19.

Aubele, M., Mattis, A., Zitzelsberger, H., Walch, A., Kremer, M., Welzl, G., Höfler, H., and Werner, M. (2000 c). Extensive ductal carcinoma *in situ* with small foci of invasive ductal carcinoma: Evidence of genetic resemblance by CGH. *Int. J. Cancer* **85**, 82–86.

Bardi, G., and Heim, S. (2009). Tumors of the digestive tract. *In* Cancer Cytogenetics (S. Heim, and F. Mitelman, eds.), 3rd ed. pp. 429–461. *Wiley-Blackwell, New Jersey.*

Bardi, G., Parada, L. A., Bomme, L., Pandis, N., Johansson, B., Willén, R., Fenger, C., Kronborg, O., Mitelman, F., and Heim, S. (1997b). Cytogenetic findings in metastases from colorectal cancer. *Int. J. Cancer* **72**, 604–607.

Bardi, G., Parada, L. A., Bomme, L., Pandis, N., Willén, R., Johansson, B., Jepsson, B., Beroukas, K., and Heim, S. (1997a). Cytogenetic comparisons of synchronous carcinomas and polyps in patients with colorectal cancer. *Br. J. Cancer* **76**, 765–769.

Bedenne, L., Faivre, J. M. C., Piard, F., Cauvin, J. M., and Hillon, P. (1992). Adenoma-carcinoma sequence or de novo carcinogenesis? A study of adenomatous remnant in a population-based series of large bowel cancers. *Cancer* **69**, 883–888.

Blanco, J. G., Edick, M. J., and Relling, M. V. (2004). Etoposide induced chimeric *Mll* gene fusions. *FASEB J.* **18**, 173–175.

Bloomfield, C. D., Archer, K. J., Mrozek, K., Lillington, D. M., Kaneko, Y., Head, D. R., Dal Cin, P., and Raimondi, S. C. (2002). 11q23 balanced chromosome aberrations in treatment-related myelodysplastic syndromes and acute leukemia: Report from an international workshop. *Genes Chromosomes Cancer* **33**, 362–378.

Bomme, L., Bardi, G., Pandis, N., Fenger, C., Kronborg, O., and Heim, S. (1996). Cytogenetic analysis of colorectal adenomas: Karyotypic comparisons of synchronous tumors. *Cancer Genet. Cytogenet.* **106**, 66–71.

Brabletz, T., Jung, A., Spaderna, S., Hlubek, F., and Kirchner, T. (2005). Opinion: Migrating cancer stem cells—an integrated concept of malignant tumour progression. *Nat. Rev. Cancer* **5**, 744–749.

Buerger, H., Otterbach, F., Simon, R., Poremba, C., Diallo, R., Decker, T., Riethdorf, L., Brinkschmidt, C., Dockhorn-Dworniczak, B., and Boecker, W. (1999). Comparative genomic hybridization of ductal carcinoma *in situ* of the breast—evidence of multiple genetic pathways. *J. Pathol.* **187**, 396–402.

Buerger, H., Simon, R., Schäfer, K. L., Diallo, R., Littmann, R., Poremba, C., van Diest, P. J., Dockhorn-Dworniczak, B., and Böcker, W. (2000). Genetic relation of lobular carcinoma *in situ*, ductal carcinoma *in situ*, and associated invasive carcinoma of the breast. *Mol. Pathol.* **53**, 118–121.

D'Arrigo, A., Belluco, C., Ambrosi, A., Digito, M., Esposito, G., Bertola, A., Fabris, M., Nofrate, V., Mammano, E., Leon, A., Nitti, D., and Lise, M. (2005). Metastatic transcriptional pattern revealed by gene expression profiling in primary colorectal carcinoma. *Int. J. Cancer* **115**, 256–262.

Diallo, R., Schaefer, K. L., Poremba, C., Shivazi, N., Willmann, V., Buerger, H., Dockhorn-Dworniczak, B., and Boecker, W. (2001). Monoclonality in normal epithelium and in hyperplastic and neoplastic lesions of the breast. *J. Pathol.* **193**, 27–32.

Fadl-Elmula, I., Bonaldi, L., Gorunova, L., Mandahl, N., Elfving, E., Mitelman, F., and Heim, S. (1998a). Cytogenetic heterogeneity in a second primary radiation-induced bladder carcinoma: Ten karyotypically unrelated clones. *Cancer Genet. Cytogenet.* **105**, 134–137.

Fadl-Elmula, I., Gorunova, L., Lundgren, R., Mandahl, N., Forsby, N., Mitelman, F., and Heim, S. (1998b). Chromosomal abnormalities in two bladder carcinomas with secondary squamous cell differentiation. *Cancer Genet. Cytogenet.* **102**, 125–130.

Fadl-Elmula, I., Gorunova, L., Mandahl, N., Elfving, E., Lundgren, R., Mitelman, F., and Heim, S. (1999b). Cytogenetic monoclonality in multifocal uroepithelial carcinomas: Evidence of intraluminal tumor seeding. *Br. J. Cancer* **81**, 6–12.

Fadl-Elmula, I., Gorunova, L., Mandahl, N., Elfving, E., Lundgren, R., Mitelman, F., and Heim, S. (2000). Karyotypic characterization of urinary bladder transitional cell carcinomas. *Genes Chromosomes Cancer* **29**, 256–265.

Fadl-Elmula, I., Gorunova, L., Mandahl, N., Elfving, E., Radomak, C., Mitelman, F., and Heim, S. (1999a). Cytogenetic analysis of upper urinary tract transitional cell carcinomas. *Cancer Genet. Cytogenet.* **115**, 123–127.

Fawole, A. S., Simpson, D. J., Rajagopal, R., Elder, J., Holland, T. A., Fryer, A., Deakin, M., Elder, J. B., and Farrell, W. E. (2002). Loss of heterozygosity on chromosome 10q is associated with earlier onset sporadic colorectal adenocarcinoma. *Int. J. Cancer* **99**, 829–833.

Fialkow, P. J. (1976). Clonal origin of human tumors. *Biochim. Biophys. Acta* **458**, 283–321.

Fioretos, T., and Johansson, B. (2009). Chronic myeloid leukemia. *In* Cancer Cytogenetics (S. Heim, and F. Mitelman, eds.), 3rd ed. pp. 179–207. *Wiley-Blackwell, New Jersey.*

Going, J. J. (2003). Epithelial carcinogenesis: Challenging monoclonality. *J. Pathol.* **200**, 1–3.

Going, J. J., Abd El-Monem, H. M., and Craft, J. A. (2001). Clonal origins of human breast cancer. *J. Pathol.* **194**, 406–412.

Gorunova, L., Höglund, M., Andren-Sandberg, A., Dawiskiba, S., Jin, Y., Mitelman, F., and Johansson, B. (1998). Cytogenetic analysis of pancreatic carcinomas: Intratumor heterogeneity and nonrandom pattern of chromosome aberrations. *Genes Chromosomes Cancer* **23**, 81–99.

Grade, M., Becker, H., Liersch, T., Ried, T., and Ghadimi, B. M. (2006). Molecular cytogenetics: Genomic imbalances in colorectal cancer and their clinical impact. *Cellular Oncol.* **28**, 71–84.

Heim, S. (1992). Is cancer cytogenetics reducible to the molecular genetics of cancer cells? *Genes Chromosomes Cancer* **5**, 188–196.

Heim, S., Mandahl, N., and Mitelman, F. (1988). Genetic convergence and divergence in tumor progression. *Cancer Res.* **48**, 5911–5916.

Heim, S., and Mitelman, F. (2009). Cancer Cytogenetics, 3rd ed. Wiley-Blackwell, New Jersey.

Heim, S., Teixeira, M. R., Dietrich, C. U., and Pandis, N. (1997). Cytogenetic polyclonality in tumors of the breast. *Cancer Genet. Cytogenet.* **95**, 16–19.

Hwang, E. S., Nyante, S. J., Yi Chen, Y., Moore, D., DeVries, S., Korkola, J. E., Esserman, L. J., and Waldman, F. M. (2004). Clonality of lobular carcinoma *in situ* and synchronous invasive lobular carcinoma. *Cancer* **100**, 2562–2572.

ISCN (2009). International System of Human Cytogenetic Nomenclature. (L. G. Shaffer, M. L. Slovak, and L. J. Campbell, Eds.). S Karger AG, Basel.

Jiang, J. K., Chen, Y. J., Lin, C. H., Yu, I. T., and Lin, J. K. (2005). Genetic changes and clonality relationship between primary colorectal cancers and their pulmonary metastases—an analysis by comparative genomic hybridisation. *Genes Chromosomes Cancer* **43**, 25–36.

Kiemeney, L. A., Witjes, J. A., Heijbroek, R. P., Verbeek, A. L., and Debruyne, F. M. (1993). Predictability of recurrent and progressive disease in individual superficial bladder cancer. *J. Urol.* **150**, 60–64.

Knosel, T., Schluns, K., Dietel, M., and Petersen, I. (2005). Chromosomal alterations in lung metastases of colorectal carcinomas: Associations with tumor specific tumor dissemination. *Clin. Exp. Metastasis* **22**, 533–538.

Knosel, T., Schluns, K., Stein, U., Schwabe, H., Schlag, P. M., Dietel, M., and Petersen, I. (2004). Chromosomal alterations during lymphatic and liver metastasis formation of colorectal cancer. *Neoplasia* **6**, 23–28.

Kuukasjarvi, T., Karhu, R., Tanner, M., Kahkonen, M., Schaffer, A., Nupponen, N., Pennanen, S., Kallioniemi, A., Kallioniemi, O. P., and Isola, J. (1997a). Genetic heterogeneity and clonal evolution underlying development of asynchronous metastasis in human breast cancer. *Cancer Res.* **57**, 1597–1604.

Kuukasjarvi, T., Tanner, M., Pennanen, S., Karhu, R., Kallioniemi, O. P., and Isola, J. (1997b). Genetic changes in intraductal breast cancer detected by comparative genomic hybridization. *Am. J. Pathol.* **150**, 1465–1471.

Leggett, B. A., Devereaux, B., Biden, K., Searle, J., Young, J., and Jass, J. (2001). Hyperplastic polyposis: Association with colorectal cancer. *Am. J. Surg. Pathol.* **25**, 177–184.

Mao, X., Hamoudi, R. A., Talbot, I. C., and Baudis, M. (2006). Allele-specific loss of heterozygosity in multiple colorectal adenomas: Toward an integrated molecular cytogenetic map II. *Cancer Genet. Cytogenet.* **167**, 1–14.

Mitelman Database of Chromosome Aberrations and Gene Fusions in Cancer (2011). (F. Mitelman, B. Johansson, and F. Mertens, Eds.), http://cgap.nci.nih.gov/Chromosomes/Mitelman.

Nishizaki, T., DeVries, S., Chew, K., Goodson III, W. H., Ljung, B. M., Thor, A., and Waldman, F. M. (1997). Genetic alterations in primary breast cancers and their metastases: Direct comparison using modified comparative genomic hybridization. *Genes Chromosomes Cancer* **19**, 267–272.

Noguchi, S., Motomura, K., Inaji, H., Imaoka, S., and Koyama, H. (1992). Clonal analysis of human breast cancer by means of the polymerase chain reaction. *Cancer Res.* **52**, 6594–6597.

Noguchi, S., Motomura, K., Inaji, H., Imaoka, S., and Koyama, H. (1994). Clonal analysis of predominantly intraductal carcinoma and pre-cancerous lesions of the breast by means of polymerase chain reaction. *Cancer Res.* **54**, 1849–1853.

Novelli, M., Cossu, A., Oukrif, D., Quaglia, A., Lakhani, S., Poulsom, R., Sasieni, P., Carta, P., Contini, M., Pasca, A., Palmieri, G., Nodmer, W., et al., (2003). X-inactivation patch size in human female tissue confounds the assessment of tumor clonality. *Proc. Natl. Acad. Sci.* **100**, 3311–3314.

Nowell, P. C. (1976). The clonal evolution of tumor cell populations. *Science* **194**, 23–28.

Nyante, S. J., Devries, S., Chen, Y. Y., and Hwang, E. (2004). Array-based comparative genomic hybridization of ductal carcinoma *in situ* and synchronous invasive lobular cancer. *Hum. Pathol.* **35**, 759–763.

Pandis, N., Bardi, G., Jin, Y., Dietrich, C., Johansson, B., Andersen, J., Mandahl, N., Mitelman, F., and Heim, S. (1994). Unbalanced 1;16-translocation as the sole karyotypic abnormality in a breast carcinoma and its lymph node metastasis. *Cancer Genet. Cytogenet.* **75**, 158–159.

Pandis, N., Heim, S., Bardi, G., Idvall, I., Mandahl, N., and Mitelman, F. (1993). Chromosome analysis of 20 breast carcinomas: Cytogenetic multiclonality and karyotypic-pathologic correlations. *Genes Chromosomes Cancer* **6**, 51–57.

Pandis, N., Jin, Y., Gorunova, L., Petersson, C., Bardi, G., Idvall, I., Johansson, B., Ingvar, C., Mandahl, N., Mitelman, F., and Heim, S. (1995a). Chromosome analysis of 97 primary carcinomas of the breast: identification of eight karyotypic subgroups. *Genes Chromosomes Cancer* **12**, 173–185.

Pandis, N., Teixeira, M. R., Adeyinka, A., Rizou, H., Bardi, G., Mertens, F., Andersen, J. A., Bondeson, L., Sfikas, K., Qvist, H., Apostolikas, N., Mitelman, F., and Heim, S. (1998). Cytogenetic comparison of primary tumors and lymph node metastases in breast cancer patients. *Genes Chromosomes Cancer* **22**, 122–129.

Pandis, N., Teixeira, M. R., Gerdes, A. M., Limon, J., Bardi, G., Andersen, J. A., Idvall, I., Mandahl, N., Mitelman, F., and Heim, S. (1995b). Chromosome abnormalities in bilateral breast carcinomas. Cytogenetic evaluation of the clonal origin of multiple primary tumors. *Cancer* **76**, 250–258.

Parada, L. A., Maranon, A., Hallen, M., Tranberg, K. -G., Stenram, U., Bardi, G., and Johansson, B. (1999). Cytogenetic analyses of secondary liver tumors reveal significant differences in genomic imbalances between primary and metastatic colon carcinomas. *Clin. Exp. Metastasis* **17**, 471–479.

Paredes-Zaglul, A., Kang, J. J., Essig, Y. P., Mao, W. G., Irby, R., Wloch, M., and Yeatman, T. J. (1998). Analysis of colorectal cancer by comparative genomic hybridization: Evidence for inclusion of the metastatic phenotype by loss of tumor suppressor genes. *Clin. Cancer Res.* **4**, 879–886.

Persson, K., Pandis, N., Mertens, F., Borg, A., Baldetorp, B., Killander, D., and Isola, J. (1999). Chromosomal aberrations in breast cancer: A comparison between cytogenetics and comparative genomic hybridization. *Genes Chromosomes Cancer* **25**, 115–122.

Schmidt, G. H., and Mead, R. (1990). On the clonal origin of tumors—lessons from studies of intestinal epithelium. *Bioassays* **12**, 37–40.

Sherbenou, D. W., and Druker, B. J. (2007). Applying the discovery of the Philadelphia chromosome. *J. Clin. Invest.* **117**, 2067–2074.

Sidransky, D., Frost, P., Von Eschenbach, A., Oyasu, R., Preisinger, A. C., and Vogelstein, B. (1992). Clonal origin bladder cancer. *N. Engl. J. Med.* **326**, 737–740.

Teixeira, M. R., Pandis, N., Bardi, G., Andersen, J. A., Bohler, P. J., Qvist, H., and Heim, S. (1997). Discrimination between multicentric and multifocal breast carcinoma by cytogenetic investigation of macroscopically distinct ipsilateral lesions. *Genes Chromosomes Cancer* **18**, 170–174.

Teixeira, M. R., Pandis, N., Bardi, G., Andersen, J. A., and Heim, S. (1996). Karyotypic comparisons of multiple tumorous and macroscopically normal surrounding tissue samples from patients with breast cancer. *Cancer Res.* **56**, 855–859.

Teixeira, M. R., Pandis, N., Bardi, G., Andersen, J. A., Mandahl, N., Mitelman, F., and Heim, S. (1994). Cytogenetic analysis of multifocal breast carcinomas: Detection of karyotypically unrelated clones as well as clonal similarities between tumour foci. *Br. J. Cancer* **70**, 922–927.

Teixeira, M. R., Pandis, N., Bardi, G., Andersen, J. A., Mitelman, F., and Heim, S. (1995). Clonal heterogeneity in breast cancer: Karyotypic comparisons of multiple intra- and extra-tumorous samples from 3 patients. *Int. J. Cancer* **63**, 63–68.

Teixeira, M. R., Pandis, N., and Heim, S. (2002). Cytogenetic clues to breast carcinogenesis. *Genes Chromosomes Cancer* **33**, 1–16.

Teixeira, M. R., Pandis, N., and Heim, S. (2009). Tumors of the breast. *In* Cancer Cytogenetics (S. Heim, and F. Mitelman, eds.), 3rd ed. pp. 493–518. *Wiley-Blackwell, New Jersey.*

Teixeira, M. R., Ribeiro, F. R., Torres, L., Pandis, N., Andersen, J. A., Lothe, R. A., and Heim, S. (2004). Assessment of clonal relationships in ipsilateral and bilateral multiple breast carcinomas by comparative genomic hybridisation and hierarchical clustering analysis. *Br. J. Cancer* **91**, 775–782.

Teixeira, M. R., Tsarouha, H., Kraggerud, S. M., Pandis, N., Dimitriadis, E., Andersen, J. A., Lothe, R. A., and Heim, S. (2001). Evaluation of breast cancer polyclonality by combined chromosome banding and comparative genomic hybridization analysis. *Neoplasia* **3**, 204–214.

Thompson, R. A., Campbell, E. W., Kramer, H. C., Jacobs, S. C., and Naslund, M. J. (1993). Late invasive recurrence despite long-term surveillance for superficial bladder cancer. *J. Urol.* **149**, 1010–1011.

Torres, L., Ribeiro, F. R., Pandis, N., Andersen, J. A., Heim, S., and Teixeira, M. R. (2007). Intratumor genomic heterogeneity in breast cancer with clonal divergence between primary carcinomas and lymph node metastases. *Breast Cancer Res. Treat.* **102**, 143–155.
Tsai, Y. C., Lu, Y., Nichols, P. W., Zlotnikov, G., Jones, P. A., and Smith, H. S. (1996). Contiguous patches of normal human mammary epithelium derived from a single stem cell: Implications for breast carcinogenesis. *Cancer Res.* **56**, 402–404.
Vogelstein, B., Fearon, E. R., Hamilton, S. R., and Feinberg, A. P. (1985). Use of restriction fragment length polymorphisms to determine the clonal origin of human tumors. *Science* **227**, 642–645.

Genotyping Techniques to Address Diversity in Tumors

David Lindgren[*], Mattias Höglund[†] and Johan Vallon-Christersson[†‡]

[*] *Center for Molecular Pathology, Department of Laboratory Medicine, Lund University, SUS Malmö, Malmö, Sweden*
[†] *Department of Oncology, Clinical Sciences, Lund University, Lund, Sweden*
[‡] *CREATE Health Strategic Center for Translational Cancer Research, Lund University, Lund, Sweden*

Array-based genotyping platforms have during recent years been established as a valuable tool for the characterization of genomic alterations in cancer. The analysis of tumor samples, however, presents challenges for data analysis and interpretation. For example, tumor samples are often admixed with nonaberrant cells that define the tumor microenvironment, such as infiltrating lymphocytes and fibroblasts, or vasculature. Furthermore, tumors often comprise subclones harboring divergent aberrations that are acquired subsequent to the tumor-initiating event. The combined analysis of both genotype and copy number status obtained by array-based genotyping platforms provide opportunities to address these challenges. In this chapter, we present the basic principles for current array-based genotyping platforms and how they can be used to infer genotype and copy number for

Advances in CANCER RESEARCH, Volume 112
Copyright 2011, Elsevier Inc. All right reserved.
0065-230X/10 $35.00
DOI: 10.1016/B978-0-12-387688-1.00006-5

acquired genomic alterations. We describe how these techniques can be used to resolve tumor ploidy, normal cell admixture, and subclonality. We also exemplify how genotyping techniques can be applied in tumor studies to elucidate the hierarchy among tumor clones, and thus, provide means to study clonal expansion and tumor evolution. © 2011 Elsevier Inc.

I. INTRODUCTION

Cancer development and tumor formation involve acquired genomic aberrations, such as sequence mutations and copy number changes. Molecular investigation of genomic alterations in tumors has traditionally been performed using methods such as loss of heterozygosity (LOH) analyses and comparative genomic hybridization (CGH). Conventional GCH, first described by Kallioniemi and coworkers (Kallioniemi *et al.*, 1992), uses differentially fluorescently labeled DNA from tumor sample and reference DNA to reveal regions of loss and gain by competitive hybridization to immobilized normal metaphase chromosomes. With the advent of array technology (Schena *et al.*, 1995), the analysis of cancer genomes advanced rapidly with greatly increased resolution and sensitivity. Array-based comparative genomic hybridization (aCGH) was first performed using gene-centered arrays originally developed for gene expression analysis, or using low-density arrays of large genomic segments cloned in bacterial artificial chromosomes (BACs) (Pollack *et al.*, 1999). Initial techniques were soon further developed for genome-wide investigation of copy number aberrations at high-resolution by tiling BAC arrays and subsequently by employing oligonucleotide probe arrays. In short, aCGH utilizes the same strategy as conventional metaphase CGH but DNA is hybridized to immobilized DNA probes mapped to known genomic locations. Current array platforms, comprising from tens of thousands up to millions of probes, allow for detection of breakpoints and copy number alterations at subgene resolution and have been widely used to screen for genomic alterations in cancer (Pinkel and Albertson, 2005). Such analyses have provided a depiction of copy number gain and loss frequencies across large tumor cohorts in a variety of cancers, highlighting recurrent alterations important during oncogenesis and tumor development (Chin *et al.*, 2006). LOH analyses have, on the other hand, been widely used in cancer research to detect regions of allelic imbalances indicating regions of genomic deletion or copy-number-neutral LOH, and have been used to identify tumor suppressor genes inactivated by mutation followed by loss of the wild-type allele. Traditionally, LOH analysis use polymorphic markers, such as nucleotide repeat regions or single nucleotide polymorphisms, to detect regions of allelic imbalance.

Whole genome genotyping (WGG) arrays based on single nucleotide polymorphisms (SNPs) (Wang *et al.*, 1998) were developed to analyze blood samples in association studies and have since its introduction successfully been used in numerous studies for identification of genetic susceptibility loci in a variety of diseases (Grant and Hakonarson, 2008). Progression of WGG arrays, or SNP arrays, has followed the identification of SNPs in the human genome derived from initiatives such as the international HapMap Project (http://www.hapmap.org), and platforms currently in use allow for genotyping of millions of SNPs simultaneously. Even though SNP arrays were not originally designed for analysis of tumor samples, it was soon demonstrated that these platforms are suitable for the analysis of cancer genomes (LaFramboise *et al.*, 2005; Lindblad-Toh *et al.*, 2000; Peiffer *et al.*, 2006; Wang *et al.*, 2004; Zhao *et al.*, 2005;). Allele-specific interrogation of tumor DNA using SNP arrays provides means to investigate the relative abundance of alleles and effectively combine the advantages of LOH analysis and aCGH analysis. Thus, SNP arrays enable researchers to detect copy-neutral events in tumors along with copy-number aberrations. SNP arrays have therefore become a valuable tool for the analysis of cancer genomes and have been used to provide detailed characterization of a variety of cancers (LaFramboise, 2009; Wang and Armstrong, 2007). The two most commonly used platforms for SNP arrays are Illumina BeadChip (Gunderson *et al.*, 2005; Steemers *et al.*, 2006) and Affymetrix (Matsuzaki *et al.*, 2004; Wang *et al.*, 1998). Experimental design and data analysis are somewhat different, although the main principles are applicable to both platforms as that they provide detection signals from individual alleles separately.

There are several inherent problems with analyses of tumor genomes. For example, a significant proportion of solid tumors are highly aneuploid and subjected to genome duplication events causing deviations from the normal diploid chromosome level (Rajagopalan and Lengauer, 2004). Therefore, a problem of determining the baseline for calling relative genomic copy-number alterations becomes apparent. In cytogenetics, the most common chromosome number in a cell population—the modal number—determines if the state of a genomic region is regarded as neutral, gained, or lost relative to a fixed ploidy status. In aCGH, the absolute copy number cannot be resolved and copy number is presented as relative to a reference point, often approximated to the mean- or median copy number, or to the predominant relative copy number (Staaf *et al.*, 2007). As SNP arrays provide assessment of allelic composition in combination with abundance, it opens up for strategies that estimate absolute copy number and ploidy (LaFramboise, 2009). SNP array platforms have also successfully been applied to address problems regarding intermixture of nonaberrant cell populations. As analysis is performed on extracted DNA rather than on individual cells *in situ*, measured copy number changes will reflect overall net changes in the cell population from which DNA is extracted. Thus, the amplitude of signal associated with a copy-number alteration is dependent on the fraction of cells harboring the alteration. When DNA from grossly dissected tumor biopsies is analyzed, presence of residual tumor-adjacent or infiltrating normal cells will reduce the dynamic range between segments of different copy number. Importantly, with aCGH data it is not straightforward to discriminate between contamination of normal genomes and varying magnitude of underlying net copy-number changes, although there have been efforts aimed at resolving this issue (Tolliver *et al.*, 2010). Traditionally, normal-contamination issues have been addressed by excluding samples with low tumor cellularity from analysis or, when feasible, by microdissection of biopsies. However, the tumor microenvironment comprising tumor cells and other cell types such as immune cells, fibroblasts, and endothelial cells is integral in tumor development and progression. The interplay between cells within the tumor microenvironment has been highlighted as important hallmarks of cancer and its composition has been shown to represent an intrinsic property of tumors (Hanahan and Weinberg, 2011). Excluding samples due to cellularity may therefore bias tumor cohort composition. In this respect, SNP arrays might offer an advantage as it has been demonstrated that interpretation of the readout of allelic imbalances can be successfully used to estimate and correct for cellularity or the fraction of cells affected by an alteration (Assie *et al.*, 2008; Nancarrow *et al.*, 2007). The presence of genetic diversity within tumor samples, that is, tumor subclonality, represents another source of sample heterogeneity that presents challenges when analyzing tumor genomes. However, the combined analysis of both genotype and copy number status obtained by SNP array analysis provide opportunities to discern subclonal alterations against the background of the predominant clone. Likewise, genotype estimates from SNP array data will provide increased possibilities to study clonal relationships between repeated tumor samples from the same individual.

Here we aim to provide the basic principles for array-based genotyping platforms and the principles of how these techniques can be used to address sample heterogeneity and subclonality in tumors. We first provide a brief description of SNP array platforms, experimental procedures, and data extraction. We then proceed to describe the calculation of *B* allele frequency (BAF) and relative copy number, and how these values are affected by underlying acquired genetic alterations. We finally discuss how these data can be used and interpreted with the aim of deducing intermixture of nonaberrant cells within tumor biopsies, as well as subclonal events and intratumor heterogeneity.

II. THE BASIC PRINCIPLES OF SNP ARRAY PLATFORMS AND SNP ARRAY DATA INTERPRETATION

A. Platforms and Probe Design

There are two SNP array platforms predominantly in use, provided by Affymetrix and Illumina, respectively. Both platforms have been extensively used for genotyping blood samples in genetic linkage studies as well as for the analysis of cancer genomes. The underlying chemistry differs between the platforms but both can be used to interrogate genotypes in a similar manner and both depend on classic base-pairing and hybridization of target DNA to nucleotide probes of complementary sequences immobilized on a solid surface (Fig. 1). The principle relies on that quantified probe intensities reflect the abundance of hybridized respective alleles. There are numerous and detailed accounts of technical aspects of how the platforms work (Matsuzaki *et al.*, 2004; Peiffer *et al.*, 2006; Steemers *et al.*, 2006) and several comparisons of how they perform (Baumbusch *et al.*, 2008; Curtis *et al.*, 2009; Gunnarsson *et al.*, 2008; Hehir-Kwa *et al.*, 2007). Here, we will confine to describe the basic principles of the platforms and highlight some of the differences between them.

Since the first SNP array platforms were presented (Wang *et al.*, 1998), array density has increased by several orders of magnitude and the current platforms comprise millions of probes in a single assay. The earlier versions of Affymetrix SNP arrays utilized comprehensive collections of probes to interrogate each individual SNP, using paired perfect-match and miss-match probes to infer genotype, a strategy adopted from Affymetrix expression array platforms. More recent versions utilize a less redundant strategy of a limited number of probes per SNP and allele. In short, 25-mer oligonucleotide probes designed to match the target DNA of interest are *in situ* synthesized on the array surface. For interrogation of SNPs, specific probes are synthesized for each of the two alleles at separate locations on the array (Fig. 1B). The overall strategy depends on preferential hybridization of perfect complementary target sequence to the probes coupled with the ability to quantify the amount of bound target. By quantification of hybridized targets to separate allele-specific probes, their individual abundance in the sample may be inferred. The Illumina SNP array platforms are more recent and have not undergone the same degree of reformation compared with Affymetrix. It shares the same basic principle of target hybridization to loci-specific probes with Affymetrix,

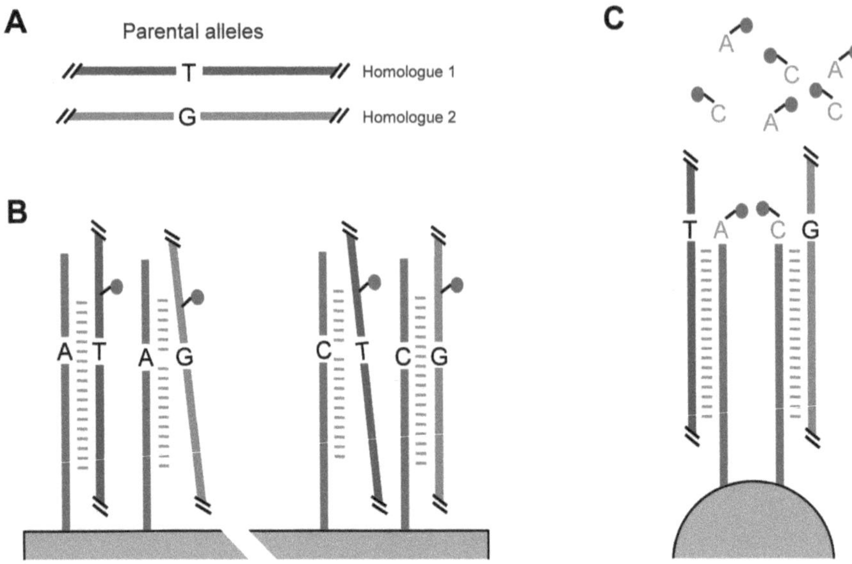

Fig. 1 Schematic illustration of the basic principles of how allele-specific intensity values are measured using the Affymetrix and Illumina assays. (**A**) Parental homologues comprising one centrally located heterozygous SNP (T/G). (**B**) The Affymetrix assay relies on multiple allele-specific probes spanning the interrogated SNP and complementary to either T or G. In the illustration, only one allele specific probe per allele is depicted. The probes are located in separated features on the array surface and will preferentially hybridize labeled target of perfect complementarity. Relative difference in allele abundance is resolved by comparing the quantified intensities from separate features harboring the respective probes. (**C**) The Illumina assay relies on a single locus-specific probe complementary to the sequence adjacent to the interrogated SNP. The probe will hybridize both parental homologues in a allele non specific manner. The hybridized target is used as template in a subsequent enzymatic single-base extension step employing differentially labeled nucleotides. The use of differentially labeled nucleotides permits dual intensity quantification in the same feature and relative difference in allele abundance is resolved by comparing the quantified intensities. (For color version of this figure, the reader is referred to the web version of this book.)

and similarly, probe density has increased to comprise millions of markers. However, there are some fundamental differences between the platforms. Illumina utilizes their BeadChip technology that permits probes to be immobilized on silica beads rather than directly onto the array surface. Probe-covered beads are then randomly distributed in microwells covering the array surface, followed by a probe location decoding procedure. Each SNP is interrogated with a single bead type covered by one unique 50-mer probe designed to target the sequence adjacent to the SNP of interest. After target hybridization, alleles are differentiated by a subsequent enzymatic single-base extension of the probe using the hybridized target as template.

Base extension results in the incorporation of differentially labeled nucleotides depending on the captured allele. The abundance of the respective SNP alleles in the sample may then be inferred by dual color quantification of signal intensities from each bead type (Fig. 1C). It should also be noted that current WGG platforms contain large numbers of probes that are not designed to interrogate SNPs. Rather these probes are solely designed to assay copy number and as such serve as aCGH probes and some are specifically designed to target copy-number polymorphic regions, that is, constitutional copy-number variations (CNVs).

B. Principles of Data Extraction and Normalization

Raw data acquisition and processing varies depending on array platform. Arrays are hybridized and labeled according to chemistry-dependent experimental procedures followed by imaging and data extraction. Raw signal measurements for the *A* and *B* alleles are preprocessed, normalized, and summarized over probe replicates or a collection of probes depending on platform. Preprocessing and normalization of probe data is performed to achieve pairs of allele-specific measurements for each SNP locus, and to this end there are various methods described (LaFramboise, 2009). Pairs of normalized allele measurements can subsequently be used to call genotypes and to infer DNA abundance and allele ratio. For calling genotype and calculating allele ratio, observed normalized intensities are related to expected values derived from collections of reference data. Transformation of intensities to relative copy-number estimates is essentially also performed by relating values to a collection of normal reference samples or to a matched control.

C. The B Allele Frequency and Relative Copy Number

The BAF, first presented using Illumina data (Peiffer *et al.*, 2006), is calculated for each SNP individually by transformation of allele intensities and represents the proportion of DNA content for allele *B* as compared to the total DNA content of *A* and *B* alleles together. The proposed transformation involves linear interpolation of allele frequencies from reference data derived from normal samples. Since BAF simply describes the total number of *B* allele copies divided by the total number of allele copies for that specific locus, a theoretical BAF can be calculated for any given genotype using the following equation:

$$\mathrm{BAF} = N_B/(N_A + N_B) \tag{1}$$

In Eq. (1), N_B is the total number of *B* allele copies and N_A is the total number of *A* allele copies.

Apart from genotyping, SNP arrays provide means for quantification of relative copy numbers at each given loci and current SNP arrays typically contain large numbers of probes that are solely designed to assay copy number and target nonsequence polymorphic loci. These probes can be used for the analysis of CNVs but many are also added to provide increased power and resolution when analyzing acquired copy-number aberrations in tumors. Relative copy-number ratio values are calculated by comparing observed normalized intensities (sum of *A* and *B*) to the expected, similarly to how BAF is derived, and is typically presented as Log2 relative ratio (LRR). Data from Affymetrix can be converted into BAF and LRR by appropriate normalization and transformation (Sun *et al.*, 2009; Wang *et al.*, 2007). Examples of expected BAF and LRR values for a normal genome and how these values are affected by acquired genetic aberrations are further discussed below.

D. Expected BAF and LRR for a Normal Genome

In a diploid genome, there are only three possible allele combinations for a given locus: homozygosity for the *A* allele (*AA*), heterozygosity (*AB*), or homozygosity for the *B* allele (*BB*). Thus, given Eq. (1), three different BAF values are possible for SNP loci in a normal diploid genome: 0 (*AA*), 0.5 (*AB*), or 1 (*BB*). The genotype status across chromosomes may conveniently be visualized using BAF plots (Fig. 2). In these plots, the BAF values for individual SNPs (*y*-axis) are plotted with respect to their genomic position (*x*-axis), similar to a copy number profile. A schematic BAF plot representation of a normal diploid genome is presented in Fig. 2A. As seen in Fig. 2A, any given SNP locus will only have one unique BAF value and this value is defined by its corresponding genotype. The design of current SNP arrays are in practice arbitrary with respect to *A* and *B* alleles, that is, when considering a large consecutive series of SNP loci, *AA*, *AB*, and *BB* genotypes will ideally appear randomly distributed. Therefore, BAF values plotted across a chromosome will give the impression of being "banded" at the macro-level. Figure 2B displays an experimentally obtained BAF plot of a chromosome from the analysis of a normal diploid genome. Three seemingly horizontal bands representing *AA*, *AB*, and *BB* genotypes are apparent, closely clustered around the theoretical BAF values of 0, 0.5, and 1, respectively (Fig. 2B). In reality, the series of consecutive BAF values across the genome shift continuously between the three states as extensive homozygous genomic segments are normally not observed. The BAF profile of a homozygous genome, for example, a haploid genome, will consequently present only two bands, restricted to theoretical BAF values 0 and 1, whereas a triploid genome will show four bands. The appearance of more than four bands is inevitably the result of

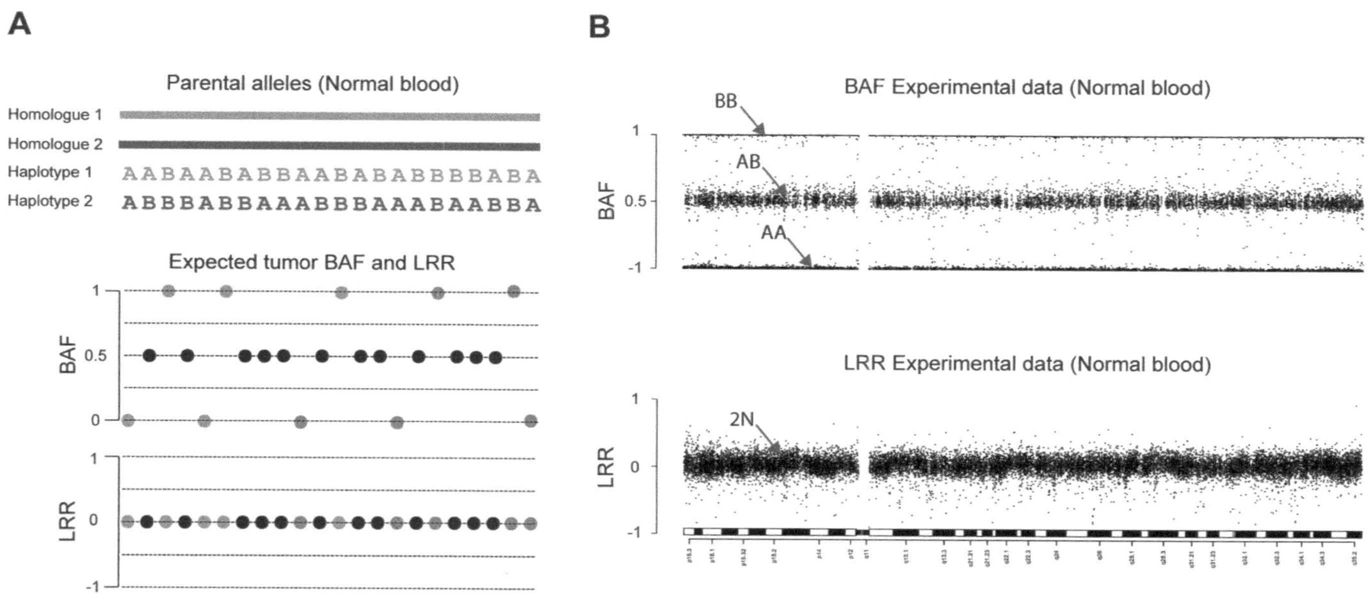

Fig. 2 SNP array analysis of a normal diploid genome. (**A**) A schematic illustration of an expected BAF and LRR plot from the analysis of a diploid genome. Each individual SNP has a BAF value of 0, 0.5, or 1, reflecting the genotype for that specific locus (*AA*, *AB*, and *BB*, respectively). BAF and LRR values for germline homozygous- and heterozygous SNPs are colored gray and black, respectively. (**B**) Experimentally obtained BAF and LRR plots of a chromosome. In the BAF plot, individual SNPs cluster close to 0, 0.5, or 1, producing three characteristic horizontal bands. At the LRR level, the majority of SNPs cluster around 0, representing the measurement of two DNA copies. (For color version of this figure, the reader is referred to the web version of this book.)

mixed samples, for example, inadvertently mixing DNA from two individuals. However, such chimeric patterns may be observed in clinical samples, for example, when analyzing recurring leukemias after the patient has undergone bone marrow transplantation (Paulsson, Lindgren, and Johansson, 2011). We will discuss BAF values in further detail in Section III, and how these values may be used for the identification of regions of genomic alteration.

In Section II.C, we described how SNP arrays estimate copy numbers for each SNP locus. By definition, a normal diploid genome has two copies of each autosome. Copy numbers are often presented as relative ratios, which are log2 transformed and centered. Therefore the copy-number profile of a normal genome is centered on 0, corresponding to two copies (Fig. 2B). However, it is worth to mention that constitutional CNVs are quite common (Iafrate *et al.*, 2004; Sebat *et al.*, 2004). Therefore, care should be taken when analyzing tumor samples—to avoid misinterpreting small copy-number gains or losses as acquired somatic alterations—especially if matched constitutional blood is unavailable for comparison (Heinrichs, Li, and Look, 2010).

III. WHOLE GENOME GENOTYPING OF TUMOR SAMPLES

Since the introduction of SNP arrays, a large number of studies have proved these platforms to be important means of analysis of acquired genomic changes. Since SNP arrays can detect chromosomal imbalances at both the copy-number level, measured as deviation of LRR, and at the genotype level, measured as deviations of BAF, the combined use of these two measurements can be used for interpretation of underlying genomic imbalances. We will here discuss the basic concept of how copy number and allelic ratios are affected by common genetic alterations such as deletions, copy-number gains, and copy-number-neutral events.

A. Changes in BAF and LLR upon Acquired Genomic Alterations

As described above, there are three possible genotypes for a given SNP locus in the normal diploid genome, either heterozygous (*AB*) or homozygous (*AA* or *BB*). Using Eq. (1) for calculating BAF, we also described that these genotypes have BAF values of 0.5 (*AB*), 0 (*AA*), and 1 (*BB*), respectively. In tumors, however, deviations from the normal diploid state are frequently observed, for example, gains of chromosomal regions harboring oncogenes or deletions of regions that harbor tumor suppressor genes. As a consequence, not only the DNA copy number will be affected, but also the balance between *A* and *B* alleles for SNPs that were constitutionally heterozygous (*AB*) within the altered region. Similarly to the normal state, the BAF formula given in Eq. (1) can be used to calculate theoretical BAF, that is, a representation of the proportion of *B* alleles to the total number of allele copies. In Table I, we list a number of possible genotypes and their corresponding BAF values. For example, a region present in three copies can have four possible genotype combinations, *AAA*, *AAB*, *ABB*, and *BBB*, which will have theoretical BAFs of 0/3 = 0, 1/3 = 0.33, 2/3 = 0.67, and 3/3 = 1, respectively. Similarly, an SNP locus with an *ABBB* genotype will have a BAF value of 0.8. Thus, the simultaneous readout of both BAF and copy number by SNP arrays provides a unique opportunity to extrapolate the actual genotype status of an altered region within a tumor genome. We will in the next paragraphs present a couple of genomic alterations and their effect on BAF and LRR.

With the simple principle of BAF in mind, let us hypothesize a scenario in which a somatic hemizygous deletion has occurred within a

Table I The Association between Genotype and BAF

Genotype	CN[a]	BAF
—	0 (HD)[b]	—
A	1	0
B	1	1
AA	2	0
AB	2	0.5
BB	2	1
AAA	3	0
AAB	3	0.33
ABB	3	0.67
BBB	3	1
AAAA	4	0
AAAB	4	0.25
AABB	4	0.5
ABBB	4	0.75
BBBB	4	1
AABBB	5	0.6
ABBBB	5	0.8
AAABBB	6	0.5
AABBBB	6	0.67
ABBBBB	6	0.83

[a]CN: total number of allele copies;
[b]HD:, homozygous deletion

diploid tumor genome leading to LOH in the affected region, that is, the possible genotypes are restricted to either *A* or *B* (Fig. 3A). Thus, BAF values for all germline heterozygous SNPs are shifted from BAF = 0.5 to either BAF = 0 or BAF = 1, depending on which chromosomal homologue has been lost. The plot will therefore display two horizontal bands of BAF values at 0 and 1. It is important to note that only SNPs heterozygous in the germline will change their respective BAF value in case of a deletion; constitutively homozygous SNP loci are by definition noninformative for studying acquired allelic imbalances at the genotype level. At the copy-number level, however, all SNPs are informative and the deletion will be detected as reduction in LRR for all measured probes within the region (Fig. 3A). In theory, the DNA content for a deletion is reduced to half of that of the normal state, and the theoretical LRR value for affected SNPs would therefore be −1 in LRR space. However, due to platform limitations the experimentally obtained response on LRR values is typically smaller than the theoretical (Peiffer *et al.*, 2006).

Similarly, a single copy-number gain will not only be reflected at the LRR level, but also introduce a shift in BAF at all SNP loci that were germline heterozygous. SNPs within a region affected by a one-copy gain caused by duplication of material from one of the homologues will have four possible allele combinations (*AAA, AAB*, *ABB*, and *BBB*), resulting in a four-banded pattern in a BAF plot (Fig. 3A). A region present in three copies may also arise through a two-copy gain of material from one homologue in combination with deletion of material from the second homologue. This will lead to complete homozygosity within the region (only *AAA* or *BBB* genotypes will be present), and the BAF pattern will be indistinguishable from that of LOH caused by deletions. The increase in LRR will however indicate that this region is present in more than two copies.

For more complex alterations involving higher allele copy numbers, multiple paired genotype combinations are possible within the gained region, again depending on which homologues are present and in what proportions. Figure 3A presents two possible scenarios of how a two-copy gain can be manifested. In the first example, two imbalanced genotypes are possible for SNPs that were germline heterozygous (*AAAB* and *ABBB*, BAF = 0.25 and BAF = 0.75, respectively). Alternatively, a net gain of two copies may arise through duplication of material from both homologues. Germline heterozygous SNPs will here remain balanced (*AABB*, BAF = 0.5) and no shift will be observed at the BAF-level. Similar to the three-copy example above, all haplotypes in the four-copy region may also be derived from the same homologue leading to complete homozygosity of the segment (*AAAA* and *BBBB*).

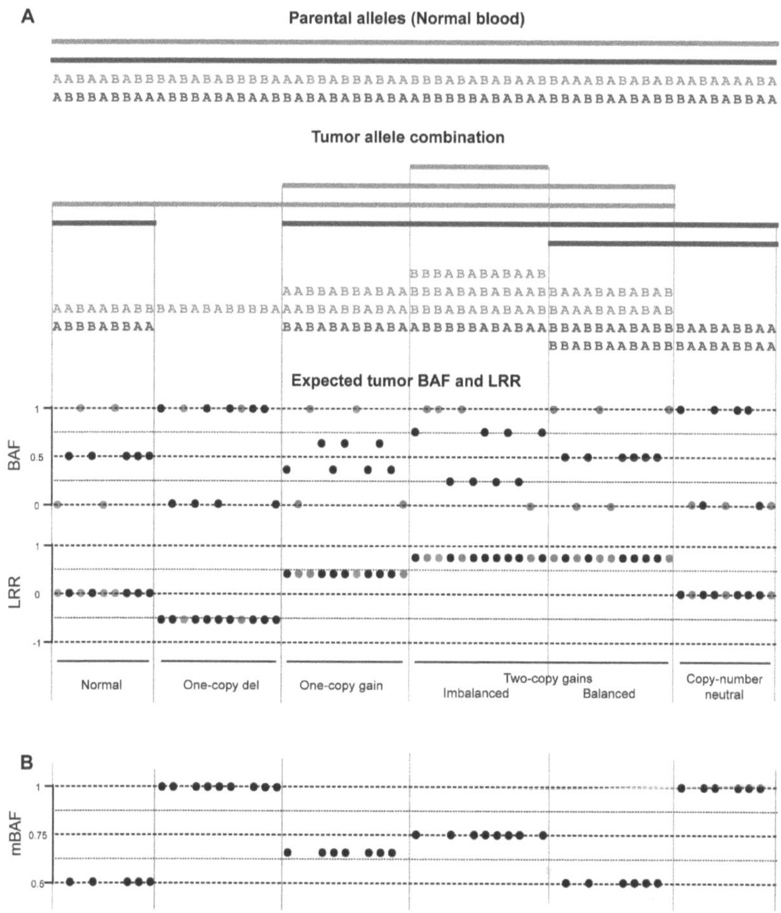

Fig. 3 Schematic examples of common genomic alterations and their expected BAF, LRR, and mBAF values. **(A)** The constitutional genotype showing the two parental homologues of a diploid genome, each with its own specific haplotype series, is shown at the top. Allele combinations for a tumor with acquired genomic alterations are shown below the parental alleles. The balance between A and B, and the total number of allele copies, will determine the BAF and LRR for each SNP locus. Thus, each alteration causes a shift in the BAF and/or the LRR profile. BAF and LRR values for germline homozygous- and heterozygous SNPs are colored gray and black, respectively. At the copy number level all SNPs are informative. Expected BAF and LRR plots for the acquired alterations are shown at the bottom. From left to right (possible genotype combinations for germline heterozygous SNPs are given within parenthesis): normal balanced genotype (*AB*), a one-copy deletion (*A* and *B* genotypes), a one-copy gain (*AAB* and *ABB* genotypes), a two-copy gain in which both surplus segments are derived from the same chromosomal homologue (*AAAB* and *ABBB* genotypes), a balanced two-copy gain (*AABB* genotypes), a segment with copy number neutral LOH (*AA* and *BB* genotypes). **(B)** Schematic mBAF transformation of the above BAF profile. The mBAF is mirrored at the 0.5 axis yielding only one possible value for SNP loci that were germline heterozygous. For example, in the *AAAB/ABBB* two copy gain segment, the *AAAB* genotypes (BAF = 0.2) will be transformed to the mirrored genotype (*BBBA*) with mBAF = 0.8. Noninformative homozygous SNPs are excluded from this plot. (For color version of this figure, the reader is referred to the web version of this book.)

Tumors may also display regions of allelic imbalance but without changes in copy number, a state often referred to as copy-number-neutral imbalance. The combination of genotype and copy-number measurements makes SNP arrays ideal for the identification of copy-number-neutral imbalances. In contrast, such aberrations are undetectable using aCGH. A simple example of a copy-number-neutral imbalance is when a chromosomal region is deleted and followed by duplication of the remaining allele (Fig. 3A). It must be stressed that definition of copy-number-neutral alterations are intimately linked to the ploidy state of the tumor. The ploidy status of tumors and its implication on BAF and LRR will, however, be discussed more in detail below. Copy-number-neutral imbalance is sometimes referred to as uniparental disomy (UPD), which is the terminology used to describe when an individual is constitutionally homozygous for a chromosomal region since both alleles are derived from a single parent. UPDs are observed as the cause of certain recessive genetic disorders and arise through meiotic segregation errors, chromosomal duplications, or mitotic recombination events during early development. Due to its narrow definition—homozygosity caused by two copies from the same parent—and close association with constitutional genetics, we will refrain from using the term UPD when discussing copy-number-neutral allelic imbalance events.

B. The Mirrored B Allele Frequency (mBAF)

In the examples above, we demonstrated how different types of acquired chromosomal alterations influence the BAFs of constitutionally heterozygous SNP loci. A consecutive series of SNP alleles (a haplotype series) on a chromosome homolog is in practice random with respect to its sequence of *A*s and *B*s. If we consider a region affected by a specific genetic alteration, we also note that BAF values for the SNPs within this region are symmetrically positioned around the 0.5 axis. A reflection of BAF data along the 0.5 axes can, therefore, be applied to obtain mirrored BAF (mBAF) values (Assie *et al.*, 2008; Staaf *et al.*, 2008). In Fig. 3B, we demonstrate this inherent symmetry for the regions of copy number and/or allelic imbalance presented in Fig. 3A. If noninformative germline homozygous SNPs are removed, an mBAF plot will display only one horizontal band reflecting the proportion between the major and minor allele for that specific region. Thus, the use of mBAF will provide similar information as for BAF but requires fewer genotype combinations to describe the genomic state, that is, the paired genotype combination for the one-copy gain presented in Fig. 3A (*AAB* and *ABB*) have BAF = 0.33 and BAF = 0.67, but mBAF = 0.67: corresponding to the BAF for the genotype that is dominated by *B* alleles (in this case *ABB*). As exemplified below, mBAF can facilitate identification of segments of allelic imbalance.

C. Delineating Regions of Genomic Imbalance

A number of computational methods have been described for the automated identification of altered regions in tumor genomes analyzed by SNP arrays. As for conventional LOH analysis, at the level of individual SNPs, a matched blood sample is needed as a reference to determine if that specific SNP is subjected to an acquired alteration or not. However, even in case of a matched normal genotype, individual SNPs are generally not sufficient for determining the genotype at a given loci due to possible technical noise. Therefore, one must make use of larger regions of consecutive SNPs to accurately predict genomic imbalances. We previously described that, when considering a larger series of SNPs, a BAF plot will appear as banded and that three bands are seen when analyzing a normal diploid genome. Through the schematic examples of genomic alterations described above (Fig. 3), we also demonstrate that most somatic alterations will introduce shifts in the BAF profile and that these shifts are a consequence of the particular haplotype combination that constitute that specific alteration. The high resolution of SNP arrays permits inference of allelic imbalance from a continuous stretch of LOH without the need of a matched normal genotype. Initially this possibility was demonstrated using hidden Markov model algorithms to infer regions of allelic imbalance (Beroukhim *et al.*, 2006; Lin *et al.*, 2004), but several and more elaborate approaches are currently available for the definition of such genomic segments (Li *et al.*, 2011; Staaf *et al.*, 2008). It should again be stressed that relatively long stretches of homozygosity may be constitutionally present, and therefore care must always be taken when inferring regions of LOH in the absence of a matched normal sample (Heinrichs *et al.*, 2010; McQuillan *et al.*, 2008;).

Use of segmentation algorithms, for example, CBS (Venkatraman and Olshen, 2007), to identify breakpoints delineating regions with a joint underlying genomic state was early adopted for aCGH data and has been repeatedly evaluated (Lai *et al.*, 2005). Segmentation-based approaches can be applied with the overall aim to describe the studied genome as a series of segments ascribed specific BAF and LLR states. Thus, any segment corresponding to a genomic event or alteration is represented by an LRR and BAF that deviate from the normal state: either by imbalanced LRR, BAF, or both. Note that when ascribing a BAF value to a segment, SNPs that are homozygous in the germline are uninformative and are disregarded.

Figure 4 displays typical BAF and mBAF patterns obtained from an SNP array analysis of a tumor and illustrates how data can be segmented in order to reduce data dimensionality. Thus, instead of describing single SNP loci, we can rather refer to the alleles and genotype of whole

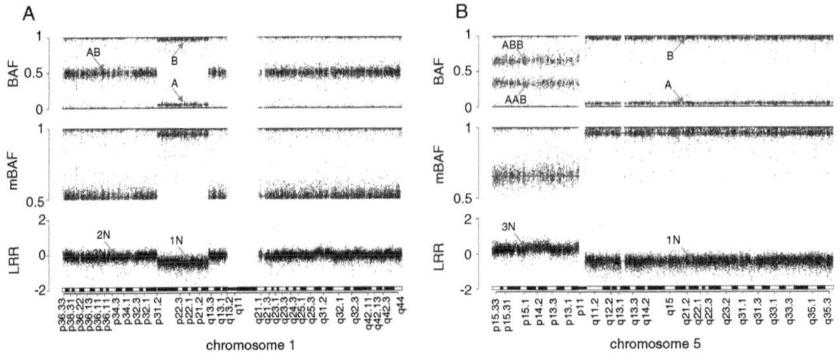

Fig. 4 Experimental BAF, mBAF, and LRR plots of two chromosomes obtained from a SNP array analysis of a tumor. Deviations from the expected BAF and LRR patterns for a normal diploid genome are observed, indicative of acquired chromosomal alterations, including (**A**) an interstitial deletion in 1p21-p31, and (**B**) gain of 5p with deletion of 5q. The mBAF profiles have been segmented, identifying breakpoints that delineate segments that constitute a discrete genomic state. Identified segments are ascribed specific mBAF and LRR values based on the SNPs between breakpoints. For example, three segments define chromosome 1, whereas chromosome 5 can be described with two separate segments. (For color version of this figure, the reader is referred to the web version of this book.)

segments, that is, we refer to haplotypes and genotypes as the collective haplotype/genotype state of one genomic region. A change in the haplotype constitution will result in a change in the banded pattern. It then becomes intuitive that most acquired alterations will introduce a shift in BAF and/LRR, and that changing from one underlying state to another will involve breakpoints in the data delineating genomic alterations (Fig. 4).

D. BAF versus LRR Plots

We have shown that SNP-array data provide both genotype and copy-number estimates for each SNP that is queried, and that these can be visually represented using mBAF and LRR profile plots. To interpret a specific genetic alteration, it is needed to take both mBAF and LRR into account, and their respective relationship can be queried by plotting LRR versus mBAF (Fig. 5). Although values from individual SNPs can be plotted, various segmentation approaches can effectively reduce the complexity of data, that is, defining regions of genomic balance or imbalance and treating these as individual events assigned representative mBAF and LRR values. When plotting segmented LRR versus mBAF (or BAF) from a tumor with a diploid chromosomal number, a characteristic pattern will emerge where genomic regions (segments)

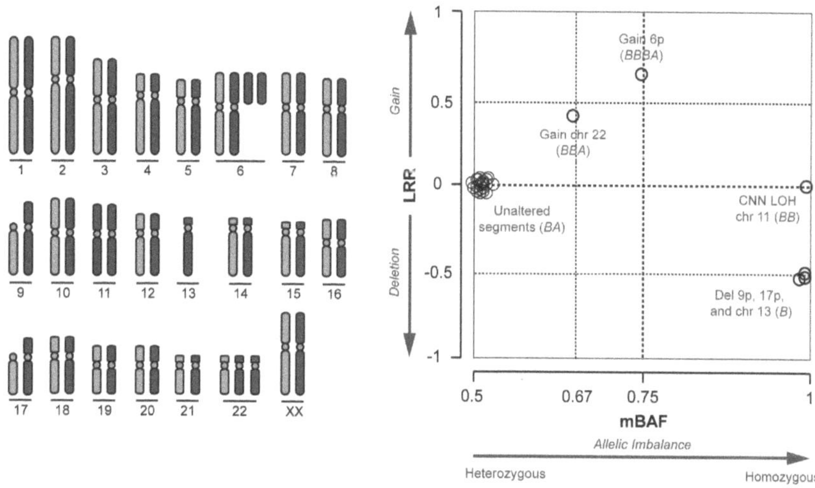

Fig. 5 A schematic karyotype of a diploid tumor and its corresponding mBAF/LRR plot. The karyotype illustrates: hemizygous deletions of chromosomes 13, 9p, and 17p; gains of chromosomes 22 (trisomy, *BBA*) and 6p (four copies, *BBBA*); and copy-number-neutral allelic imbalance of chromosome 11 (*BB*). The parental chromosomal homologues are colored in yellow or blue, respectively. For the mBAF/LRR plot, each individual circle in the figure represents a continuous chromosomal segment with identical LRR and BAF as given by the tumor karyotype. For example, chromosome 6 is represented by two separate segments: 6p (mBAF = 0.75 and LRR = 0.65) and the unaffected 6q (BAF = 0.5 and LRR = 0). Chromosome 11 has no intrachromosomal breakpoints and is only represented by one segment (BAF = 1 and LRR = 0). (See Page 6 in Color Section at the back of the book.)

with identical allele combinations (genotypes) will appear close to each other within the mBAF/LRR space (Fig. 5).

For example, segments of one-copy gain (*BBA*) will appear together as a cluster of values with elevated LRR and mBAF, approaching their theoretical values of mBAF = 0.67 and LRR = 0.58. Correspondingly, regions representing copy-number loss (*B*) will cluster around their theoretical values of mBAF = 1 and LRR = (−1), whereas copy–number-neutral LOH (*BB* genotypes) will in this example be positioned at the same mBAF as losses (mBAF = 1) but at LRR = 0 (Fig. 5). All unaltered segments (*AB*) will form a dense cluster at mBAF = 0.5 and LRR = 0. Hence, each individual tumor will demonstrate a characteristic mBAF/LRR pattern depending on what specific alterations (genotype combinations) have been acquired. Popova and coauthors (Popova *et al.*, 2009) termed this BAF/LRR pattern as the "genomic alteration print" (GAP) of a tumor. Pattern-recognition strategies have been applied on similar representations of SNP-array data to resolve tumor cellularity, underlying ploidy of the tumor, as well as intratumor heterogeneity (Attiyeh *et al.*, 2009; Popova *et al.*, 2009).

IV. WGG ANALYSES OF COMPLEX AND HETEROGENEOUS CELL POPULATIONS

We have so far discussed relatively simple examples of alterations affecting one homogenous population of tumor cells. In practice, however, WGG analyses are often performed on heterogeneous tumor samples that contain more than one distinct population of cells. For example, primary tumor samples are often admixed with cells without somatic alterations. These nonaberrant cells can include cells that define the tumor microenvironment, such as infiltrating lymphocytes and fibroblasts, or vasculature. Normal cells can also be present due to sampling procedures leading to inclusion of varying amounts of tumor-adjacent non-neoplastic tissue. Thus, the proportion of nonaberrant cells will vary from sample to sample. Regardless of the cause and nature of included nonaberrant cells, the presence of normal diploid cells within a tumor sample can cause problems in downstream analyses and subsequent interpretations of the data. Moreover, cancers may to varying degrees be composed of multiple clones harboring divergent aberrations that are acquired subsequent to the tumor-initiating event. Collectively, the presence of heterogeneity in tumor samples imposes challenges on data analysis and interpretation. By providing the combination of genotype and copy-number information, SNP-array data can however be used for resolving some of this complexity and thereby increase our possibilities to study the mechanisms and actions underlying cancer.

A. Tumor Ploidy

Tumor genomes are often highly aneuploid and may reach near-triploid, tetraploid, or even higher ploidy states. Such deviations from the normal diploid state will have direct implications on both BAF and LRR and the underlying aneuploidy has to be taken into account when making assumptions of the genotype status of altered regions. In our examples so far, we have only dealt with simple chromosomal alterations on a diploid background. If we instead consider a tumor with a near triploid genome, we will expect most chromosomes to be in allelic imbalance. Chromosomal regions without copy number alterations relative to the modal chromosome number can, for example, have two copies of one allele and one copy of the other allele—resulting in a characteristic four-banded pattern in the

BAF profile (Fig. 6A). Deletions in a triploid *ABB* background will thus be seen as a shift toward either homozygosity, that is, two identical alleles are retained (*BB*) (Fig. 6A), or toward allelic balance, that is, one copy of each allele is retained (*AB*).

A tetraploid tumor, on the other hand, will typically display a BAF profile where most chromosomes have a balanced genotype (*AABB*) (Fig. 6B). Such a scenario could entail tetraploidization from a diploid genome through incomplete cytokinesis, endoreduplication, or cell fusion. In theory, BAF cannot be used to discriminate strictly tetraploid cells from diploid cells after duplication of the genome. In reality though, the complexity of the tumor genotype is typically such that tetraploidy can be discerned since additional alterations are acquired after tetraploidization. A one-copy deletion on a tetraploid background is shown in Fig. 6B.

Deviations from the diploid state will also give rise to highly characteristic patterns in the mBAF/LRR plot. This is exemplified in Fig. 7A by a near triploid tumor karyotype and its corresponding mBAF/LRR plot. Note that the majority of chromosomal segments cluster at LRR = 0 with mBAF = 0.67, indicative of an imbalanced *BBA* genotype. Segments representing deletions, and thus ascribed negative LRR, are seen at either

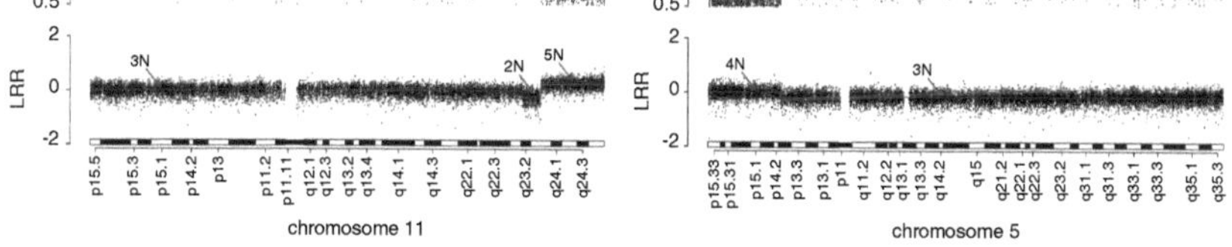

Fig. 6 Experimental BAF, mBAF, and LRR plots obtained from two tumors with increased ploidy number. (**A**) Representation of chromosome 11 for a near triploid tumor. Most of the chromosome displays a four-banded pattern characteristic for a trisomy (mBAF = 0.67 and LRR = 0). A deletion is however observed in 11q23 (mBAF = 0.97 and LRR = –0.22), and a gain in 11q24 (mBAF = 0.58 and mBAF = 0.26). The deletion possibly indicates one-copy loss resulting in a *BB* genotype. The gain possibly indicates duplication of material from both homologues relative to the trisomic state (*AABBB*). (**B**) Chromosome 5 for a near tetraploid tumor. A small segment of the chromosome (5p14-p15) is present in four copies (mBAF = 0.51, LRR = 0). A net loss of one copy for the remainder of the chromosome results in an unbalanced mBAF value and a negative LRR value, closely matching an *ABB* genotype. (For color version of this figure, the reader is referred to the web version of this book.)

mBAF = 0.5 or mBAF = 1, depending on which homolog is lost (*AB* or *BB* genotype, respectively). In Fig. 7B, an example of a near tetraploid karyotype and its corresponding mBAF/LRR plot is given. For this karyotype segments without relative copy-number alterations are located at

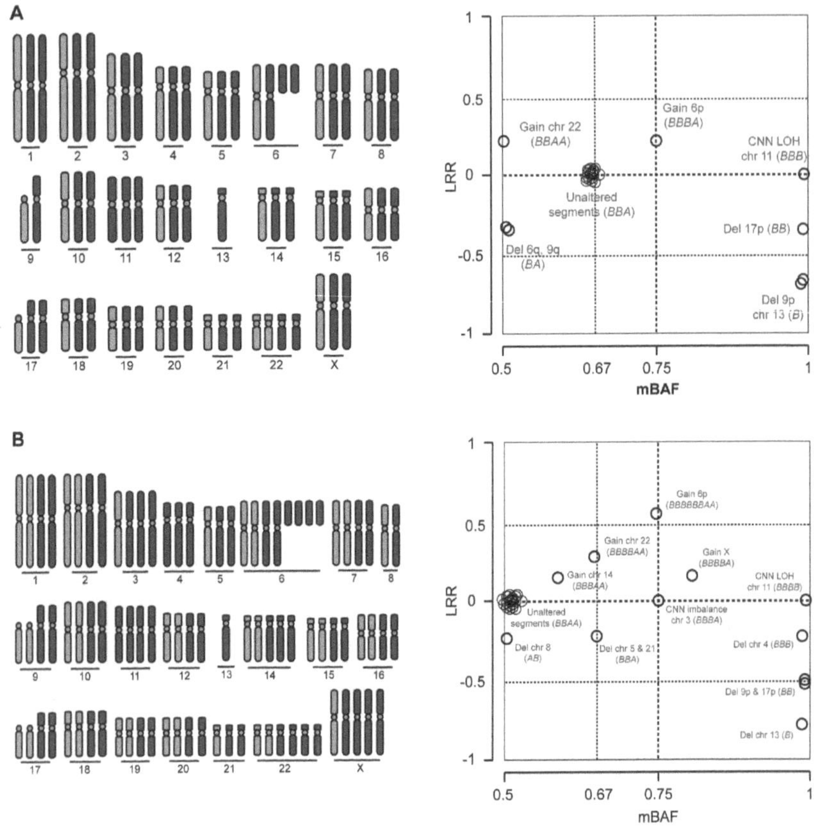

Fig. 7 Karyotypes of two aneuploid tumors and their corresponding theoretical representations in mBAF/LRR plots. (**A**) A near-triploid tumor. The parental homologues are colored in yellow and blue, respectively. Relative copy-number gains are observed for chromosome 22 and 6p. Net losses of material are seen for 9p, 9q, 17p, 6q, and chromosome and 13. Copy number neutral LOH is seen for chromosome 11 of which one homologue is present in three copies. In the mBAF/LRR plot the majority of segments are in allelic imbalance (*BBA*) and located at mBAF = 0.67 and LRR = 0. Deletions and gains are shifted towards either heterozygosity (*AB* or *AABB*) or towards increased allelic imbalance (*B*, *BB*, or *BBBA*). (**B**) A near-tetraploid tumor karyotype and its representation in mBAF/LRR space. Most segments are located at mBAF = 0.5 and LRR = 0. A number of different genotypes are seen for regions with negative LRR (e.g., *BBA*, *BB*, BA, and *B*). Similarly, a variety of genotypes representing net gain of material are observed (e.g., *BBBAA*, *BBBBA*, and *BBBBAA*), each with its' specific expected LRR and mBAF values. (See Page 7 in Color Section at the back of the book.)

mBAF = 0.5 (*AABB*). In contrast to the mBAF/LRR of diploid tumors, a tetraploid background will allow for a variety of possible genotypes for regions subjected to deletions, for example, *BBA*, *BB*, and *B*.

B. BAF and LRR in an Admixture of Tumor and Normal Cells

The above theoretical examples have focused on situations when there is only one clone present within the sample, that is, all analyzed cells have identical genotypes. Given that the LRR copy number reflects the DNA content, increasing proportions of cells with a normal karyotype will cause the LRR for a genomic alteration to converge toward that of the normal cells. In much the same way, the BAF patterns of a tumor will be affected by the presence of nonaberrant cells within the sample. However, whereas the change in LRR for imbalanced regions is linearly proportional to the fraction of present diploid cells, the effect on expected BAF is not always linear. Instead the effect on BAF for a given alteration will depend on its specific genotype. A simple example to illustrate how BAF for a genomic alteration is influenced by the presence of normal cells is given in Fig. 8A. Here, a schematic representation of a sample with eight tumor and two normal cells is displayed, that is, 20% normal cells are present within the sample. Let us hypothesize that the tumor cells carry a one-copy deletion leading to LOH in the affected region. Since BAF simply describes the frequency of *B* alleles in a given region, Eq. (1) can be used to calculate expected BAF to 0.83 (Fig. 8A). Figure 8B displays expected BAF and LRR plots for a hemizygous deletion in a sample with 20% normal cell admixture.

Equation (1) can, with some minor modifications, be used to calculate BAF values for any given locus in case of heterogeneous samples. A general formula to describe the relationship between BAF and fraction of normal cells is given by the following equation:

$$\text{BAF} = (x + N_B(1 - x))/(2x + N_A(1 - x) + N_B(1 - x)) \qquad (2)$$

In Eq. (2), x is the fraction of cells with a normal karyotype, and N_A and N_B denote the number of *A* and *B* allele copies for the specific aberration genotype, respectively. As previously stated, BAF is not necessarily linearly affected by the proportion of normal cells. In Fig. 8C, we use Eq. (2) to plot theoretical mBAF for a number of different chromosomal states as defined by their genotypes and show how these vary with increasing normal cell admixture. For example, mBAF for an aberration with genotype *B*, corresponding to a hemizygous deletion in a diploid tumor, will shift from 1 to 0.56 with increasing fraction of normal cells from 0% to 80%. As shown in Fig. 8C, the relationship between mBAF and normal

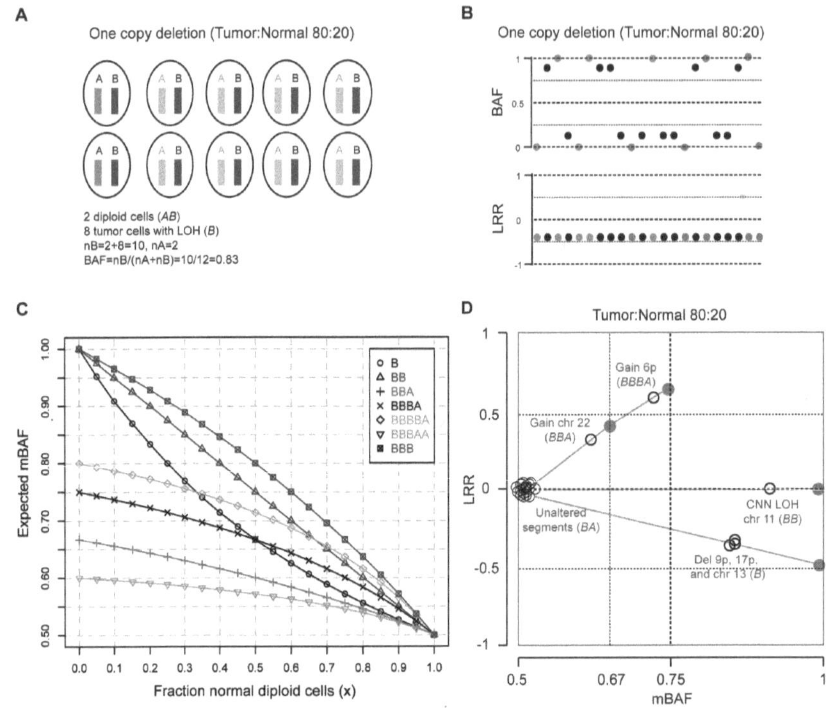

Fig. 8 BAF and LRR in case of tumor and normal cell admixture. (**A**) Schematic example of a sample containing two normal diploid cells and eight tumor cells with an acquired deletion (the lost allele is grayed out). The tumor will contribute with eight *B* alleles, whereas the normal cells contribute with two *A* and two *B* alleles. The expected mBAF is calculated to 0.83. (**B**) Schematic BAF plot illustrating a hemizygous deletion in a sample with 80% tumor cells and 20% normal cells. Germline heterozygous SNPs within the deleted region will not reach their expected BAF values (0 or 1) since the background of normal cells will contribute with both *A* and *B* alleles. (**C**) Line-plots of expected mBAF for a number of different genotypes as a function of the fraction of normal diploid cells present within the sample. For example, a tumor segment with an *AAB* genotype has an expected mBAF of 0.6 if intermixed with 50% normal diploid cells. (**D**) Schematic mBAF/LRR plot of a sample with 80% tumor and 20% normal cells. The tumor karyotype is identical to the karyotype presented in Fig. 5. Since normal diploid cells (*AB*) are present in the sample, the mBAF and LRR for the respective alterations are shifted toward mBAF = 0.5 and LRR = 0 along theoretical lines (gray). Solid gray circles indicate expected LRR and mBAF values in case of no normal cells present. (For color version of this figure, the reader is referred to the web version of this book.)

cell admixture is not linear for genotype *B*. Given Eq. (2), a linear relationship is, however, seen for genotype *BB*. The behavior of LRR and BAF in response to normal contamination has been extensively described (Nancarrow *et al.*, 2007), and experimentally corroborated using serial dilution experiments of tumor cell lines and matched normal blood (Assie *et al.*, 2008; Staaf *et al.*, 2008; Van Loo *et al.*, 2010). As a consequence, it is possible to use experimental array SNP data to infer the fraction of non-aberrant cells present within a sample. Several studies have successfully demonstrated this using tumor biopsies by comparing BAF-derived estimates with cellularity scores from histological examination (Assie *et al.*, 2008; Nancarrow *et al.*, 2007; Sun *et al.*, 2009). It follows that BAF of an altered region is in fact reflecting the fraction of cells harboring the alteration and not only the fraction of normal cells. Thus, any deviation from the expected value can be caused by clonal heterogeneity rather that normal contamination. The principles of estimating the fraction of normal cells can be illustrated using a simple example (Fig. 8D). The figure illustrates an expected mBAF/LRR pattern for the example given in Fig. 5, but in this case on a background of 20% normal cells. As demonstrated using well-characterized CLL samples (Staaf *et al.*, 2008), once the cellularity of a sample is resolved it is possible to also estimate the fraction of tumor cells carrying individual alterations. However, the combination of normal contamination and increased clonal heterogeneity can rapidly increase the complexity of the data and thereby reduce the possibility to resolve underlying genotype status.

C. Tumor Subclonality

The presence of genetic variation between different subclones within a tumor mass is a well-known phenomenon. Even though tumor cells generally are clonally related and show identical alterations at some loci, subclonal differences are often observed. Subclonal genetic alterations may readily be identified at the individual cell level by conventional cytogenetics or fluorescence *in situ* hybridization. Current molecular analyses of bulk samples will however only give an average estimate of all imbalances. For SNP arrays, the effect on BAF and LRR of subclonal alterations will in practice follow the same line of reasoning as discussed in the examples above about nonaberrant cell involvement. If we further expand our example of a sample of 80% tumor cells and 20% normal diploid cells (Figs. 5 and 8D) and hypothesize that 50% of the tumor cells carry some additional alterations, we can simply calculate expected mBAF for these using Eq. (2). A deletion (*B*) present in 40% of the cells will display mBAF of 0.625. If we instead consider a late copy number gain (*BBA*), mBAF for the altered region will be 0.583. Subclonal events will in this respect behave as regular alterations occurring in a proportion the cells and BAF will be affected as if normally contaminated. When plotting segmented LRR versus mBAF values, subclones are readily discernable as segment values deviating from the expected pattern set by the percentage of normal contamination (Fig. 9, arrows). Subclonal alterations in a region

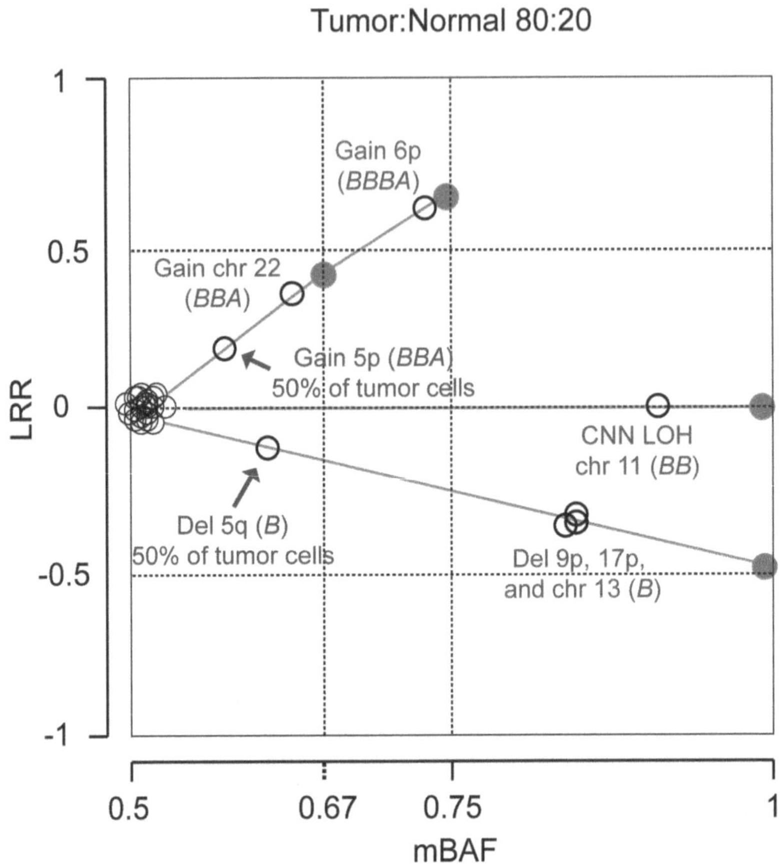

Fig. 9 Schematic mBAF/LRR plot of a sample with a 20% diploid cell background (80% tumor cells) and two subclonal events within the tumor cell population. The tumor karyotype is identical to the karyotype presented in Fig. 5, although 50% of the tumor cells also have acquired a gain of 5p (*BBA*) and a loss of 5q (*B*). Since only 40% of the cells in the sample carry these alterations, the respective mBAF and LRR for these segments will deviate (arrows) from the pattern observed for the alterations present in all tumor cells. (For color version of this figure, the reader is referred to the web version of this book.)

already affected by an earlier alteration will be much harder to detect. Although it is possible to propose plausible models to explain any observed pattern, subsequent validation is necessary to definitively resolve the underlying states. For instance, it is not possible to distinguish between a case including two clones present in a 50:50 relationship that do not share alterations from a homogenous tumor population harboring the union of these alterations but with 50% normal admixture. Likewise, cell populations comprising highly rearranged genomes and mixed ploidy will add complexity beyond examples presented here. Nonetheless, for regions that do not follow expected patterns one can at least assume the presence of subclonal events. Thus, SNP arrays may provide means for the detection of subclonal events and also propose a likely genotype that will explain the observed pattern.

D. Tracing Clonal Relationships Using SNP Arrays

Depiction of copy-number gain and loss frequencies across large tumor cohorts highlights recurrent alterations and can be used to classify tumors into groups with related karyotypes (Russnes *et al.*, 2010). Although providing clues to genetic events important for tumor initiation, progression, and metastasis, studies on nonrelated individuals will never be able to "recapitulate" clonal evolution and expansion events *per se*. To be able to discern and model the underlying chronology of events, repeated samples from the same individual have to be studied. Unfortunately such studies are quite rare, most probably since availability of multiple tumor specimens from single individuals is scarce. Certain cancers are, however, more permissive in studying clonal evolution. One example includes urothelial carcinomas where sampling from multifocal and highly recurrent tumors through noninvasive cystoscopies is possible (Höglund, 2007). The limited availability of multiple samples from individual patients can be circumvented by macro- or microdissection (Navin *et al.*, 2010) or cell-sorting procedures followed by expansion in animal models (Navin *et al.*, 2011), effectively performing multiple samplings of the same tumor. Interestingly, a number of studies have shown that the bulk of tumor cells at different time points, although sharing some common alterations, differs with respect to their array of genomic alterations. In many cases, one must assume an ancestral clone that the tumors are derived from, that is, there is a clonal relationship but not a strict linear evolution (Höglund, 2007; Mullighan *et al.*, 2008).

As described above, SNP arrays provide opportunities to investigate tumor heterogeneity. The possibility to reveal allelic imbalances and infer genotypes of acquired alterations facilitates elucidation of clonal relationships between tumors. Various models to study cancer evolution by LOH, karyotype, and CGH data have been proposed (Höglund *et al.*, 2005; Letouzé *et al.*, 2010; Navin and Hicks, 2010). For example, TuMult uses a computational approach, tracing breakpoints, for unraveling the succession of genomic alterations that has occurred during the process of carcinogenesis (Letouzé *et al.*, 2010). Given the high resolution of current arrays, the presence of multiple identical breakpoints in tumors is highly indicative of a shared origin. Investigating shared copy number alterations and mapping breakpoints may be supplemented by genotype information provided by SNP-array analysis.

We will here present some hypothetical examples of how SNP-array data can be used to analyze multiple tumors from the same patient in order to investigate clonal expansion, chronology of events, and divergence in clonal evolution. We first return to our example describing a sample of 80% tumor cells and 20% normal diploid cells in which we demonstrated how intratumor heterogeneity could be readily discerned (Figs. 5, 8D, and 9). In our example, the presence of a subclone was indicated by segmented mBAF/LRR values signifying an acquired deletion. The observed BAF for the alteration corresponds to that 50% of the tumor cells carry the deletion, and we can look for the deletion in other tumor samples from the same patient. For instance, if a metastasis or recurrence is available from the same patient we can investigate whether it too carries the identified deletion. The estimated proportion of cells that carry the deletion can yield information on whether the recurrence or metastasis represents an expansion of the specific subclone identified in the primary tumor, for example, if the deletion is present in the majority of the cells. Recall that the example from Fig. 9 included an additional late alteration: a one-copy gain estimated to be present in 50% of the tumor cells. When analyzing the recurrence, the identified gain might be estimated to be present in the same proportion as the deletion, indicating that both alterations were confined to the same subclone in the primary. Alternatively, one might fail to detect the gain in the recurrence altogether. The latter scenario will suggest that the two alterations were in fact confined to separate subclones in the primary. In line with this simplified example, numerous paired analyses of tumor samples can be imagined that aim to describe plausible relationships between tumors from the same individual.

Apart from discerning possible subclonal expansions, as exemplified above, the inherent properties of SNP arrays provide additional possibilities for tracing clonal hierarchies. It is of importance to stress one obvious, but fundamental, principle in a clonal evolution model; a subsequent clone cannot reacquire an allele that has been lost, that is, a clone that is heterozygous for a given locus cannot be a direct descendant from a clone that is homozygous for that locus. Such a situation is exemplified in Fig. 10A in which one tumor clone (C1) carries only one homologue of chromosome 9 and the other clone (C2) has retained both homologues. In this example, we can conclude that C2 cannot be directly descending from C1; however, the opposite is of course possible. Analogously, homozygous deletions are ideal to discern clonal relationships since complete loss of a locus also represents a state that cannot be reversed.

With this simple principle in mind, we will introduce the concept of "imbalance haplotype" (IH). That is, for any region of allelic imbalance, it

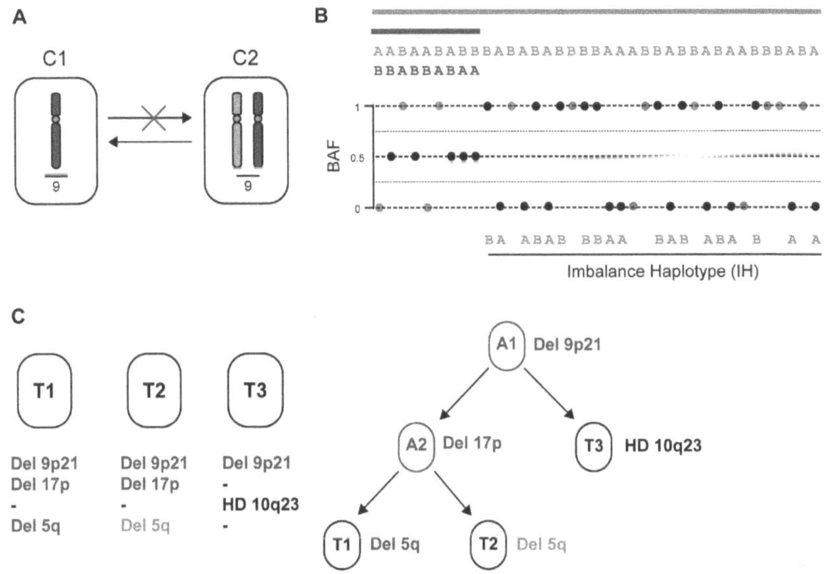

Fig. 10 Addressing clonal hierarchy using SNP array data. (**A**) Schematic example of two tumor clones (C1 and C2). The C2 clone has lost one homologue of chromosome 9. Therefore, subclone C2 cannot be a descendant from C1 (crossed arrow). The reverse is possible (arrow). (**B**) Definition of imbalance haplotype (IH). The IH is defined by the series of alleles that are in excess for a given genomic alteration. In this example, a loss of a chromosomal segment is illustrated. Each SNP within the IH is called from its respective BAF. Noninformative germline homozygous SNPs (gray circles) are not considered in the IH sequence. (**C**) Schematic example of three tumors derived from the same individual. Identified alterations in each tumor are listed to the left. Incompatible IHs for the 5q deletion is indicated by separate colors. A hierarchical tree describing clonal relationships can be deduced from the given alterations (right). Two obligate ancestral clones (A1 and A2, respectively) must be assumed as intermediate steps to describe the clonal evolution in this example. See main text for details. (For color version of this figure, the reader is referred to the web version of this book.)

is possible to determine the dominating haplotype, that is, the consecutive series of SNP alleles that are in abundance. We exemplify this for a deletion in which BAF is used to infer the complete haplotype sequences of the parental alleles (Fig. 10B). Importantly, if the actual haplotype is known, it can be used to query an alteration of the same region but in a separate sample from the same patient. By extrapolating the actual haplotype series from BAF values, we therefore can conclude if alterations in tumors from the same patient can be ascribed to the same chromosomal homologue or not.

We will now use this line of reasoning in a hypothetical example aimed to model an underlying hierarchy among tumor clones (Fig. 10C). In the example given we use three tumors (T1, T2, and T3) that are obtained from the same individual, but at separate time points. From the list of alterations, we can identify a focal deletion at 9p21 that is present in all tumors. The IHs for this region are also identical and we can therefore assume that all three tumors stem from a shared cell of origin and that this deletion is an early event. Deletion of the same homologue of chromosome arm 17p is shared by tumors T1 and T2, but not by T3. Thus, T3 cannot be a descendant of either T1 or T2. Neither can T1 and T2 be linearly derived from T3 since this tumor harbors a homozygous deletion at 10q23. Furthermore, the T1 and T2 tumors both display heterozygous deletions of 5q; however, different haplotypes are lost in the respective tumors (incompatible IHs). Thus, neither of these clones can directly have given rise to one another. This simple way of deducing clonal relationships, based on compatible and incompatible events, thus, provide us with an opportunity to connect the tumors hierarchically (Fig. 10C, right). We can conclude that the tumors share a common origin; moreover, we can reject a straightforward linear model of clonal evolution. We can, in addition, infer nodes of lineage deviation representing obligate ancestral clones.

Even though the above example may be overly simplified, it still conveys the basal concept of how SNP arrays may be used to address issues of clonality and tumor evolution. There are, to our knowledge, no reports that use SNP arrays to infer IHs and that take these into consideration when assessing clonal relationships between tumors. Nonetheless, the same conceptual thinking, that is, demonstrating loss of incompatible genotypes, has been applied in earlier studies using LOH analyses, demonstrating its feasibility (Lindgren *et al.*, 2008; van Tilborg *et al.*, 2000;).

V. CONCLUDING REMARKS

Throughout recent years, molecular techniques to study cancer have progressed in terms of resolution and sensitivity, but also with respect to accessibility due to decreased cost. Microarray-based platforms have evolved from proof-of-concepts—presented little more than a decade ago—to highly standardized off-the-shelf assays for genome-wide analysis of gene expression, DNA copy number, and genotypes. Undoubtedly, technologies will continue to evolve and much of what is considered at the forefront today will be superseded tomorrow. We have aimed to present some basic concepts pertaining to the analysis of tumor-heterogeneity using genotyping techniques. In doing so, we have also tried to give a brief account of currently available and standardized platforms for genome-wide genotyping. However, we have refrained from discussing in depth any particular analysis methods inherently tied to the mentioned platforms. Much of what has been presented in terms of data interpretation can in theory be applied to genotype and copy number data in general. Recent developments in high-throughput sequencing techniques lend promise to resolving some of the limitations of current array-based technology in the analysis of tumor-heterogeneity. Mainly, in terms of sensitivity, array-based analysis may fail to detect alterations confined to minor subpopulations. Nonetheless, current techniques have their merits and will undoubtedly continue to contribute to our understanding of tumor heterogeneity, development, and progression.

REFERENCES

Assie, G., LaFramboise, T., Platzer, P., Bertherat, J., Stratakis, C. A., and Eng, C. (2008). SNP arrays in heterogeneous tissue: Highly accurate collection of both germline and somatic genetic information from unpaired single tumor samples. *Am. J. Hum. Genet.* **82**, 903–915.

Attiyeh, E. F., Diskin, S. J., Attiyeh, M. A., Mosse, Y. P., Hou, C., Jackson, E. M., Kim, C., Glessner, J., Hakonarson, H., Biegel, J. A., and Maris, J. M. (2009). Genomic copy number determination in cancer cells from single nucleotide polymorphism microarrays based on quantitative genotyping corrected for aneuploidy. *Genome Res.* **19**, 276–283.

Baumbusch, L. O., Aaroe, J., Johansen, F. E., Hicks, J., Sun, H., Bruhn, L., Gunderson, K., Naume, B., Kristensen, V. N., Liestol, K., Borresen-Dale, A. L., and Lingjaerde, O. C. (2008). Comparison of the Agilent, ROMA/NimbleGen and Illumina platforms for classification of copy number alterations in human breast tumors. *BMC Genomics* **9**, 379.

Beroukhim, R., Lin, M., Park, Y., Hao, K., Zhao, X., Garraway, L. A., Fox, E. A., Hochberg, E. P., Mellinghoff, I. K., Hofer, M. D., Descazeaud, A., Rubin, M. A., et al., (2006). Inferring loss-of-heterozygosity from unpaired tumors using high-density oligonucleotide SNP arrays. *PLoS Comput. Biol.* **2**, e41.

Chin, K., DeVries, S., Fridlyand, J., Spellman, P. T., Roydasgupta, R., Kuo, W. L., Lapuk, A., Neve, R. M., Qian, Z., Ryder, T., Chen, F., Feiler, H., et al., (2006). Genomic and transcriptional aberrations linked to breast cancer pathophysiologies. *Cancer Cell* **10**, 529–541.

Curtis, C., Lynch, A. G., Dunning, M. J., Spiteri, I., Marioni, J. C., Hadfield, J., Chin, S. F., Brenton, J. D., Tavare, S., and Caldas, C. (2009). The pitfalls of platform comparison: DNA copy number array technologies assessed. *BMC Genomics* **10**, 588.

Grant, S. F. A., and Hakonarson, H. (2008). Microarray technology and applications in the arena of genome-wide association. *Clin Chem* **54**, 1116–1124.

Gunderson, K. L., Steemers, F. J., Lee, G., Mendoza, L. G., and Chee, M. S. (2005). A genome-wide scalable SNP genotyping assay using microarray technology. *Nat. Genet.* **37**, 549–554.

Gunnarsson, R., Staaf, J., Jansson, M., Ottesen, A. M., Göransson, H., Liljedahl, U., Ralfkiaer, U., Mansouri, M., Buhl, A. M., Smedby, K. E., Hjalgrim, H., Syvanen, A. C., et al., (2008). Screening for copy-number alterations and loss of heterozygosity in chronic lymphocytic leukemia—a comparative study of four differently designed, high resolution microarray platforms. *Genes Chromosomes Cancer* **47**, 697–711.

Hanahan, D., and Weinberg, R. A. (2011). Hallmarks of cancer: The next generation. *Cell* **144**, 646–674.

Hehir-Kwa, J. Y., Egmont-Petersen, M., Janssen, I. M., Smeets, D., van Kessel, A. G., and Veltman, J. A. (2007). Genome-wide copy number profiling on high-density bacterial artificial chromosomes, single-nucleotide polymorphisms, and oligonucleotide microarrays: A platform comparison based on statistical power analysis. *DNA Res.* **14**, 1–11.

Heinrichs, S., Li, C., and Look, A. T. (2010). SNP array analysis in hematologic malignancies: Avoiding false discoveries. *Blood* **115**, 4157–4161.

Höglund, M. (2007). On the origin of syn- and metachronous urothelial carcinomas. *Eur. Urol.* **51**, 1185–1193 discussion 1193.

Höglund, M., Frigyesi, A., Säll, T., Gisselsson, D., and Mitelman, F. (2005). Statistical behavior of complex cancer karyotypes. *Genes Chromosomes Cancer* **42**, 327–341.

Iafrate, A. J., Feuk, L., Rivera, M. N., Listewnik, M. L., Donahoe, P. K., Qi, Y., Scherer, S. W., and Lee, C. (2004). Detection of large-scale variation in the human genome. *Nat. Genet.* **36**, 949–951.

Kallioniemi, A., Kallioniemi, O. P., Sudar, D., Rutovitz, D., Gray, J. W., Waldman, F., and Pinkel, D. (1992). Comparative genomic hybridization for molecular cytogenetic analysis of solid tumors. *Science* **258**, 818–821.

LaFramboise, T. (2009). Single nucleotide polymorphism arrays: A decade of biological, computational and technological advances. *Nucleic Acids Res.* **37**, 4181–4193.

LaFramboise, T., Weir, B. A., Zhao, X., Beroukhim, R., Li, C., Harrington, D., Sellers, W. R., and Meyerson, M. (2005). Allele-specific amplification in cancer revealed by SNP array analysis. *PLoS Comput. Biol.* **1**, e65.

Lai, W. R., Johnson, M. D., Kucherlapati, R., and Park, P. J. (2005). Comparative analysis of algorithms for identifying amplifications and deletions in array CGH data. *Bioinformatics* **21**, 3763–3770.

Letouzé, E., Allory, Y., Bollet, M. A., Radvanyi, F., and Guyon, F. (2010). Analysis of the copy number profiles of several tumor samples from the same patient reveals the successive steps in tumorigenesis. *Genome Biol.* **11**, R76.

Li, A., Liu, Z., Lezon-Geyda, K., Sarkar, S., Lannin, D., Schulz, V., Krop, I., Winer, E., Harris, L., and Tuck, D. (2011). GPHMM: An integrated hidden Markov model for identification of copy number alteration and loss of heterozygosity in complex tumor samples using whole genome SNP arrays. *Nucleic Acids Res.* **39**(12), 4928–4941.

Lin, M., Wei, L. -J., Sellers, W. R., Lieberfarb, M., Wong, W. H., and Li, C. (2004). dChipSNP: Significance curve and clustering of SNP-array-based loss-of-heterozygosity data. *Bioinformatics* **20**, 1233–1240.

Lindblad-Toh, K., Tanenbaum, D. M., Daly, M. J., Winchester, E., Lui, W. O., Villapakkam, A., Stanton, S. E., Larsson, C., Hudson, T. J., Johnson, B. E., Lander, E. S., and Meyerson, M. (2000). Loss-of-heterozygosity analysis of small-cell lung carcinomas using single-nucleotide polymorphism arrays. *Nat. Biotechnol.* **18**, 1001–1005.

Lindgren, D., Gudjonsson, S., Jee, K. J., Liedberg, F., Aits, S., Andersson, A., Chebil, G., Borg, A., Knuutila, S., Fioretos, T., Månsson, W., and Höglund, M. (2008). Recurrent and multiple bladder tumors show conserved expression profiles. *BMC Cancer.* **8**, 183.

Matsuzaki, H., Dong, S., Loi, H., Di, X., Liu, G., Hubbell, E., Law, J., Berntsen, T., Chadha, M., Hui, H., Yang, G., Kennedy, G. C., et al., (2004). Genotyping over 100,000 SNPs on a pair of oligonucleotide arrays. *Nat. Method.* **1**, 109–111.

McQuillan, R., Leutenegger, A. L., Abdel-Rahman, R., Franklin, C. S., Pericic, M., Barac-Lauc, L., Smolej-Narancic, N., Janicijevic, B., Polasek, O., Tenesa, A., Macleod, A. K., Farrington, S. M., et al., (2008). Runs of homozygosity in European populations. *Am. J. Hum. Genet.* **83**, 359–372.

Mullighan, C. G., Phillips, L. A., Su, X., Ma, J., Miller, C. B., Shurtleff, S. A., and Downing, J. R. (2008). Genomic analysis of the clonal origins of relapsed acute lymphoblastic leukemia. *Science* **322**, 1377–1380.

Nancarrow, D. J., Handoko, H. Y., Stark, M. S., Whiteman, D. C., and Hayward, N. K. (2007). SiDCoN: A tool to aid scoring of DNA copy number changes in SNP chip data. *PLoS ONE* **2**, e1093.

Navin, N., Kendall, J., Troge, J., Andrews, P., Rodgers, L., McIndoo, J., Cook, K., Stepansky, A., Levy, D., Esposito, D., Muthuswamy, L., Krasnitz, A., et al., (2011). Tumour evolution inferred by single-cell sequencing. *Nature* **472**(7341), 90–94.

Navin, N., Krasnitz, A., Rodgers, L., Cook, K., Meth, J., Kendall, J., Riggs, M., Eberling, Y., Troge, J., Grubor, V., Levy, D., Lundin, P., et al., (2010). Inferring tumor progression from genomic heterogeneity. *Genome Res.* **20**, 68–80.

Navin, N. E., and Hicks, J. (2010). Tracing the tumor lineage. *Mol. Oncol.* **4**, 267–283.

Paulsson, K., Lindgren, D., and Johansson, B. (2011). SNP array analysis of leukemic relapse samples after allogeneic hematopoietic stem cell transplantation with a sibling donor identifies meiotic recombination spots and reveals possible correlation with the breakpoints of acquired genetic aberrations. *Leukemia.* [Epub ahead of print]

Peiffer, D. A., Le, J. M., Steemers, F. J., Chang, W., Jenniges, T., Garcia, F., Haden, K., Li, J., Shaw, C. A., Belmont, J., Cheung, S. W., Shen, R. M., et al., (2006). High-resolution genomic profiling of chromosomal aberrations using Infinium whole genome genotyping. *Genome Res.* **16**, 1136–1148.

Pinkel, D., and Albertson, D. G. (2005). Array comparative genomic hybridization and its applications in cancer. *Nat. Genet.* **37**, S11–S17.

Pollack, J. R., Perou, C. M., Alizadeh, A. A., Eisen, M. B., Pergamenschikov, A., Williams, C. F., Jeffrey, S. S., Botstein, D., and Brown, P. O. (1999). Genome-wide analysis of DNA copy-number changes using cDNA microarrays. *Nat. Genet.* **23**, 41–46.

Popova, T., Manié, E., Stoppa-Lyonnet, D., Rigaill, G., Barillot, E., and Stern, M. H. (2009). Genome Alteration Print (GAP): A tool to visualize and mine complex cancer genomic profiles obtained by SNP arrays. *Genome Biol.* **10**, R128.

Rajagopalan, H., and Lengauer, C. (2004). Aneuploidy and cancer. *Nature* **432**, 338–341.

Russnes, H. G., Vollan, H. K., Lingjaerde, O. C., Krasnitz, A., Lundin, P., Naume, B., Sorlie, T., Borgen, E., Rye, I. H., Langerod, A., Chin, S. F., Teschendorff, A. E., et al., (2010). Genomic architecture characterizes tumor progression paths and fate in breast cancer patients. *Sci. Transl. Med.* **2**(38), ra47.

Schena, M., Shalon, D., Davis, R. W., and Brown, P. O. (1995). Quantitative monitoring of gene expression patterns with a complementary DNA microarray. *Science* **270**, 467–470.

Sebat, J., Lakshmi, B., Troge, J., Alexander, J., Young, J., Lundin, P., Maner, S., Massa, H., Walker, M., Chi, M., Navin, N., Lucito, R., et al., (2004). Large-scale copy number polymorphism in the human genome. *Science* **305**, 525–528.

Staaf, J., Jönsson, G., Ringnér, M., and Vallon-Christersson, J. (2007). Normalization of array-CGH data: Influence of copy number imbalances. *BMC Genomics* **8**, 382.

Staaf, J., Lindgren, D., Vallon-Christersson, J., Isaksson, A., Göransson, H., Juliusson, G., Rosenquist, R., Höglund, M., Borg, A., and Ringnér, M. (2008). Segmentation-based detection of allelic imbalance and loss-of-heterozygosity in cancer cells using whole genome SNP arrays. *Genome Biol.* **9**, R136.

Steemers, F. J., Chang, W., Lee, G., Barker, D. L., Shen, R., and Gunderson, K. L. (2006). Whole-genome genotyping with the single-base extension assay. *Nat. Method.* **3**, 31–33.

Sun, W., Wright, F. A., Tang, Z., Nordgard, S. H., Van Loo, P., Yu, T., Kristensen, V. N., and Perou, C. M. (2009). Integrated study of copy number states and genotype calls using high-density SNP arrays. *Nucleic Acid. Res.* **37**, 5365–5377.

Tolliver, D., Tsourakakis, C., Subramanian, A., Shackney, S., and Schwartz, R. (2010). Robust unmixing of tumor states in array comparative genomic hybridization data. *Bioinformatics* **26**, i106–i114.

Van Loo, P., Nordgard, S. H., Lingjærde, O. C., Russnes, H. G., Rye, I. H., Sun, W., Weigman, V. J., Marynen, P., Zetterberg, A., Naume, B., Perou, C. M., Børresen-Dale, A. -L., et al., (2010). Allele-specific copy number analysis of tumors. *Proc. Natl. Acad. Sci. USA* **107**, 16910–16915.

van Tilborg, A. A., de Vries, A., de Bont, M., Groenfeld, L. E., van der Kwast, T. H., and Zwarthoff, E. C. (2000). Molecular evolution of multiple recurrent cancers of the bladder. *Hum. Mol. Genet.* **9**, 2973–2980.

Venkatraman, E. S., and Olshen, A. B. (2007). A faster circular binary segmentation algorithm for the analysis of array CGH data. *Bioinformatics* **23**, 657–663.

Wang, D. G., Fan, J. B., Siao, C. J., Berno, A., Young, P., Sapolsky, R., Ghandour, G., Perkins, N., Winchester, E., Spencer, J., Kruglyak, L., Stein, L., et al., (1998). Large-scale identification, mapping, and genotyping of single-nucleotide polymorphisms in the human genome. *Science* **280**, 1077–1082.

Wang, K., Li, M., Hadley, D., Liu, R., Glessner, J., Grant, S. F., Hakonarson, H., and Bucan, M. (2007). PennCNV: An integrated hidden Markov model designed for high-resolution copy number variation detection in whole-genome SNP genotyping data. *Genome Res.* **17**, 1665–1674.

Wang, Y., and Armstrong, S. A. (2007). Genome-wide SNP analysis in cancer: Leukemia shows the way. *Cancer Cell.* **11**, 308–309.

Wang, Z. C., Lin, M., Wei, L. -J., Li, C., Miron, A., Lodeiro, G., Harris, L., Ramaswamy, S., Tanenbaum, D. M., Meyerson, M., Iglehart, J. D., and Richardson, A. (2004). Loss of heterozygosity and its correlation with expression profiles in subclasses of invasive breast cancers. *Cancer Res.* **64**, 64–71.

Zhao, X., Weir, B. A., LaFramboise, T., Lin, M., Beroukhim, R., Garraway, L., Beheshti, J., Lee, J. C., Naoki, K., Richards, W. G., Sugarbaker, D., Chen, F., et al., (2005). Homozygous deletions and chromosome amplifications in human lung carcinomas revealed by single nucleotide polymorphism array analysis. *Cancer Res.* **65**, 5561–5570.

Prostate Cancer as a Model System for Genetic Diversity in Tumors

Jeremy A. Squire*, Paul C. Park*, Maisa Yoshimoto*, Jennifer Alami*, Julia L. Williams*, Andrew Evans† and Anthony M. Joshua‡,§,¶

* *Department of Pathology and Molecular Medicine, Queen's University, Kingston, ON, Canada*
† *Department of Pathology, University Health Network, Toronto, Canada*
‡ *Department of Medical Biophysics, University of Toronto, Toronto, Canada*
§ *Division of Applied Molecular Oncology, Ontario Cancer Institute, Princess Margaret Hospital, Toronto, Canada*
¶ *Department of Medical Oncology, Princess Margaret Hospital, Toronto, Canada*

This chapter will summarize novel understandings of the early molecular events in prostatic carcinogenesis that may underlie both the genetic and clinical heterogeneity. Areas covered include preneoplasia, stem cell concepts, telomere abnormalities, and the nature of tumor–stromal interactions. The oncogenomics of prostate cancer is reviewed with emphasis on androgen signaling, *ETS* gene family aberrations, and

Advances in CANCER RESEARCH, Volume 112
Copyright 2011, Elsevier Inc. All right reserved.
0065-230X/10 $35.00
DOI: 10.1016/B978-0-12-387688-1.00007-7

PTEN deletion. The notion that "field cancerization," coupled with genomic instability may explain both the occurrence of multifocal disease, and the recent observations of genetic diversity of *ERG* alteration in individual tumors are discussed. Collectively, genomic studies are rapidly moving human prostate cancer closer to the promise of personalized medicine, so that specific genetic profiles of individual tumors will determine the best therapeutic approaches.
© 2011 Elsevier Inc.

I. INCIDENCE OF PROSTATE CANCER

Prostate cancer is the most commonly diagnosed malignancy in men in North America, with ~220,000 new cases expected to be diagnosed in 2011 (The American Cancer Society, 2011). The cancer is characterized by heterogeneity with regard to pathologic development, molecular abnormalities, and clinical course. There is also wide geographic variation in the incidence of clinical prostate cancer, with higher rates in the United States and Canada in comparison to Asia. These increased rates are likely due in part to the widespread use of transurethral resection of the prostate and prostate-specific antigen (*PSA*) testing, while recent increases in the Asian countries may be related to the westernized influences of diet and life style (Dijkman and Debruyne, 1996; Hsing *et al.*, 2000). Compounding this increasing incidence rate is the demographic trend of longer life spans and the implementation of, as yet unproven, public health prostate cancer surveillance in older men undergoing the *PSA* testing. The health problem created by *PSA* screening in the post second world war "baby-boomer" generation of men is now enormous. In 2007, there were 2.2 million men alive in the USA who had a diagnosis of prostate cancer (Altekruse *et al.*, 2009). The increasing incidences around the world underscore the need to improve our understanding of the molecular basis of genetic heterogeneity in this common disease. The focus for this review will therefore be on the common genetic and genomic changes that take place in tumor tissues that may explain the clinical diversity of this disease. Other aspects of the general genetic and epigenetic aetiology of prostate cancer initiation and progression have recently been reviewed in this series (Knudsen and Vasioukhin, 2010).

II. CLINICAL HETEROGENEITY OF PROSTATE CANCER

Understanding the molecular basis for this clinically heterogeneous disease is of fundamental importance because if disease is detected while still organ-confined, patients can be potentially cured by early surgical intervention or radiotherapy. Conversely, there is increasing awareness that prostate cancer may be overtreated and a "watchful-waiting" approach is appropriate. For example, in western populations, one in six men is diagnosed with prostate cancer whilst only one in 34 die of metastatic disease (Jemal *et al.*, 2007). However, following definitive management, approximately one-third of patients relapse with biochemical relapse detected as an increase in *PSA* serum levels post-treatment. Whilst first-line treatment involves androgen, ablation therapies there are now various types of surgical, chemotherapeutic, and radiation-based therapies available. However, most treatments eventually fail and prostate cancer still ranks second amongst male cancer-related deaths in USA (Altekruse SF *et al.*, 2009). Further research aimed at understanding the basis and interactions between the clinical variables and the molecular genetic diversity within a prostate cancer tumor itself is essential in designing treatment protocols tailored to the specific prognosis for the individual patient's health.

III. PATHOLOGY OF THE NORMAL PROSTATE AND PROSTATE CANCER

The prostate gland, heterogeneous at the cellular level is supported by a stroma and extracellular matrix composed of smooth muscle cells, fibroblasts, lymphocytes, and neurovascular tissue. The prostatic epithelium is made up of four distinct cell types: secretory, basal, transient amplifying (an intermediate between the previous two cell types) (Isaacs and Coffey, 1989), and neuroendocrine cells. The secretory cells that line the prostatic glands and ducts are responsible for the physiological secretions of the gland. The basal cells are much less abundant and rest on the basement membrane adjacent to the secretory cells. Basal cells exhibit many stem cell characteristics, including their relatively undifferentiated state, high proliferative capacity, protection from apoptosis, and a long life span (Bonkhoff, Stein, and Remberger, 1994; De Marzo *et al.*, 1998; Foster *et al.*, 2002; Potten and Loeffler, 1990). Neuroendocrine cells are scattered throughout the gland and are involved in the regulation of prostatic secretory activity and cell growth. There are three anatomical zones in the gland: peripheral, transitional, and central. Morphological changes in both epithelial and stromal cell compartments take place during the normal aging process in the gland of older males leading to benign prostatic hyperplasia (BPH). This lesion is not considered to be the precursor to prostate cancer (Kristal *et al.*, 2010). While the number of luminal epithelial cells increases in BPH, the columnar appearance and nuclear cytology remain normal. This lesion is very common in older males with up to

90% of men affected by BPH as they reach the age of 80 (Bushman, 2009). Other histologic lesions of the epithelial cell compartment such as proliferative inflammatory atrophy (PIA) and high-grade prostatic intraepithelial neoplasia (HPIN) are considered to be precursor lesions to prostate cancer (discussed below). HPIN usually appears in the peripheral regions of the prostate gland and is characterized by dysplasia of prostatic epithelial cells. In PIN, luminal epithelial cells accumulate and show nuclear enlargement with prominent nucleoli (Ayala and Ro, 2007). Prostatic adenocarcinomas, which account for 98% of all primary prostate cancer usually presents in the peripheral zones, but tumors may also arise in the transitional zone. Prostate cancer can present as multifocal lesions, which may also express a range of differentiation phenotypes at the histologic level.

The Gleason score is one of the most accepted and important predictors of prognosis and behavior of prostate cancer (Allsbrook *et al.*, 1999). The Gleason grade is a scoring system used by pathologists to establish the relative degree of tumor differentiation and is based on both the architectural pattern and the tumor histology. For example, Gleason grades 1–3 will form organized ductal structures, surrounded by distinct stromal borders (Fig. 1). However, Gleason grades 4–5 possess severely disrupted glandular architecture without a lumen and may exhibit gland fusion and cribriform morphology independent of stromal architecture. In addition, Gleason grade is often heterogeneous within or between different tumor foci, so the enumeration system involves obtaining an overall score as the sum of two most prevalent growth patterns in the gland.

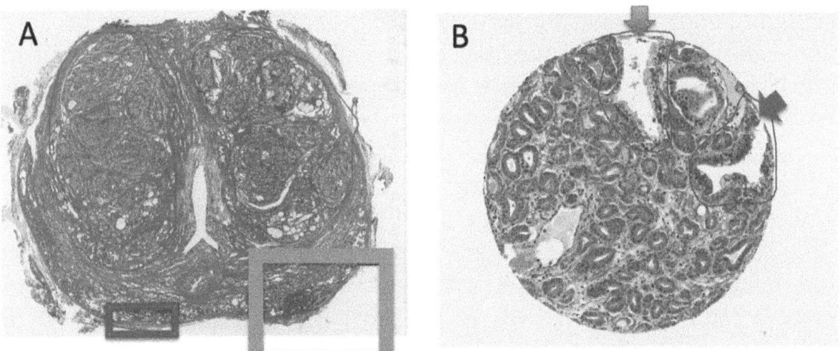

Fig. 1 Histopathology of the prostate. (A) Whole-mount hematoxylin (H and E) and eosin section of the prostate showing an outlined small area of prostate cancer within the grey rectangle and an area containing high-grade prostatic intraepithelial neoplasia (HPIN) within the green rectangle. (B) An H and E stained tissue microarray core illustrates prostate cancer comprising Gleason pattern 3. At the top, the blue arrow identifies normal glandular elements and the green arrow to the right identifies HPIN. (See Page 7 in Color Section at the back of the book.)

IV. PROSTATE CANCER ONCOGENOMICS

Advances in cytogenetics and genomics facilitated the characterization of common genomic alterations in prostate cancer, which are predominantly characterized by deletions (10q, *PTEN*; 13q, *RB1*; 8p, *NKX3.1* (a prostate-specific tumor suppressor); 5q; 2q; 17p; and less commonly 6q; 7p; 16q; 18q) with a smaller number of recurrent gains (8q, *MYC*; and chromosome 7). Of the gains, *MYC* at cytoband 8q24 is the most commonly observed, being reported in approximately 30–40% of primary prostate tumors, and 90% of metastatic prostate cancers (Ishkanian *et al.*, 2009; Jenkins *et al.*, 1997; Sato *et al.*, 2006) and *MYC* gain is associated with poor patient outcome (Sato *et al.* 2006). Other more complex patterns as well as an accumulation in the number of genomic gains and amplifications (Xq11.2-q12, *AR*) emerge in more advanced disease (Table I). Genomic rearrangements leading to the formation of

Table I Fusion Gene Five Prime Partners and Classes Reported to Date.

5′ partner	Class[b]	3′ partner
TMPRSS2	I	*ERG*
TMPRSS2	I	*ETV1*
TMPRSS2	I	*ETV4*
HERV-K_22q11.23	II	*ETV1*
SLC45A3	II	*ETV1*
C15orf21	III	*ETV1*
HNRPA2B1	IV	*ETV1*
KLK2	II	*ETV4*
CANT1	II	*ETV4*
ACSL3	II	*ETV1*
SLC45A3	II	*ERG*
FLJ35294	II	*ETV1*
DDX5	IV	*ETV1*
TMPRSS2	I	*ETV5*
SLC45A3	II	*ETV5*
EST14	II	*ETV1*
HERVK17	II	*ETV1*
FOXP1	II	*ETV1*
SLC45A3	II	*ELK4*
NDRG1	II	*ERG*
SLC45A3	*ETS* neg	*BRAF*[a]
ESRP1	*ETS* neg	*RAF1*

[a]Only 3′ partners identified to date that are not a member of the *ETS* transcription factor family.
[b]Tomlins *et al.* (2007).

TMPRSS2:ETS gene fusions and deletion of the *PTEN* tumor suppressor are the two most frequent alterations observed in prostate cancer (Tomlins *et al.*, 2005; Yoshimoto *et al.*, 2007, 2008). The *TMPRSS2:ERG* gene fusion is the principal genomic alteration and a characteristic signature in approximately half of prostatic malignancies.

V. DISCOVERY OF TMPRSS2:ETS FUSION GENES IN PROSTATE CANCER

The discovery of the *TMPRSS2:ETS* gene fusions exemplifies the current shift in the strategy in cancer genomics from experimental to bioinformatics approaches. By surveying 132 gene-expression prostate cancer datasets from the Oncomine database (www.oncomine.org, Compendia Bioscience) using a data transformation algorithm, cancer outlier profile analysis (COPA), Chinnaiyan and colleagues identified genes with aberrant expression profiles in a subset of samples (Tomlins *et al.*, 2005) which appeared to be cancer-related genes known to participate in chromosomal rearrangements in hematological malignancies and sarcomas.

Two mutually exclusive Erythroblastosis virus E26 transformation-specific (*ETS*) transcription factors, *ETS variant 1* (*ETV1*, 7p21.3) and *ETS-related gene* (*ERG*, 21q22.2), were identified as high-ranking outliers in several independent gene-expression profiling datasets. Exon-walking quantitative PCR of patient samples found the 3′ regions of *ERG* and *ETV1* to be overexpressed but the corresponding 5′ regions were absent. 5′ RNA ligase-mediated rapid amplification of cDNA ends identified the 5′ end of these transcripts as the promoter sequences belonging to the prostate-specific, androgen-regulated *transmembrane protease, serine II* (*TMPRSS2*, 21q22.3).

VI. FREQUENCY OF TMPRSS2:ETS GENE FUSIONS IN PROSTATE CANCER

This breakthrough study reported that >70% of radical prostatectomy samples harbored a fusion of the 5′ untranslated region (UTR) of *TMPRSS2* with the coding sequences of either *ERG* or *ETV1* (Tomlins *et al.*, 2005). Interestingly, only months prior, Petrovics and colleagues (Petrovics *et al.*, 2005) reported that *ERG* was the most commonly overexpressed oncogene in prostate cancer by qPCR analysis. Subsequent studies showed that the frequencies of the *TMPRSS2:ERG* fusion gene (or *ERG* rearrangement) are exceptionally variable and inconsistent in the literature, ranging from 27% to 79% in radical prostatectomy and biopsy samples, generally from *PSA* screened cohorts (Barwick *et al.*, 2010; Magi-Galluzzi *et al.*, 2011; Mehra *et al.*, 2007; Tomlins *et al.*, 2005; Watson *et al.*, 2009; Yoshimoto *et al.*, 2008).

Some of the discrepancies in frequency relate to differences in the patient cohort (race as well as global geographical location or *PSA*-screened vs. population based), the type of specimen examined, as well as the technique used to detect the fusion gene. This concept is clearly illustrated when comparing population-based cohorts, which have a lower reported frequency of 15–35%, as compared to radical prostatectomy cohorts (Attard *et al.*, 2008; Demichelis *et al.*, 2007; FitzGerald *et al.*, 2008). The *TMPRSS2:ERG* gene fusion is found in approximately half of Caucasian patients, with a lower reported frequency in African-American men and less common in Asian cohorts (Lee *et al.*, 2010; Magi-Galluzzi *et al.*, 2011; Miyagi *et al.*, 2010; Mosquera *et al.*, 2009). An excellent example of discrepancies based on patient populations was demonstrated in a study that compared UK and Chinese patients, wherein 41.3% of the UK patients but only 7.5% of Chinese cohort were found to harbor *ERG* rearrangements detected by FISH (Mao *et al.*, 2010). More recently, a contradicting report stating that 90% of the Chinese radical prostatectomy specimens harbored *ETS* rearrangements further emphasizes the multifactorial nature of the variability in the frequency of this gene fusion (Sun *et al.*, 2010).

Hormone refractory and/or metastatic prostate cancer exhibits less variability in the occurrence of *TMPRSS2:ERG* rearrangements with reported frequencies ranging from 29–59% (Attard *et al.*, 2009; Boormans *et al.*, 2010; Gopalan *et al.*, 2009; Han *et al.*, 2009; Mehra *et al.*, 2008; Perner *et al.*, 2007; Saramäki *et al.*, 2008; Stott *et al.*, 2010). Additionally, minute prostatic adenocarcinoma, a form of prostate cancer that is considered less clinically significant, harbors *ETS* fusions in approximately half of reported cases (Albadine *et al.*, 2009).

Investigators have also evaluated this rearrangement in the peripheral and transitional zones of the prostate. Nearly half of the peripheral tumors (43.3%) derived from radical prostatectomy samples were *ERG* rearrangement positive, whereas all 30 corresponding transitional tumors displayed a normal *ERG* locus by FISH (Guo *et al.*, 2009). In contrast, other studies found that *TMPRSS2:ERG* (13.3%) and *ERG* rearrangements (11.9%) are present in transitional zone tumors, despite being identified at a lower rate compared to peripheral zone lesions (Bismar and Trpkov, 2010). BPH and normal epithelium are negative for *ERG* rearrangements and fusion transcripts, with the exception of the report by Clark and colleagues (Cerveira *et al.*, 2006; Clark *et al.*,

2007; Dai *et al.*, 2008; Han *et al.*, 2009; Lu *et al.*, 2009; Mosquera *et al.*, 2009; Perner *et al.*, 2007; Rajput *et al.*, 2007; Sun *et al.*, 2010; Wang *et al.*, 2006; Zhang *et al.*, 2010). This study found that eight of 17 (47%) normal epithelial samples adjacent to fusion positive prostate cancer were also positive for *TMPRSS2:ERG* rearrangements and found a 6% fusion-positive rate in BPH samples (Clark *et al.*, 2007). These data illustrate the enormous variability in the frequency of *TMPRSS2:ERG* fusion positivity with respect to the cohort, disease stage, type and quality of the tissue sample, as well as the method of detection.

VII. ANDROGEN DRIVEN TMPRSS2 GENE FUSIONS

TMPRSS2 (21q22.3) is a 70-kDa serine protease family member that is associated with physiological and pathological processes such as digestion, tissue remodelling, blood coagulation, fertility, inflammatory responses, tumor cell invasion, and apoptosis. The normal function of this protein is not yet known but it is composed of a type II transmembrane domain, a receptor class A low-density lipoprotein domain, a scavenger receptor cysteine-rich domain, a protease domain, and a cytoplasmic domain (Paoloni-Giacobino *et al.*, 1997). The 32-kDa serine protease domain undergoes autocleaveage, secretion into the prostate epithelia and interacts with cell surface proteins, the extracellular matrix and proteins of neighboring cells (Afar *et al.*, 2001; Vaarala *et al.*, 2001; Wilson *et al.*, 2005). The *TMPRSS2* gene is composed of 14 exons, with protein coding sequences only in the latter half. Importantly, *TMPRSS2* harbors androgen responsive elements in its 5′ UTR. Prostate epithelial cells express TMPRSS2 at higher levels relative to other tissues, while *TMPRSS2* gene expression is further elevated in prostate cancer relative to BPH and normal prostatic epithelium (Afar *et al.*, 2001). Androgens and the *AR* are essential to the growth and development of the prostate gland but also play an important role in the initiation and progression of prostate cancer (Balk and Knudsen, 2008). It is well established that despite chemical castration the *AR* continues to function to drive the prostate growth, likely through a variety of mechanisms such as hypersensitivity to low levels of androgens, activation in the absence of, or via unconventional ligands, or amplification of the *AR* gene locus, resulting in elevated levels of *AR* protein. All of these mechanisms could potentially allow *AR* activation in the presence of low androgen concentration (Ai *et al.*, 2009; Kawata *et al.*, 2010; Vis and Schröder, 2009). Consequently, *TMPRSS2* in turn plays an important role in prostate cancer progression in spite of hormonal ablation regimens as its promoter region drives the expression of the fused *ETS* gene.

VIII. Ets TRANSCRIPTION FACTORS

Twenty-seven human *ETS* transcription factor family members have been identified, all of which share a conserved DNA binding domain that recognizes unique sequences containing GGA(A/T) (Nye *et al.*, 1992). *ERG* (21q22.2) is the *ETS* transcription factor most commonly known to participate in prostate cancer gene fusions. The *ERG* gene contains 11 exons, with the transcriptional start site in exon 3. The *ERG* protein can interact with *ETS* members as well as other transcription factors, such as *JUN* and *FOS*, through its protein–protein interacting domain, SAM-PNT (Basuyaux *et al.*, 1997; Carrère *et al.*, 1998; Verger *et al.*, 2001). The conserved *ETS* DNA binding domain permits binding to purine-rich DNA sequences (Reddy and Rao, 1991; Reddy *et al.*, 1987), and in this manner, exerts its effects in numerous cellular processes including membrane remodelling, angiogenesis, differentiation, proliferation, and tumorigenesis (Birdsey *et al.*, 2008; Carver *et al.*, 2009; Ellett *et al.*, 2009; Kruse *et al.*, 2009; McLaughlin *et al.*, 2001; Oikawa and Yamada, 2003; Randi *et al.*, 2009; Sato, 2001). Emerging evidence suggests that formation of the fusion gene may promote prostatic tumorigenesis, progression, and invasive disease and may be associated with prostate cancer-related mortality (Demichelis *et al.*, 2007; Hawksworth *et al.*, 2010; Klezovitch *et al.*, 2008; Wang *et al.*, 2008). Functionally, *ERG* overexpression in prostate cancer is highly implicated in promoting motility and invasiveness (Demichelis *et al.*, 2007; Schulz *et al.*, 2010; Singh *et al.*, 2002; Sreekumar *et al.*, 2009; Tomlins *et al.*, 2008; Trojanowska, 2000). In prostate cancer, *ERG* expression has been associated with a variety of targets including elevated levels of histone deactylase 1 (HDAC1) and subsequent down regulation of HDAC1 target genes, upregulation of WNT pathway proteins, and inhibition of apoptotic signaling (Iljin *et al.*, 2006). HDAC1 upregulation is common in prostate cancer, but was found to be uniformly increased in *ERG* rearranged tumors (Björkman *et al.*, 2008; Iljin *et al.*, 2006). Activation of the WNT pathway leads to transcription of numerous genes involved in tumorigenesis, including *AR, MYC, JUN, CCND1, BMP4,* and *MMP7* (Schweizer *et al.*, 2008; Terry *et al.*, 2006). Highlighting the importance of WNT pathway activation in fusion-positive prostate cancer is the resultant increase in *AR* transcription and expression levels. Consequently transcription of the fusion gene is increased, further amplifying *ERG* expression. Moreover, β-catenin and the *AR* interact in an androgen-dependent manner to regulate *AR* target genes, whereas in androgen-insensitive tumors, both β-catenin and *AR* target genes are expressed (Schweizer *et al.*, 2008).

ERG overexpression has recently been shown to activate *MYC* and result in its overexpression. These experiments revealed that *MYC* is activated by *ERG* and together their co-overexpression results in the suppression of prostate-epithelial differentiation genes and shorter time to biochemical recurrence (Hawksworth *et al.*, 2010; Sun *et al.*, 2008). *ERG* can also induce *ICAM2* expression, resulting in *AKT* activation via PDK1 and subsequent inhibition of BAD leading to suppression of apoptotic signals (McLaughlin *et al.*, 2001). In addition, *ERG* can bind BRCA1, a co-activator of the *AR*, and together were shown to regulate *IGFR* expression (Chai *et al.*, 2001; Schayek *et al.*, 2009). *IGFR* expression eventually leads to activation of *AKT* upon IGFR ligand binding. *ERG* regulates matrix metalloproteinases thus influencing extracellular matrix remodelling and the invasive potential of the cell (Ellett *et al.*, 2009; Hawksworth *et al.*, 2010; Schulz *et al.*, 2010; Singh *et al.*, 2002).

The *TMPRSS2:ERG* fusion gene constitutes the majority (>85%) of *ETS* rearrangements in prostate cancer, likely owing to their close proximity (2.7 Mb) and identical orientation on chromosome 21. Numerous alternative *ETS* members also can fuse to *TMPRSS2*, albeit at a much lower frequency. Similarly, variability in 5′ partners have also been identified (Table I). An intensive study examining *ETV1* rearrangements found that only nine of 23 samples had previously identified 5′ partners, demonstrating the extensive variability in fusion partners pairing with *ETS* transcription factors other than *ERG* (Attard *et al.*, 2008). The combined frequency of the remaining fusion variants accounts for approximately 10% of cases.

IX. DETECTION OF FUSION GENE REARRANGEMENTS BY FISH

Our laboratory uses differentially labeled bacterial artificial chromosome (BAC) clones as probes for the 3′ (RP11-476D17) and 5′ (RP11-95I21) segments of *ERG* with an additional BAC probe specific for the 5′ region of *TMPRSS2* (RP11-535H11) or for the transcriptional regulatory sequences (telomeric) of *TMPRSS2* (RP11-35C4, RP11-891L10; RP11-260O11) (Yoshimoto *et al.*, 2006). The advantage of this method is that the mechanism of rearrangement can be deduced (Yoshimoto *et al.*, 2006, 2007). This probe configuration can distinguish between fusion gene formation through deletion (loss of the 3′ *TMPRSS2* signals; Fig. 2C),

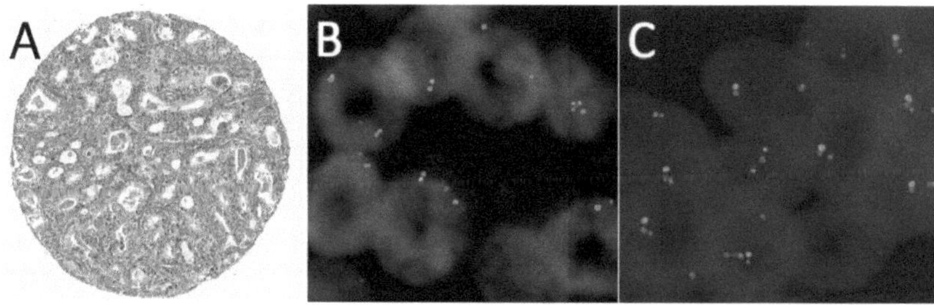

Fig. 2 Identifying genomic alterations of *PTEN* and *TMPRSS2–ERG* in prostate cancer. (A) H and E tissue from a tissue microarray core to illustrate prostate cancer comprising Gleason pattern 3. (B) Adjacent section comprising DAPI stained (blue) nuclei showing hemizygous deletion (one copy loss) of the *PTEN* gene. Only one orange *PTEN*-specific signal is present but both green and blue flanking probes are present. (C) Adjacent section with nuclei showing presence of *TMPRSS2–ERG* fusion rearrangement taking place by genomic deletion of the green 5′ *ERG* probe. (See Page 8 in Color Section at the back of the book.)

or through translocation (split of 5′ and 3′ *TMPRSS2* signals). In addition to identifying the fusion status and type, the inherent cell-by-cell analysis addresses the heterogeneous nature of the disease, and provides a biological context for such information. Characterization of the rearrangement mechanism is essential as differential clinical impacts may be observed with the various mechanisms resulting in *ETS* gene fusions. Moreover, with this FISH assay any intratumor or intertumor heterogeneity can be documented at the cellular level to better define differences in genctic diversity in multifocal disease (discussed below).

X. LOSS OF PTEN IN PROSTATE CANCER

Phosphatase and tensin homologue deleted on chromosome 10 (*PTEN*) is among the most commonly mutated tumor suppressor genes in human cancer (Zhang and Yu, 2010). *PTEN* targets proteins in signaling pathways that regulate cell growth, survival, and modulate genome stability. Inactivation of *PTEN* in human prostate cancer by diverse genomic mechanisms is well documented and attests to its crucial role on prostate cancer oncogenesis.

Early studies using loss of heterozygosity (LOH) analyses of deletions of 10q showed that 35–58% of *PTEN* LOH is present in advanced prostate cancer (Feilotter *et al.*, 1998; Müller *et al.*, 2000; Verhagen *et al.*, 2006). Submicroscopic deletions of *PTEN* have been characterized more extensively in radical prostatectomy tissue samples using four color FISH assays with high specificity (Fig. 1B). *PTEN* deletions are found in ~40% of localized prostate cancer (Bednarz *et al.*, 2010; Bismar *et al.*, 2011; Hermans *et al.*, 2004; Ishkanian *et al.*, 2009; Lapointe *et al.*, 2007; Liu *et al.*, 2009; Reid *et al.*, 2010; Sircar *et al.*, 2009; Taylor *et al.*, 2010; Verhagen *et al.*, 2006; Yoshimoto *et al.*, 2006, 2007, 2008). In advanced disease, FISH has shown that both hemizygous (one copy lost) and homozygous (both copies lost) *PTEN* deletions have been observed (Liu *et al.*, 2009; McMenamin *et al.*, 1999; Reid *et al.*, 2010; Schmitz *et al.*, 2007; Verhagen *et al.*, 2006), with the incidence of *PTEN* deletion being much higher, approaching 70–80% of tumors (Bismar *et al.*, 2011; Holcomb *et al.*, 2009; Liu *et al.*, 2009; Sircar *et al.*, 2009). It has been shown that hemizygous *PTEN* deletions are associated with earlier biochemical relapse, and that the homozygous deletions creating "*PTEN* null" tumors are strongly linked to metastasis and androgen independent progression (Bismar *et al.*, 2011; Carver *et al.*, 2009; Reid *et al.*, 2010; Sarker *et al.*, 2009; Sircar *et al.*, 2009; Squire, 2009). Genomic losses and inactivations of the *PTEN* gene most likely lead to an adverse clinical course through haploinsufficiency and concomitant reduced levels of the *PTEN* protein (Koksal *et al.*, 2004). Loss of the *PTEN* leads to upregulation of PI3K/*AKT*, which is a strongly associated progression pathway in prostate cancer (Ayala *et al.*, 2004; Ohigashi *et al.*, 2005; Sarker *et al.*, 2009; Sircar *et al.*, 2009; Wang *et al.*, 2010).

PTEN loss also appears to play an important role in the induction of prostate cancer in mouse models of the disease. It has been shown that deletion of both alleles of *PTEN* leads to developmental defects and early death, *PTEN* heterozygous mice developed HPIN, but without progression to invasive cancer (Di Cristofano *et al.*, 1998; Kwabi-Addo *et al.*, 2001). King *et al.* (2009) developed an *ERG* transgenic mouse that expressed the most common human variant form of the fusion gene and found that these mice failed to develop HPIN preneoplasia (discussed below) or acquire prostate tumors. When they crossed this fusion transgene with mice that were either *PTEN*-deficient or had prostate-specific high *AKT* activity, they obtained offspring that developed HPIN, but did not show prostate cancer progression. Similarly, the *ERG* overexpressing transgenic mouse of Carver *et al.* (2009) also developed HPIN, but only when crossed with *PTEN*-deficient mice. However, in this strain, progression to prostate cancer was a consistent feature in offspring. It has also been shown in the mouse that reduced levels of the *PTEN* protein can lead to genomic instability (Alimonti *et al.*, 2010). It has therefore been suggested that loss of genome stability may enhance the ability of cells bearing *PTEN* loss to acquire additional chromosomal aberrations such as the *TMPRSS2–ERG* fusion in human prostate cancer (Bismar *et al.*, 2011). Moreover, both *PTEN* loss and the presence of *TMPRSS2–ERG* fusion genes interact to lead to a less favorable clinical outcome in human prostate cancer (Reid *et al.*, 2010; Yoshimoto *et al.*, 2008). An example of the detection by FISH of both *PTEN* loss and concomitant *TMPRSS2–ERG* fusion in the same tumor is presented in Fig. 2. Collectively both the mouse and human data suggest that downstream pathways of both *PTEN/AKT* and androgen-dependent overexpressed *ETS* targets may act as drivers of prostate cancer progression.

In vitro studies (Carver *et al.*, 2009) have attempted to demonstrate the role of downstream *ETS* targets in prostate cancer progression by showing that *PTEN* haploinsufficiency or *AKT* overexpression conferred a proliferative advantage in prostate cancer cell lines, whereas *ERG* expression increased migration. Two proteins with known functions in cell migration—the chemokine receptor CXCR4 and the metalloproteinase ADAMTS1—were identified as direct targets of *ERG*. Small interfering (si)RNA-mediated knockdown of CXCR4 blocked migration in prostate cancer cell lines overexpressing *ERG*. These data suggest that the *ERG*-mediated increase in migration cooperates with PI(3)K pathway-driven proliferation to promote rapid tumorigenesis. It will be important to uncover other transcriptional targets of *ERG*, and to determine if inhibiting *ERG* or one of its targets can block the initiation of adenocarcinoma in HPINs. Collectively, both the mouse and human data (summarized in Fig. 3) suggest that future therapeutic strategies

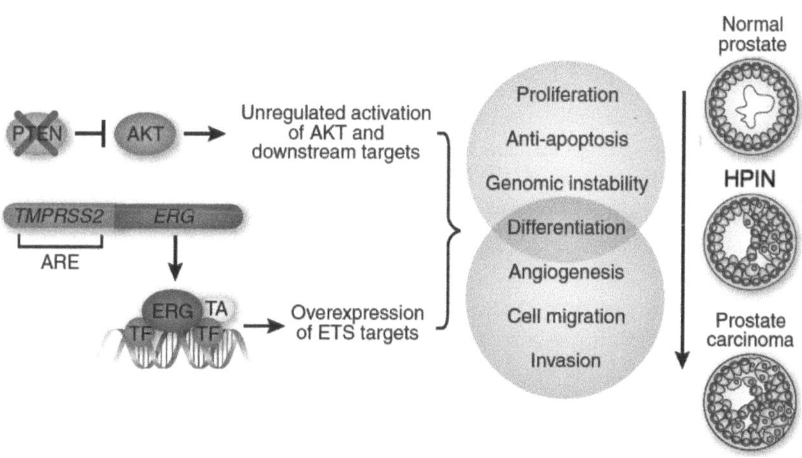

Fig. 3 Interactions between *PTEN* loss and *TMPRSS2–ERG* fusion in prostate cancer. Loss of *PTEN* and concomitant activation of *AKT* could act in partnership with the *TMPRSS2–ERG* fusion protein to promote progression to prostate cancer through downstream pathways that increase the selective advantage of premalignant high-grade prostatic intraepithelial neoplasia (HPIN) cells. ARE: androgen response elements; TA: transcriptional activator; TF: transcription factor, [based on Squire (2009)]. (For color version of this figure, the reader is referred to the web version of this book.)

must identify effectors within both of the *PTEN/AKT* and ETS signaling pathways to improve treatment response in human prostate cancer.

XI. GENOMIC ALTERATIONS IN PROSTATE CANCER PRENEOPLASIA

The evolution of prostate cancer from precursor cells of the normal epithelium is characterized by a number of genomic abnormalities that are predominantly generated in the preneoplastic phase. HPIN (Fig. 1B) is considered most likely to represent a preneoplastic forerunner to prostate cancer on the basis of pathological (Qian *et al.*, 1997), epidemiological (Sakr *et al.*, 1993, 1994) and cytogenetic (Beheshti *et al.*, 2002) evidence. For example, analyses of DNA isolated by laser capture microdissection of HPIN detected with prostate cancer have demonstrated similar chromosomal abnormalities such as the recurrent loss of chromosome arm 8p and the gain of 8q, chromosome 7, 10q, and Xq (Hughes *et al.*, 2006; Nupponen and T Visakorpi, 2000).

The extraordinary sensitivity of FISH to detect genomic losses of *PTEN* and *TMPSSS2–ERG* fusion events at the cellular level has meant that it has been possible to document the presence of genetic biomarkers associated with prostate cancer in subsets of HPIN (Ribeiro *et al.*, 2006; Yoshimoto *et al.*, 2006). *TMPRSS2:ERG* gene fusions are reported in 10–21% of HPIN lesions, but are identified almost exclusively in HPIN lesions adjacent to fusion-positive prostate cancer (Carver *et al.*, 2009; Cerveira *et al.*, 2006; Han *et al.*, 2009; Mosquera *et al.*, 2008; Perner *et al.*, 2007; Zhang *et al.*, 2010). Hemizygous deletions of *PTEN* are similarly found in HPIN that are in close proximity to *PTEN* deletion-positive cancer. It is therefore difficult to determine whether the presence of these genetic markers in HPIN is directly associated with transition from HPIN to cancer, or whether the HPIN lesions adjacent to cancer foci are actually infiltrating tumor cells that are presenting a histologic "HPIN-like" appearance.

Another proposed precursor lesion, PIA, is comprised of discrete foci of proliferative glandular epithelium with the morphological appearance of simple atrophy (Ruska *et al.*, 1998) or postatrophic hyperplasia (Cheville and Bostwick, 1995) in association with inflammation. The key features of these lesions are the presence of two distinct cell layers, mononuclear and/or polymorphonuclear inflammatory cells in both the epithelial and stromal compartments. The information linking PIA to prostate cancer is less obvious than the genetic and clinical relationship between HPIN and prostate cancer. PIA is thought to be a lesion arising as a proliferative response to environmental damage such as inflammation and accumulative oxidative stress in the lifetime of the gland in elderly men. In contrast, HPIN and early prostate cancer lesions are thought to arise as a consequence of genomic damage (De Marzo *et al.*, 2004, 2007). Whilst the results of several pathological studies in this area have not been definitive (Anton *et al.*, 1999; Billis, 1998; Billis and Magna, 2003; Faith *et al.*, 2005; Franks, 1954; Liavåg, 1968; Putzi and De Marzo, 2000), more recent analyses of PIA indicate that the lesion does express molecular signals of early neoplastic transformation. Observations such as GSTP1 hypermethylation (Nakayama *et al.*, 2003), increased GSTA1 and COX-2 (Parsons *et al.*, 2001; Zha *et al.*, 2001), reduced NKX3.1 expression (Bethel *et al.*, 2006), early chromosomal abnormalities (Shah *et al.*, 2001), elevated Bcl-2 (De Marzo *et al.*, 1999) and telomere shortening (Perner *et al.*, 2007) are all consistent with PIA lesions being analogous to HPIN as an early prostate cancer precursor. PIA may also represent the intermediate luminal cell type that has been proposed as a target of neoplastic transformation in prostate cancer (van Leenders *et al.*, 2003). Collectively, these etiological and pathological observations suggest the involvement of both PIA and HPIN as early prostatic lesions that may undergo additional genomic alterations to progress to prostate cancer.

XII. GENETIC DIVERSITY GENERATED BY TELOMERE LOSS IN PRENEOPLASIA

One potential mechanism that could initiate chromosomal rearrangements characteristic of prostate cancer and preneoplasia is telomere attrition (Meeker *et al.*, 2002; Vukovic *et al.*, 2003). Telomeres stabilize and protect the ends of chromosomes, but shorten because of cell division in the absence of telomerase and/or oxidative stress or other microenvironmental influences (Zvereva *et al.*, 2010). FISH analysis of prostate cancer, HPIN, and nondysplastic prostate epithelium has demonstrated a significant decrease in telomere length in both HPIN and prostate cancer in comparison with normal epithelium (Vukovic *et al.*, 2003). In addition, elevated rates of aneusomy were observed suggesting that increased levels of chromosomal aberrations were associated with decreased telomere length. This reduction in telomere length was more evident in the histologic regions of the prostate containing unifocal prostate cancer, suggesting that the microenvironment of the gland where cancer first occurred may have been subjected to genotoxic stress, local proliferation and concomitant telomere attrition (Cattan *et al.*, 2008; Houben *et al.*, 2008; Richter and Proctor, 2007). More recently, it has been shown that telomere attrition in the prostate gland is associated with particular genomic aberrations that may contribute to the chromosomal instability that characterizes prostate cancer carcinogenesis (Joshua *et al.*, 2010).

XIII. CHROMOSOMAL INSTABILITY AND GENETIC DIVERSITY IN PROSTATE CANCER

Genomic instability is a general term to describe the overall processes that increase the rate of mutation, thus enabling cells to develop new and aggressive phenotypes. Instability mechanisms fall into two broad groups: microsatellite or chromosomal. Chromosomal instability describes the cellular processes that increases the rate at which whole chromosomes, or large portions of chromosomes are gained, lost or rearranged in tumors (Rajagopalan *et al.*, 2003). This process appears to be more important in prostate cancer.

Analysis of the cytogenetics and genomics of prostate cancer and precursor lesions such as PIN have shown that specific chromosomal alterations are recurrent and take place early on in the malignant process (Hughes *et al.*, 2006; Ribeiro *et al.*, 2006). Previous studies have suggested several genomic loci that are gained in a high proportion of both HPIN and prostate cancer (7q22, 8q21-qter, 16p12-13) or lost in a high proportion (6q16, 8q21-qter, 10q22-23, 13q13-26, 16q22-qter). Recently reviewed by Saramaki and Visakorpi (2007) a few trends from these analyses are evident. (i) Losses tend to predominant earlier in carcinogenesis, suggesting haploinsufficency as a likely phenomenon (Chaib *et al.*, 2003; Li *et al.*, 2005). (ii) The rate of evolution of genomic instability is higher in HPIN than in more advanced stages of prostate cancer, although the tumor certainly continues to mutate. This is consistent, although not exclusively, with a mechanism of telomere generated chromosomal instability (Chin *et al.*, 2004) although other mechanisms such as centrosome defects (Pihan *et al.*, 2001, 2003) and DNA repair defects (Fan *et al.*, 2004; Meng *et al.*, 2005) in combination with checkpoint defects (Kiviharju-af Hällström *et al.*, 2007) found in prostate cells are certainly possible (Löbrich and Jeggo, 2007). (iii) The aberrations found give insight into genetic loci that have functional importance in the cancer phenotype. For example, loss of *NKX3.1* (8p21.2), *PTEN* (10q23.3), *CDH1* (*E-cadherin*; 16q22.1) and gain of *EZHZ2* (7q36.1), and *AR* (Xq12) have all been subsequently verified on the basis of initial comparative genomic hybridization studies. Novel loci identified on the basis of higher resolution technologies are also of increasing interest such as losses of *MXI1* (10q25.2, a *MYC* antagonist), *FOXO1A* (13q14.11, an inhibitor of *AR* signaling), *ATBF1* (16q22.3, transactivator of the

CDKN1A cyclin-dependent kinase inhibitor) and gains of *MCM7* (7q21.3, part of a complex that binds DNA origins and prepares them for initiation of replication).

XIV. MULTIFOCAL GENOTYPIC HETEROGENEITY OF PROSTATE CANCER

Multifocal prostate cancer, (reviewed in Andreoiu and Cheng, 2010) generally consists of a dominant (or index) tumor and one or multiple separate smaller tumors (Arora *et al.*, 2004; Bostwick *et al.*, 1998; Ruijter *et al.*, 1996; Wise *et al.*, 2002). Moreover, multifocal prostate cancer frequently shows histologic heterogeneity among different tumor foci exhibiting different Gleason grades (Arora *et al.*, 2004; Mehra *et al.*, 2007; Ruijter *et al.*, 1996). Collectively, the histologic and biological heterogeneity of multifocal prostate cancer suggests the tumor foci arise independently within the same gland. The emerging genetic data suggests that an earlier unknown carcinogenic insult such as increased oxidative stress may have taken place in the peripheral zone of part of the gland and could be associated with a predisposition to elevated level of genetic instability (perhaps driven by telomere attrition), that may underlie acquisition of multifocal prostate cancer (see Fig. 4).

"Field cancerization" was a term adopted to explain multifocal oral cancers in which clinically occult multifocal preneoplastic foci emerge within the epithelium of an anatomic region exposed to the same carcinogen(s) (Slaughter *et al.*, 1953). Such preneoplastic lesions may not be apparent on histological examination but molecular analyses have indicated clonal genomic changes such as p53 loss, loss of heterozygosity and microsatellite instability in a variety of epithelial neoplasms (Braakhuis *et al.*, 2003; Garcia *et al.*, 1999).

Surprisingly there has been a little consideration of the concept of field cancerization in the context of prostatic carcinogenesis, despite the observation that 80% of radical prostatectomy specimens demonstrate more than one neoplastic focus (Greene *et al.*, 1991). Furthermore, the occurrence of multicentric HPIN lesions in approximately 70% of radical prostatectomies (Qian *et al.*, 1997) also draws attention to the idea that a more generalized carcinogenic process may be operating. Genetic evidence has been accumulating that this phenomena is more widespread than has been traditionally appreciated.

Bostwick *et al.* (1998) and others (Cheng *et al.*, 1998; Macintosh *et al.*, 1998; Ruijter *et al.*, 1999) have analyzed markers of allelic imbalance in radical prostatectomy samples and found evidence suggesting

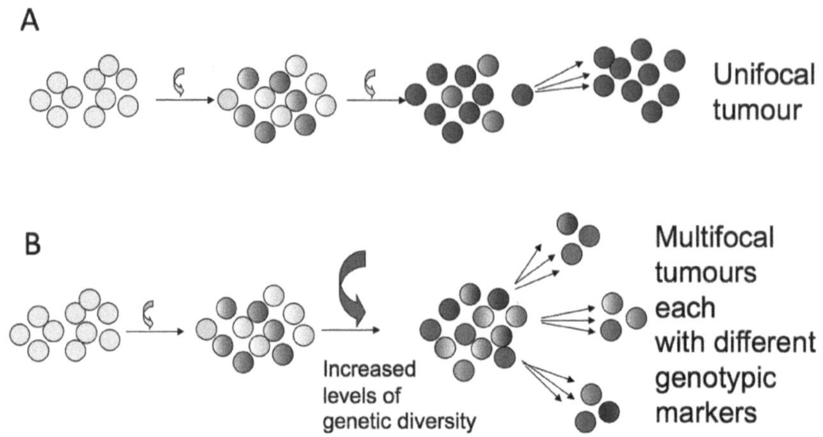

Fig. 4 Models to explain varying genetic diversity present in unifocal and multifocal prostate cancer. (A) Example of a tumor presenting as a single genetically clonal focus of cancer on radical prostatectomy. In this model, a single progenitor cell develops genomic instability (varying tones of red colored circles) and selection favors growth of one genomically homogeneous abnormal clone. (B) In this model, an early carcinogenic insult affects multiple cells and generates more extensive instability in a region of the gland. Within this region, progenitors can acquire different genomic aberrations (multicolored circles) under the influence of selection, leading to multiple independent genetically diverse tumors. (See Page 8 in Color Section at the back of the book.)

that the multiple foci of prostate cancer arise independently within the same prostate. This concept was confirmed with analyses of *TMPRSS2-ETS* fusion genes in whole mount prostatectomy specimens (Gao *et al.*, 2007). Similarly, FISH analysis classifying the different types of *TMPRSS2* rearrangement present in different foci from multifocal prostate cancers also suggested that individual tumors have distinct genetic rearrangements (Mehra *et al.*, 2007). Overall, 70% of the tumors in the study showed *TMPRSS2* rearrangement, including 63% through deletion (loss of the 3′ *TMPRSS2* signal), 27% through translocation (separation of 5′ and 3′ *TMPRSS2* signals), and 10% through both mechanisms in different tumor foci. Of the 30 *TMPRSS2* rearranged cases, 30% showed concordance in all tumor foci in an individual gland, whereas 70% from the same gland were discordant in at least one focus. Similar studies have documented the diversities fusion *TMPRSS2* rearrangements in benign glands, HPIN and multifocal cancers (Furusato *et al.*, 2008). Taken together these findings demonstrate that multifocal prostate cancer appears to be a heterogeneous disease at the genetic level, arising from multiple, independent clonal expansions within the gland (summarized in Fig. 5).

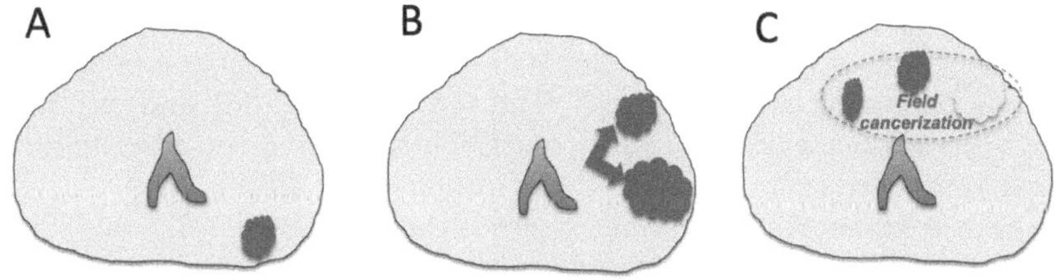

Fig. 5 Schematic depictions of whole mount sections and genomic clonality in unifocal and multifocal prostate cancer. The three panels are interpretative representations in pink of whole mount sections of prostate containing focal prostate cancer (see H and E whole mount in Fig. 1A). The tumor foci are colored to indicate different clonality based on genomic analyses. (A) In this whole mount, section one focal region of cancer is depicted in red. (B) Schematic section showing two foci of tumor growth. Both are depicted as red because they each had the same pattern of clonal genomic change suggesting a unifocal tumor with intraductal migration (depicted as red arrows) from a common progenitor. (C) Schematic section showing three genomically distinct clonal presentations shown as green, red, and yellow tumors. As discussed in the text, it is envisaged that a more generalized carcinogenic process or "field cancerization" may have affected a large section (delineated by dashed ellipse). Within this area three foci of tumors developed and the differing pattern of genetic change within each focus of cancer implies an independent origin for each tumor focus. (See Page 8 in Color Section at the back of the book.)

XV. INVOLVEMENT OF THE STROMA IN THE DISEASE PROCESS

Recent genetic approaches suggest that alterations may extend further from a preneoplastic or a neoplastic focus than can be appreciated by traditional histologic methods. Biomarkers associated with neoplastic progression in prostate cancer such as AMACR (Ananthanarayanan *et al.*, 2005; Yu *et al.*, 2004), EPCA (Dhir *et al.*, 2004), *AKT*-1 (Ayala *et al.*, 2004), GST-Pi and telomerase (Montironi *et al.*, 1999), as well as altered proliferation/apoptosis (Ananthanarayanan *et al.*, 2006) and nuclear morphology (Bartels *et al.*, 1998; Montironi *et al.*, 1999) have all been found to be altered in normal epithelium surrounding a neoplastic focus. Indeed, expression microarray studies have found 70% similarity in the genetic profile between tumor samples and prostatic tissue adjoining the tumors (Yu *et al.*, 2004) although it should be appreciated that obtaining homogeneous RNA from patient samples with intermingling of normal and tumor tissues remains technically challenging.

Traditionally, field carcinogenesis referred only to the epithelial component of the relevant organ. There is emerging evidence that the prostatic stroma also is affected by the tissue microenvironment either as a primary or a secondary phenomenon; hypermethylation of *GSTP1* and *RARB2* were found in a subset of histologically normal appearing stroma from radical prostatectomy specimens (Hanson *et al.*, 2006) and telomere length in stroma adjacent to foci of HPIN was found to correlate with the telomere length in the HPIN foci (Joshua *et al.*, 2007).

Taken together, these findings suggests that the peripheral zone of the prostate is likely subject to etiological factors that have epigenetic and/or genetic alterations throughout this zone, leading to the emergence of multiple tumors (Fig. 5). The selective nature of the carcinogenic insult that leads to effects predominantly in the peripheral zone remain unclear; however regional differences in sex steroid metabolism (Cavalieri *et al.*, 2002), dietary carcinogen and inflammation localization (De Marzo *et al.*, 2007) or a differential between the proliferative compartment and apoptotic indices have been suggested (Laczkó *et al.*, 2005).

XVI. STEM CELL ORIGINS FOR PROSTATE CANCER

Self-renewing cancer "stem cells" or cancer initiating cells (CICs) are not to be confused with normal adult stem cells, which are organ-specific cells that have the biologic property of self-renewal and with each division produce both progenitor cells and at least one offspring that maintains the stem cell phenotype in a regulated manner. Although stem cell research has focused predominately in hematologic malignancies, impetus for research in solid malignancies has increased with the identification of putative CICs in various malignancies (Al-Hajj *et al.*, 2003; O'Brien *et al.*, 2007; Singh *et al.*, 2003).

There have been two current, theoretical "leading contenders" for the cell type in the prostate that leads to the adenocarcinoma. Basically, CICs exist in both basal and luminal cell populations and are postulated to comprise of intermediate cells that possess characteristics of both cell types. Wang *et al.* have argued in favor of the CICs being derived from multipotent luminal cells (prostate stem cells known as "castration-resistant Nkx3.1-positive" cells) whereas in contrast, Goldstein *et al.*, have proposed that the prostate cancer "cell of origin" is a basal cell with low levels or the complete absence of androgen receptors (Goldstein *et al.*, 2010; Wang *et al.*, 2009).

Various subpopulations have been delineated as putative CICs, on the basis of expression of "stemness" cell surface markers (Signoretti and Loda, 2006). For example, because most prostate tumors immunophenotypically resemble luminal cells ($CK8^{+}/18^{+}AR^{+}CD44^{-}p63^{-}$) it was proposed that prostate cancer CICs are due to the dedifferentiation and transformation of luminal cells (Nagle *et al.*, 1987; Liu *et al.*, 1999). Equally, other candidates including the transit-amplifying cell (De Marzo *et al.*, 1998; van Leenders and Schalken, 2001; Verhagen *et al.*, 1992) and basal cell (Bonkhoff and Remberger, 1996) have been proposed to be CICs based on their keratin profile and intrinsic androgen independence respectively (Litvinov *et al.*, 2003). Recent efforts to characterize the prostatic CICs has been most promising with a basal cell subpopulation of $CD44^{+}\alpha_2\beta_1CD133+$ cells that possess a high *in vitro* proliferative potential and can reconstitute prostatic-like acini in immune-compromised male nude mice with concomitant expression of differentiation markers, such as *AR*, PAP, and CK18 (Collins and Maitland, 2006; Richardson *et al.*, 2004). Several lines of evidence suggest that while the normal prostatic stem cell lies within the basal compartment (Tang *et al.*, 2007), putative candidates for prostate CICs remain more contentious.

An emerging key regulator in stem cell self-renewal and cancer cell proliferation, the polycomb group genes, specifically the polycomb-group transcriptional repressor BMi-1 has recently been shown to be overexpressed in prostate cancer such that tumors exhibiting elevated Gleason scores (8 plus) have significant upregulation of Bmi-1 (Lukacs *et al.*, 2010). Conversely, presence of Bmi-1 in lower grade tumors is highly predictive for *PSA* recurrence (van Leenders *et al.*, 2007). Through *in vitro* prostate sphere forming assay along with mouse tissue regeneration systems, the role of Bmi-1 has been elucidated in regulating normal mouse prostate stem cell (PrSC) self-renewal and cancer initiation. The loss of Bmi-1 completely prevented the formation of prostate hyperplasia in the FGF10 model and subsequently attenuated the primary outgrowth of prostate cancer induced by *PTEN* deletion. Bmi-1 was shown to inhibit prostate cells from FGF10-driven hyperplasia and slowed the growth of aggressive *PTEN*-deletion-induced prostate cancer. Furthermore, cell fractionation studies showed localization of Bmi-1 expression in the prostate basal/stem cell enriched fraction (Lukacs *et al.*, 2010).

The search for markers of CICs in human prostate cancer remains ongoing. Several laboratories have successfully isolated a subpopulation of cells that have enhanced tumorigenicity *in vitro* and/or *in vivo*, while using cells from either established prostate cancer cell lines or xenografts (Sun and Gao, 2011). Ultimately, the characterization of CICs and the nature of their involvement in prostatic carcinogenesis remain promising areas of investigation (reviewed in Collins *et al.*, 2006; Lawson and Witte, 2007; Maitland and Collins, 2005)) leading to potential targeted therapy for this disease.

XVII. PROSPECTIVE: EMERGING INSIGHTS FROM PROSTATE CANCER GENOMIC SEQUENCING

The discovery of recurrent gene fusions between *TMPRSS* and *ETS* transcription factors, and the recognition that there is a high frequency of *PTEN* deletions in clinically aggressive prostate cancer, suggests that genomic rearrangements may provide a major mechanism driving prostate oncogenesis. However, both *PTEN* and *ETS* have been the subject of intense study for many years, and it is likely that novel somatic mutations will be discovered by deep sequencing that might provide further insights concerning the genetic and clinical diversity of this disease. Recently the complete sequence of seven primary human prostate cancers and their paired normal counterparts was published (Berger *et al.*, 2011). In this study, the *TMPRSS2–ERG* fusion positive tumors were frequently rearranged near open chromatin, androgen receptor, and *ERG* DNA binding sites. However, tumors that did not have an *ETS* fusion failed to express such alterations, suggesting there may be a mechanistic link between chromatin or transcriptional regulation and the onset of genomic aberrations in prostate cancer oncogenesis. Collectively genomic discoveries are rapidly moving the study of human prostate cancer closer to the promise of personalized medicine; so that specific genetic profiles of tumors will determine the best treatment course for the individual patient. Genomic analyses coupled with targeted functional studies in appropriate model systems that mirror cellular contexts of the human prostate, will help to identify critical signaling pathways that are causally involved in prostate cancer initiation, and may explain the clinical diversity that characterizes this disease.

ACKNOWLEDGMENT

We thank the Canadian Cancer Society (Grant 018124) for support associated with this work and all members of the Squire laboratory in Toronto and now in Kingston for their excellent contributions in this area.

REFERENCES

Afar, D. E., Vivanco, I., Hubert, R. S., Kuo, J., et al., (2001). Catalytic cleavage of the androgen-regulated *TMPRSS2* protease results in its secretion by prostate and prostate cancer epithelia. *Cancer Res.* **61**(4), 1686–1992.

Ai, J., Wang, Y., Dar, J. A., Liu, J., et al., (2009). HDAC6 regulates androgen receptor hypersensitivity and nuclear localization via modulating Hsp90 acetylation in castration-resistant prostate cancer. *Mol. Endocrinolo* **23**(12), 1963–1972.

Al-Hajj, M., Wicha, M. S., Benito-Hernandez, A., Morrison, S. J., and Clarke, M. F. (2003). Prospective identification of tumorigenic breast cancer cells. *Proc. Nat. Acad. Sci. USA* **100**(7), 3983–3988.

Albadine, R., Latour, M., Toubaji, A., Haffner, M., et al., (2009). *TMPRSS2–ERG* gene fusion status in minute (minimal) prostatic adenocarcinoma. *Mod. Pathol.* **22**(11), 1415–1422.

Alimonti, A., Carracedo, A., Clohessy, J. G., Trotman, L. C., et al., (2010). Subtle variations in *Pten* dose determine cancer susceptibility. *Nat. Genet.* **42**(5), 454–458.

Allsbrook, W. C., Mangold, K. A., Yang, X., and Epstein, J. I. (1999). The Gleason grading system: an overview. *J. Urol. Pathol.* **10**, 141–157.

Altekruse, S. F., Kosary, C. L., Krapcho, M., Neyman, N., Aminou, R., Waldron, W., Ruhl, J., Howlader, N., Tatalovich, Z., Cho, H., Mariotto, A., Eisner, M. P., Lewis, D. R., Cronin, K., Chen, H. S., Feuer, E. J., Stinchcomb, D. G, and Edwards, B. K, (eds.), 2009. *SEER Cancer Statistics Review, 1975–2007*, Bethesda, MD.

Ananthanarayanan, V., Deaton, R. J., Yang, X. J., Pins, M. R., and Gann, P. H. (2005). Alpha-methylacyl-CoA racemase (AMACR) expression in normal prostatic glands and high-grade prostatic intraepithelial neoplasia (HGPIN): association with diagnosis of prostate cancer. *Prostate* **63**(4), 341–346.

Ananthanarayanan, V., Deaton, R. J., Yang, X. J., Pins, M. R., and Gann, P. H. (2006). Alteration of proliferation and apoptotic markers in normal and premalignant tissue associated with prostate cancer. *BMC Cancer* **6**, 73.

Andreoiu, M., and Cheng, L. (2010). Multifocal prostate cancer: biologic, prognostic, and therapeutic implications. *Human Pathology* **41**, 781–793.

Anton, R. C., Kattan, M. W., Chakraborty, S., and Wheeler, T. M. (1999). Postatrophic hyperplasia of the prostate: lack of association with prostate cancer. *Amer. J. Surg. Pathol.* **23**(8), 932–936.

Arora, R., Koch, M. O., Eble, J. N., Ulbright, T. M., et al., (2004). Heterogeneity of Gleason grade in multifocal adenocarcinoma of the prostate. *Cancer* **100**(11), 2362–2366.

Attard, G., Clark, J., Ambroisine, L., Fisher, G., et al., (2008a). Duplication of the fusion of *TMPRSS2* to *ERG* sequences identifies fatal human prostate cancer. *Oncogene* **27**(3), 253–263.

Attard, G., Clark, J., Ambroisine, L., Mills, I. G., et al., (2008b). Heterogeneity and clinical significance of ETV1 translocations in human prostate cancer. *Br. J. Cancer.* **99**(2), 314–320.

Attard, G., Jameson, C., Moreira, J., Flohr, P., et al., (2009). Hormone-sensitive prostate cancer: a case of *ETS* gene fusion heterogeneity. *J. Clin. Pathol.* **62**(4), 373–376.

Ayala, A. G., and Ro, J. Y. (2007). Prostatic intraepithelial neoplasia: recent advances. *Arch. Path. Lab. Med.* **131**(8), 1257–1266.

Ayala, G., Thompson, T., Yang, G., Frolov, A., et al., (2004). High levels of phosphorylated form of *Akt*-1 in prostate cancer and non-neoplastic prostate tissues are strong predictors of biochemical recurrence. *Clin. Cancer Res.* **10**(19), 6572–6578.

Balk, S. P., and Knudsen, K. E. (2008). *AR*, the cell cycle, and prostate cancer. *Nucl. Receptor Signaling* **6**, e001.

Bartels, P. H., Silva, V. D., da Montironi, R., Hamilton, P. W., et al., (1998). Chromatin texture signatures in nuclei from prostate lesions. *Anal. Quant. Cytology Histology.* **20**(5), 407–416.

Barwick, B., Abramovitz, M., and Kodani, M. (2010). Prostate cancer genes associated with *TMPRSS2–ERG* gene fusion and prognostic of biochemical recurrence in multiple cohorts. *J Cancer.* **102**(3), 570–576.

Basuyaux, J. P., Ferreira, E., Stéhelin, D., and Butticè, G. (1997). The ETS transcription factors interact with each other and with the c-Fos/c-JUN complex via distinct protein domains in a DNA-dependent and -independent manner. *J Biol. Chem.* **272**(42), 26188–26195.

Bednarz, N., Eltze, E., Semjonow, A., Rink, M., et al., (2010). BRCA1 loss preexisting in small subpopulations of prostate cancer is associated with advanced disease and metastatic spread to lymph nodes and peripheral blood. *Clin. Cancer Res.* **16**(13), 3340–3348.

Beheshti, B., Vukovic, B., Marrano, P., Squire, J. A., and Park, P. C. (2002). Resolution of genotypic heterogeneity in prostate tumors using polymerase chain reaction and comparative genomic hybridization on microdissected carcinoma and prostatic intraepithelial neoplasia foci. *Cancer Gen. Cytogen.* **137**(1), 15–22.

Berger, M. F., Lawrence, M. S., Demichelis, F., Drier, Y., et al., (2011). The genomic complexity of primary human prostate cancer. *Nature.* **470**(7333), 214–220.

Bethel, C. R., Faith, D., Li, X., Guan, B., et al., (2006). Decreased NKX3.1 protein expression in focal prostatic atrophy, prostatic intraepithelial neoplasia, and adenocarcinoma: association with gleason score and chromosome 8p deletion. *Cancer Res.* **66**(22), 10683–10690.

Billis, A. (1998). Prostatic atrophy: an autopsy study of a histologic mimic of adenocarcinoma. *Mod. Pathol.* **11**(1), 47–54.

Billis, A., and Magna, L. A. (2003). Inflammatory atrophy of the prostate, Prevalence and significance. *Arch. Pathol. Lab. Med.* **127**(7), 840–844.

Birdsey, G. M., Dryden, N. H., Amsellem, V., Gebhardt, F., et al., (2008). Transcription factor Erg regulates angiogenesis and endothelial apoptosis through VE-cadherin. *Blood* **111**(7), 3498–3506.

Bismar, T. A., and Trpkov, K. (2010). *TMPRSS2–ERG* gene fusion in transition zone prostate cancer. *Mod. Pathol.* **23**(7), 1040–1041.

Bismar, T. A., Yoshimoto, M., Vollmer, R. T., Duan, Q., et al., (2011). *PTEN* genomic deletion is an early event associated with *ERG* gene rearrangements in prostate cancer. *BJU Int.* **107**(3), 477–485.

Björkman, M., Iljin, K., Halonen, P., Sara, H., et al., (2008). Defining the molecular action of HDAC inhibitors and synergism with androgen deprivation in *ERG*-positive prostate cancer. *Int. J Cancer.* **123**(12), 2774–2781.

Bonkhoff, H., and Remberger, K. (1996). Differentiation pathways and histogenetic aspects of normal and abnormal prostatic growth: a stem cell model. *Prostate.* **28**(2), 98–106.

Bonkhoff, H., Stein, U., and Remberger, K. (1994). The proliferative function of basal cells in the normal and hyperplastic human prostate. *Prostate.* **24**(3), 114–118.

Boormans, J. L., Hermans, K. G., Made, A. C. JZ-van der., Leenders, G. J. H. Lvan., et al., (2010). Expression of the androgen-regulated fusion gene *TMPRSS2–ERG* does not predict response to endocrine treatment in hormone-naïve, node-positive prostate cancer. *Eur. Urol.* **57**(5), 830–835.

Bostwick, D. G., Shan, A., Qian, J., Darson, M., et al., (1998). Independent origin of multiple foci of prostatic intraepithelial neoplasia: comparison with matched foci of prostate carcinoma. *Cancer.* **83**(9), 1995–2002.

Braakhuis, B. J. M., Tabor, M. P., Kummer, J. A., Leemans, C. R., and Brakenhoff, R. H. (2003). A genetic explanation of Slaughter's concept of field cancerization: evidence and clinical implications. *Cancer Res.* **63**(8), 1727–1730.

Bushman, W. (2009). Etiology, epidemiology, and natural history of benign prostatic hyperplasia. *Urol. Clin. North Am.* **36**(4), 403–415.

Carrère, S., Verger, A., Flourens, A., Stehelin, D., and Duterque-Coquillaud, M. (1998). Erg proteins, transcription factors of the ETS family, form homo, heterodimers and ternary complexes via two distinct domains. *Oncogene.* **16**(25), 3261–3268.

Carver, B. S., Tran, J., Gopalan, A., Chen, Z., et al., (2009). Aberrant *ERG* expression cooperates with loss of *PTEN* to promote cancer progression in the prostate. *Nat. Gen.* **41**(5), 619–624.

Cattan, V., Mercier, N., Gardner, J. P., Regnault, V., et al., (2008). Chronic oxidative stress induces a tissue-specific reduction in telomere length in CAST/Ei mice. *Free Rad. Biol. Med.* **44**(8), 1592–5198.

Cavalieri, E. L., Devanesan, P., Bosland, M. C., Badawi, A. F., and Rogan, E. G. (2002). Catechol estrogen metabolites and conjugates in different regions of the prostate of Noble rats treated with 4-hydroxyestradiol: implications for estrogen-induced initiation of prostate cancer. *Carcinogenesis* **23**, 329–333.

Cerveira, N., Ribeiro, F. R., Peixoto, A., Costa, V., et al., (2006). *TMPRSS2–ERG* gene fusion causing *ERG* overexpression precedes chromosome copy number changes in prostate carcinomas and paired HGPIN lesions. *Neoplasia* **8**(10), 826–832.

Chai, Y., Chipitsyna, G., Cui, J., Liao, B., et al., (2001). c-Fos oncogene regulator Elk-1 interacts with BRCA1 splice variants BRCA1a/1b and enhances BRCA1a/1b-mediated growth suppression in breast cancer cells. *Oncogene.* **20**(11), 1357–1367.

Chaib, H., MacDonald, J. W., Vessella, R. L., Washburn, J. G., et al., (2003). Haploinsufficiency and reduced expression of genes localized to the 8p chromosomal region in human prostate tumors. *Genes Chromosomes Cancer* **37**(3), 306–313.

Cheng, L., Song, S. Y., Pretlow, T. G., Abdul-Karim, F. W., et al., (1998). Evidence of independent origin of multiple tumors from patients with prostate cancer. *J. Nat. Cancer Inst.* **90**(3), 233–237.

Cheville, J. C., and Bostwick, D. G. (1995). Postatrophic hyperplasia of the prostate. A histologic mimic of prostatic adenocarcinoma. *Am. J Surg. Pathol.* **19**(9), 1068–1076.

Chin, K., Solorzano, C. O., de Knowles, D., Jones, A., et al., (2004). *In situ* analyses of genome instability in breast cancer. *Nat. Gen.* **36**(9), 984–988.

Clark, J., Merson, S., Jhavar, S., Flohr, P., et al., (2007). Diversity of *TMPRSS2–ERG* fusion transcripts in the human prostate. *Oncogene* **26**(18), 2667–2673.

Collins, A. T., and Maitland, N. J. (2006). Prostate cancer stem cells. *Eur. J Cancer.* **42**(9), 1213–1218.

Dai, M. -J., Chen, L. -L., Zheng, Y. -B., Chen, W., et al., (2008). Frequency and transcript variant analysis of gene fusions between *TMPRSS2* and *ETS* transcription factor genes in prostate cancer. *Zhonghua yi xue za zhi.* **88**(10), 669–673.

De Marzo, A. M., DeWeese, T. L., Platz, E. A., and Meeker, A. K. (2004). Pathological and molecular mechanisms of prostate carcinogenesis: implications for diagnosis, detection, prevention, and treatment. *J. Cellular Biochem.* **91**(3), 459–477.

De Marzo, A. M., Marchi, V. L., Epstein, J. I., and Nelson, W. G. (1999). Proliferative inflammatory atrophy of the prostate: implications for prostatic carcinogenesis. *Am. J Pathol.* **155**(6), 1985–1992.

De Marzo, A. M., Nelson, W. G., Meeker, A. K., and Coffey, D. S. (1998). Stem cell features of benign and malignant prostate epithelial cells. *J. Urol.* **160**(6 Pt 2), 2381–2392.

De Marzo, A. M., Platz, E. A., Sutcliffe, S., Xu, J., et al., (2007). Inflammation in prostate carcinogenesis. *Nat. Rev. Cancer.* **7**(4), 256–269.

Demichelis, F., Fall, K., Perner, S., Andrén, O., et al., (2007). *TMPRSS2:ERG* gene fusion associated with lethal prostate cancer in a watchful waiting cohort. *Oncogene.* **26**(31), 4596–4599.

Dhir, R., Vietmeier, B., Arlotti, J., Acquafondata, M., et al., (2004). Early identification of individuals with prostate cancer in negative biopsies. *J. Urol.* **171**(4), 1419–1423.

Di Cristofano, A., Pesce, B., Cordon-Cardo, C., and Pandolfi, P. P. (1998). *Pten* is essential for embryonic development and tumour suppression. *Nat. Gen.* **19**(4), 348–355.

Dijkman, G. A., and Debruyne, F. M. (1996). Epidemiology of prostate cancer. *Eur. Urol.* **30** (3), 281–295.

Ellett, F., Kile, B. T., and Lieschke, G. J. (2009). The role of the *ETS* factor erg in zebrafish vasculogenesis. *Mech. Dev.* **126**(3–4), 220–229.

Faith, D., Han, S., Lee, D. K., Friedl, A., et al., (2005). P16 Is upregulated in proliferative inflammatory atrophy of the prostate. *Prostate.* **65**(1), 73–82.

Fan, R., Kumaravel, T. S., Jalali, F., Marrano, P., et al., (2004). Defective DNA strand break repair after DNA damage in prostate cancer cells: implications for genetic instability and prostate cancer progression. *Cancer Res.* **64**(23), 8526–8533.

Feilotter, H. E., Nagai, M. A., Boag, A. H., Eng, C., and Mulligan, L. M. (1998). Analysis of *PTEN* and the 10q23 region in primary prostate carcinomas. *Oncogene.* **16**(13), 1743–1748.

FitzGerald, L. M., Agalliu, I., Johnson, K., Miller, M. A., et al., (2008). Association of *TMPRSS2–ERG* gene fusion with clinical characteristics and outcomes: results from a population-based study of prostate cancer. *BMC Cancer.* **8**, 230.

Forusato, B., Gao, C. -L., Ravindranath, L., Chen, Y., McLeod, D., Dobi, A., Srivastava, S., Petrovics, G., and Sesterhenn, I. A. (2008). Mapping of *TMPRSS2–ERG* fusions in the context of miultifocal prostate cancer. *Mod. Pathol.* **21**, 67–75.

Foster, C. S., Dodson, A., Karavana, V., Smith, P. H., and Ke, Y. (2002). Prostatic stem cells. *J. Pathol.* **197**(4), 551–565.

Franks, L. M. (1954). Atrophy and hyperplasia in the prostate proper. *J. Pathol. Bacteriol.* **68**, 617–621.

Gao, C., Furusato, B., McLeod, D. G., Srivastava, S., et al., (2007). Mapping of TMPRSS2-ERG fusion in Multi-Focal Prostate Cancer. American Urological Association, Anaheim, California.

Garcia, S. B., Park, H. S., Novelli, M., and Wright, N. A. (1999). Field cancerization, clonality, and epithelial stem cells: the spread of mutated clones in epithelial sheets. *J. Pathol.* **187**(1), 61–81.

Goldstein, A. S., Huang, J., Guo, C., Garraway, I. P., and Witte, O. N. (2010). Identification of a cell of origin for human prostate cancer. *Science.* **329**(5991), 568–571.

Gopalan, A., Leversha, M. A., Satagopan, J. M., Zhou, Q., et al., (2009). *TMPRSS2–ERG* gene fusion is not associated with outcome in patients treated by prostatectomy. *Cancer Res.* **69**(4), 1400–1406.

Greene, D. R., Wheeler, T. M., Egawa, S., Dunn, J. K., and Scardino, P. T. (1991). A comparison of the morphological features of cancer arising in the transition zone and in the peripheral zone of the prostate. *J. Urol.* **146**(4), 1069–1076.

Guo, C. C., Zuo, G., Cao, D., Troncoso, P., and Czerniak, B. A. (2009). Prostate cancer of transition zone origin lacks *TMPRSS2–ERG* gene fusion. *Mod. Pathol.* **22**(7), 866–871.

Han, B., Mehra, R., Lonigro, R. J., Wang, L., et al., (2009). Fluorescence *in situ* hybridization study shows association of *PTEN* deletion with *ERG* rearrangement during prostate cancer progression. *Mod. Pathol.* **22**(8), 1083–1093.

Hanson, J. A., Gillespie, J. W., Grover, A., Tangrea, M. A., et al., (2006). Gene promoter methylation in prostate tumor-associated stromal cells. *J. Nat. Cancer Inst.* **98**(4), 255–261.

Hawksworth, D., Ravindranath, L., Chen, Y., Furusato, B., et al., (2010). Overexpression of *C-MYC* oncogene in prostate cancer predicts biochemical recurrence. *Prostate Cancer Prostatic Dis.* **13**(4), 311–315.

Hermans, K. G., Alewijk, D. C., van Veltman, J. A., Weerden, W., van., et al., (2004). Loss of a small region around the *PTEN* locus is a major chromosome 10 alteration in prostate cancer xenografts and cell lines. *Genes Chromosomes Cancer.* **39**(3), 171–184.

Holcomb, I. N., Young, J. M., Coleman, I. M., Salari, K., et al., (2009). Comparative analyses of chromosome alterations in soft-tissue metastases within and across patients with castration-resistant prostate cancer. *Cancer Res.* **69**(19), 7793–7802.

Houben, J. M. J., Moonen, H. J. J., Schooten van, F. J., and Hageman, G. J. (2008). Telomere length assessment: biomarker of chronic oxidative stress? *Free Rad. Biol. Med.* **44**(3), 235–246.

Hsing, A. W., Tsao, L., and Devesa, S. S. (2000). International trends and patterns of prostate cancer incidence and mortality. *Int. J. Cancer.* **85**(1), 60–67.

Hughes, S., Yoshimoto, M., Beheshti, B., Houlston, R. S., et al., (2006). The use of whole genome amplification to study chromosomal changes in prostate cancer: insights into genome-wide signature of preneoplasia associated with cancer progression. *BMC Genomics.* **7**, 65.

Iljin, K., Wolf, M., Edgren, H., Gupta, S., et al., (2006). *TMPRSS2* fusions with oncogenic *ETS* factors in prostate cancer involve unbalanced genomic rearrangements and are associated with HDAC1 and epigenetic reprogramming. *Cancer Res.* **66**(21), 10242–10246.

Isaacs, J. T., and Coffey, D. S. (1989). Etiology and disease process of benign prostatic hyperplasia. *Prostate.* **2**, 33–50.

Ishkanian, A. S., Mallof, C. A., Ho, J., Meng, A., et al., (2009). High-resolution array CGH identifies novel regions of genomic alteration in intermediate-risk prostate cancer. *Prostate.* **69**(10), 1091–1100.

Jemal, A., Siegel, R., Ward, E., Murray, T., et al., (2007). Cancer statistics, 2007. *Cancer J. Clinicians* **57**(1), 43–66.

Jenkins, R. B., Qian, J., Lieber, M. M., and Bostwick, D. G. (1997). Detection of c-MYC oncogene amplification and chromosomal anomalies in metastatic prostatic carcinoma by fluorescence *in situ* hybridization. *Cancer Res.* **57**(3), 524–531.

Joshua, A. M., Shen, E., Yoshimoto, M., Marrano, P., et al., (2011). Topographical analysis of telomere length and correlation with genomic instability in whole mount prostatectomies. *Prostate.* **71**(7), 778–790.

Joshua, A. M., Vukovic, B., Braude, I., Hussein, S., et al., (2007). Telomere attrition in isolated high-grade prostatic intraepithelial neoplasia and surrounding stroma is predictive of prostate cancer. *Neoplasia.* **9**(1), 81–89.

Kawata, H., Ishikura, N., Watanabe, M., Nishimoto, A., et al., (2010). Prolonged treatment with bicalutamide induces androgen receptor overexpression and androgen hypersensitivity. *Prostate.* **70**(7), 745–754.

King, J. C., Xu, J., Wongvipat, J., Hieronymus, H., et al., (2009). Cooperativity of *TMPRSS2–ERG* with PI3-kinase pathway activation in prostate oncogenesis. *Nat. Gen.* **41**(5), 524–526.

Kiviharju-af Hällström, T. M., Jäämaa, S., Mönkkönen, M., Peltonen, K., et al., (2007). Human prostate epithelium lacks Wee1A-mediated DNA damage-induced checkpoint enforcement. *Proc. Nat. Acad. Sci. USA.* **104**(17), 7211–7216.

Klezovitch, O., Risk, M., Coleman, I., Lucas, J. M., et al., (2008). A causal role for *ERG* in neoplastic transformation of prostate epithelium. *Proc. Nat. Acad. Sci. USA.* **105**(6), 2105–2110.

Knudsen, B. S., and Vasioukhin, V. (2010). Mechanisms of prostate cancer initiation and progression. *Adv. Cancer Res.* **109**, 1–50.

Koksal, I. T., Dirice, E., Yasar, D., Sanlioglu, A. D., et al. (2004). The assessment of PTEN tumor suppressor gene in combination with Gleason scoring and serum PSA to evaluate progression of prostate carcinoma. 22, 4, 307–312.

Kristal, A. R., Price, D. K., Till, C., Schenk, J. M., et al., (2010). Androgen receptor CAG repeat length is not associated with the risk of incident symptomatic benign prostatic hyperplasia: results from the prostate cancer prevention trial. *Prostate.* 70(6), 584–590.

Kruse, E. A., Loughran, S. J., Baldwin, T. M., Josefsson, E. C., et al., (2009). Dual requirement for the *ETS* transcription factors Fli-1 and Erg in hematopoietic stem cells and the megakaryocyte lineage. *Proc. Nat. Acad. Sci. USA.* **106**(33), 13814–13819.

Kwabi-Addo, B., Giri, D., Schmidt, K., Podsypanina, K., et al., (2001). Haploinsufficiency of the *Pten* tumor suppressor gene promotes prostate cancer progression. *Proc. Nat. Acad. Sci. USA.* **98**(20), 11563–11568.

Laczkó, I., Hudson, D. L., Freeman, A., Feneley, M. R., and Masters, J. R. (2005). Comparison of the zones of the human prostate with the seminal vesicle: morphology, immunohistochemistry, and cell kinetics. *Prostate.* **62**(3), 260–266.

Lapointe, J., Li, C., Giacomini, C. P., Salari, K., et al., (2007). Genomic profiling reveals alternative genetic pathways of prostate tumorigenesis. *Cancer Res.* **67**(18), 8504–8510.

Lawson, D. A., and Witte, O. N. (2007). Stem cells in prostate cancer initiation and progression. *J. Clin. Invest.* **117**(8), 2044–2050.

Lee, K., Chae, J. Y., Kwak, C., Ku, J. H., and Moon, K. C. (2010). *TMPRSS2–ERG* gene fusion and clinicopathologic characteristics of Korean prostate cancer patients. *Urology.* **76**(5), 1268 e7–1268.e13.

van Leenders, G. J., and Schalken, J. A. (2001). Stem cell differentiation within the human prostate epithelium: implications for prostate carcinogenesis. *BJU Int.* **88**, 35–42.

van Leenders, G. J. L. H., Dukers, D., Hessels, D., Kieboom vanden, S. W. M., et al., (2007). Polycomb-group oncogenes EZH2, BMI1, and RING1 are overexpressed in prostate cancer with adverse pathologic and clinical features. *Eur. Urol.* **52**(2), 455–463.

Leenders, G. J. L. H., Gage, W. R., Hicks, J. L., Balken, B., van., et al., (2003). Intermediate cells in human prostate epithelium are enriched in proliferative inflammatory atrophy. *Amer. J. Pathol.* **162**(5), 1529–1537.

Li, L. -C., Carroll, P. R., and Dahiya, R. (2005). Epigenetic changes in prostate cancer: implication for diagnosis and treatment. *J. Natl. Cancer Inst.* **97**, 103–115.

Liavåg, I. (1968). The localization of prostatic carcinoma. An autopsy study. *Scandinavian J. Urol. Nephrology.* **2**(2), 65–71.

Litvinov, I. V., De Marzo, A. M., and Isaacs, J. T. (2003). Is the Achilles' heel for prostate cancer therapy a gain of function in androgen receptor signaling? *J. Clin. Endocrinol. Metabol.* **88**(7), 2972–2982.

Liu, A. Y., True, L. D., LaTray, L., Ellis, W. J., et al., (1999). Analysis and sorting of prostate cancer cell types by flow cytometry. *Prostate.* **40**(3), 192–199.

Liu, W., Laitinen, S., Khan, S., Vihinen, M., et al., (2009). Copy number analysis indicates monoclonal origin of lethal metastatic prostate cancer. *Nat. Med.* **15**(5), 559–565.

Lu, B., Maqsodi, B., Yang, W., McMaster, G. K., et al., (2009). Detection of *TMPRSS2–ERG* fusion gene expression in prostate cancer specimens by a novel assay using branched DNA. *Urology.* **74**(5), 1156–1161.

Lukacs, R. U., Memarzadeh, S., Wu, H., and Witte, O. N. (2010). Bmi-1 is a crucial regulator of prostate stem cell self-renewal and malignant transformation. *Cell Stem Cell.* 7(6), 682–693.

Löbrich, M., and Jeggo, P. A. (2007). The impact of a negligent G2/M checkpoint on genomic instability and cancer induction. *Nat. Rev. Cancer.* 7(11), 861–869.

Macintosh, C. A., Stower, M., Reid, N., and Maitland, N. J. (1998). Precise microdissection of human prostate cancers reveals genotypic heterogeneity. *Cancer Res.* **58**(1), 23–28.

Magi-Galluzzi, C., Tsusuki, T., Elson, P., Simmerman, K., et al., (2011). *TMPRSS2–ERG* gene fusion prevalence and class are significantly different in prostate cancer of Caucasian, African-American and Japanese patients. *Prostate.* **71**(5), 489–497.

Maitland, N. J., and Collins, A. (2005). A tumour stem cell hypothesis for the origins of prostate cancer. *BJU Int.* **96**(9), 1219–1223.

Mao, X., Yu, Y., Boyd, L. K., Ren, G., et al., (2010). Distinct genomic alterations in prostate cancers in Chinese and Western populations suggest alternative pathways of prostate carcinogenesis. *Cancer Res.* **70**(13), 5207–5212.

McLaughlin, F., Ludbrook, V. J., Cox, J., Carlowitz, Ivon., et al., (2001). Combined genomic and antisense analysis reveals that the transcription factor Erg is implicated in endothelial cell differentiation. *Blood.* **98**(12), 3332–3339.

McMenamin, M. E., Soung, P., Perera, S., Kaplan, I., et al., (1999). Loss of *PTEN* expression in paraffin-embedded primary prostate cancer correlates with high Gleason score and advanced stage. *Cancer Res.* **59**(17), 4291–4296.

Meeker, A. K., Hicks, J. L., Platz, E. A., March, G. E., et al., (2002). Telomere shortening is an early somatic DNA alteration in human prostate tumorigenesis. *Cancer Res.* **62**(22), 6405–6409.

Mehra, R., Han, B., Tomlins, S. A., Wang, L., et al., (2007). Heterogeneity of *TMPRSS2* gene rearrangements in multifocal prostate adenocarcinoma: molecular evidence for an independent group of diseases. *Cancer Res.* **67**(17), 7991–7995.

Mehra, R., Tomlins, S. A., Yu, J., Cao, X., et al., (2008). Characterization of TMPRSS2–ETS gene aberrations in androgen-independent metastatic prostate cancer. *Cancer Res.* **68**(10), 3584–3590.

Meng, A. X., Jalali, F., Cuddihy, A., Chan, N., et al., (2005). Hypoxia down-regulates DNA double strand break repair gene expression in prostate cancer cells. *Radiother. Oncol* **76** (2), 168–176.

Miyagi, Y., Sasaki, T., Fujinami, K., Sano, J., et al., (2010). *ETS* family-associated gene fusions in Japanese prostate cancer: analysis of 194 radical prostatectomy samples. *Mod. Pathol.* **23**(11), 1492–1498.

Montironi, R., Hamilton, P. W., Scarpelli, M., Thompson, D., and Bartels, P. H. (1999). Subtle morphological and molecular changes in normal-looking epithelium in prostates with prostatic intraepithelial neoplasia or cancer. *Eur. Urol.* **35**(5–6), 468–473.

Mosquera, J. -M., Mehra, R., Regan, M. M., Perner, S., et al., (2009). Prevalence of *TMPRSS2–ERG* fusion prostate cancer among men undergoing prostate biopsy in the United States. *Clin. Cancer Res.* **15**(14), 4706–4711.

Mosquera, J. -M., Perner, S., Genega, E. M., Sanda, M., et al., (2008). Characterization of *TMPRSS2–ERG* fusion high-grade prostatic intraepithelial neoplasia and potential clinical implications. *Clin. Cancer Res.* **14**(11), 3380–3385.

Müller, M., Rink, K., Krause, H., and Miller, K. (2000). *PTEN*/MMAC1 mutations in prostate cancer. *Prostate Cancer Prostatic Dis.* **3**, S1–S32.

Nagle, R. B., Ahmann, F. R., McDaniel, K. M., Paquin, M. L., et al., (1987). Cytokeratin characterization of human prostatic carcinoma and its derived cell lines. *Cancer Res.* **47** (1), 281–286.

Nakayama, M., Bennett, C. J., Hicks, J. L., Epstein, J. I., et al., (2003). Hypermethylation of the human glutathione S-transferase-pi gene (GSTP1) CpG island is present in a subset of proliferative inflammatory atrophy lesions but not in normal or hyperplastic epithelium of the prostate: a detailed study using laser-capture micr. *Am. J. Pathol.* **163**(3), 923–933.

Nupponen, N. N., and Visakorpi, T. (2000). Molecular cytogenetics of prostate cancer. *Microsc. Res. Tech.* **51**(5), 456–463.

Nye, J. A., Petersen, J. M., Gunther, C. V., Jonsen, M. D., and Graves, B. J. (1992). Interaction of murine *ETS*-1 with GGA-binding sites establishes the *ETS* domain as a new DNA-binding motif. *Genes Dev.* **6**(6), 975–990.

Ohigashi, T., Mizuno, R., Nakashima, J., Marumo, K., and Murai, M. (2005). Inhibition of Wnt signaling downregulates *AKT* activity and induces chemosensitivity in *PTEN*-mutated prostate cancer cells. *Prostate.* **62**(1), 61–68.

Oikawa, T., and Yamada, T. (2003). Molecular biology of the *ETS* family of transcription factors. *Gene.* **303**, 11–34.

O'Brien, C. A., Pollett, A., Gallinger, S., and Dick, J. E. (2007). "A human colon cancer cell capable of initiating tumour growth in immunodeficient mice". *Nature.* **445**(7123), 106–110.

Paoloni-Giacobino, A., Chen, H., Peitsch, M. C., Rossier, C., and Antonarakis, S. E. (1997). Cloning of the *TMPRSS2* gene, which encodes a novel serine protease with transmembrane, LDLRA, and SRCR domains and maps to 21q22.3. *Genomics* **44**(3), 309–320.

Parsons, J. K., Nelson, C. P., Gage, W. R., Nelson, W. G., et al., (2001). GSTA1 expression in normal, preneoplastic, and neoplastic human prostate tissue. *Prostate.* **49**(1), 30–37.

Perner, S., Mosquera, J. -M., Demichelis, F., Hofer, M. D., et al., (2007). *TMPRSS2–ERG* fusion prostate cancer: an early molecular event associated with invasion. *Am. J. Surg. Pathol.* **31**(6), 882–888.

Petrovics, G., Liu, A., Shaheduzzaman, S., Furusato, B., et al., (2005). Frequent overexpression of ETS-related gene-1 (ERG1) in prostate cancer transcriptome. *Oncogene* **24**(23), 3847–3852.

Pihan, G. A., Purohit, A., Wallace, J., Malhotra, R., et al., (2001). Centrosome defects can account for cellular and genetic changes that characterize prostate cancer progression. *Cancer Res.* **61**(5), 2212–2219.

Pihan, G. A., Wallace, J., Zhou, Y., and Doxsey, S. J. (2003). Centrosome abnormalities and chromosome instability occur together in pre-invasive carcinomas. *Cancer Res.* **63**(6), 1398–1404.

Potten, C. S., and Loeffler, M. (1990). Stem cells: attributes, cycles, spirals, pitfalls and uncertainties. Lessons for and from the crypt. *Development (Cambridge, England)* **110** (4), 1001–1020.

Putzi, M. J., and De Marzo, A. M. (2000). Morphologic transitions between proliferative inflammatory atrophy and high-grade prostatic intraepithelial neoplasia. *Urology* **56**(5), 828–832.

Qian, J., Wollan, P., and Bostwick, D. G. (1997). The extent and multicentricity of high-grade prostatic intraepithelial neoplasia in clinically localized prostatic adenocarcinoma. *Human Pathol.* **28**(2), 143–148.

Rajagopalan, H., Nowak, M. A., Vogelstein, B., and Lengauer, C. (2003). The significance of unstable chromosomes in colorectal cancer. *Nat. Rev. Cancer.* **3**(9), 695–701.

Rajput, A. B., Miller, M. A., De Luca, A., Boyd, N., et al., (2007). Frequency of the *TMPRSS2–ERG* gene fusion is increased in moderate to poorly differentiated prostate cancers. *J. Clin. Pathol.* **60**(11), 1238–1243.

Randi, A. M., Sperone, A., Dryden, N. H., and Birdsey, G. M. (2009). Regulation of angiogenesis by *ETS* transcription factors. *Biochem. Soc. Trans.* **37**(6), 1248–1253.

Reddy, E. S., and Rao, V. N. (1991). ERG, an *ETS*-related gene, codes for sequence-specific transcriptional activators. *Oncogene* **6**(12), 2285–2289.

Reddy, E. S., Rao, V. N., and Papas, T. S. (1987). The erg gene: a human gene related to the *ETS* oncogene. *Proc. Nat. Acad. Sci. USA* **84**(17), 6131–6135.

Reid, A. H. M., Attard, G., Ambroisine, L., Fisher, G., et al., (2010). Molecular characterisation of *ERG. ETV1 and PTEN gene loci identifies patients at low and high risk of death from prostate cancer. Brit. J. Cancer* **102**(4), 678–684.

Ribeiro, F. R., Diep, C. B., Jerónimo, C., Henrique, R., et al., (2006). *Statistical dissection of genetic pathways involved in prostate carcinogenesis. Genes Chromosomes Cancer* **45**(2), 154–163.

Richardson, G. D., Robson, C. N., Lang, S. H., Neal, D. E., et al., (2004). CD133, a novel marker for human prostatic epithelial stem cells. *J. Cell Sci.*3539–3545

Richter, T., and Proctor, C. (2007). The role of intracellular peroxide levels on the development and maintenance of telomere-dependent senescence. *Exp. Gerontol.* **42**(11), 1043–1052.

Ruijter, E. T., Kaa vande, C. A., Schalken, J. A., Debruyne, F. M., and Ruiter, D. J. (1996). Histological grade heterogeneity in multifocal prostate cancer. Biological and clinical implications. *J. Pathol.* **180**(3), 295–299.

Ruijter, E. T., Miller, G. J., Kaa, C. A., van de Bokhoven, Avan., et al., (1999). Molecular analysis of multifocal prostate cancer lesions. *J. Pathol.* **188**(3), 271–277.

Ruska, K. M., Sauvageot, J., and Epstein, J. I. (1998). Histology and cellular kinetics of prostatic atrophy. *Am. J. Surg. Pathol.* **22**(9), 1073–1077.

Sakr, W. A., Grignon, D. J., Crissman, J. D., Heilbrun, L. K., et al., (1994). High grade prostatic intraepithelial neoplasia (HGPIN) and prostatic adenocarcinoma between the ages of 20–69: an autopsy study of 249 cases. *In vivo (Athens, Greece)* **8**(3), 439–443.

Sakr, W. A., Haas, G. P., Cassin, B. F., Pontes, J. E., and Crissman, J. D. (1993). The frequency of carcinoma and intraepithelial neoplasia of the prostate in young male patients. *J. Urol.* **150**(2 Pt 1), 379–385.

Saramäki, O. R., Harjula, A. E., Martikainen, P. M., Vessella, R. L., et al., (2008). *TMPRSS2–ERG* fusion identifies a subgroup of prostate cancers with a favorable prognosis. *Clin. Cancer Res.* **14**(11), 3395–3400.

Saramaki, O., and Visakorpi, T. (2007). Chromosomal aberrations in prostate cancer. *Frontiers Biosc.* **12**, 3287–3301.

Sarker, D., Reid, A. H. M., Yap, T. A., and Bono, J. S de. (2009). Targeting the PI3K/*AKT* pathway for the treatment of prostate cancer. *Clin. Cancer Res.* **15**(15), 4799–4805.

Sato, H., Minei, S., Hachiya, T., Yoshida, T., and Takimoto, Y. (2006). Fluorescence *in situ* hybridization analysis of c-*MYC* amplification in stage TNM prostate cancer in Japanese patients. *Int. J. Urol.* **13**(6), 761–766.

Sato, Y. (2001). Role of *ETS* family transcription factors in vascular development and angiogenesis. *Cell Struct. Funct.* **26**(1), 19–24.

Schayek, H., Haugk, K., Sun, S., True, L. D., et al., (2009). Tumor suppressor BRCA1 is expressed in prostate cancer and controls insulin-like growth factor I receptor (IGF-IR) gene transcription in an androgen receptor-dependent manner. *Clin. Cancer Res.* **15**(5), 1558–1565.

Schmitz, M., Grignard, G., Margue, C., Dippel, W., et al., (2007). Complete loss of *PTEN* expression as a possible early prognostic marker for prostate cancer metastasis. *Int. J. Cancer* **120**(6), 1284–1292.

Schulz, W. A., Ingenwerth, M., Djuidje, C. E., Hader, C., et al., (2010). Changes in cortical cytoskeletal and extracellular matrix gene expression in prostate cancer are related to oncogenic *ERG* deregulation. *BMC Cancer* **10**, 505.

Schweizer, L., Rizzo, C. A., Spires, T. E., Platero, J. S., et al., (2008). The androgen receptor can signal through Wnt/beta-Catenin in prostate cancer cells as an adaptation mechanism to castration levels of androgens. *BMC Cell Biol.* **9**, 4.

Shah, R., Mucci, N. R., Amin, A., Macoska, J. A., and Rubin, M. A. (2001). Postatrophic hyperplasia of the prostate gland: neoplastic precursor or innocent bystander? *Am. J. Pathol.* **158**(5), 1767–1773.

Signoretti, S., and Loda, M. (2006). Defining cell lineages in the prostate epithelium. *Cell Cycle (Georgetown, Tex.)* 5(2), 138–141.

Singh, S., Barrett, J., Sakata, K., Tozer, R. G., and Singh, G. (2002). *ETS* proteins and *MMPs*: partners in invasion and metastasis. *Current Drug Targ.* **3**(5), 359–367.

Singh, S. K., Clarke, I. D., Terasaki, M., Bonn, V. E., et al., (2003). Identification of a cancer stem cell in human brain tumors. *Cancer Res.* **63**(18), 5821–5828.

Sircar, K., Yoshimoto, M., Monzon, F. A., Koumakpayi, I. H., et al., (2009). *PTEN* genomic deletion is associated with p-*AKT* and *AR* signalling in poorer outcome, hormone refractory prostate cancer. *J. Pathol.* **218**(4), 505–513.

Slaughter, D. P., Southwick, H. W., and Smejkal, W. (1953). Field cancerization in oral stratified squamous epithelium; clinical implications of multicentric origin. *Cancer* **6**(5), 963–968.

Squire, J. A. (2009). *TMPRSS2–ERG* and *PTEN* loss in prostate cancer. *Nat. Genet.* **41**(5), 509–510.

Sreekumar, A., Poisson, L. M., Rajendiran, T. M., Khan, A. P., et al., (2009). Metabolomic profiles delineate potential role for sarcosine in prostate cancer progression. *Nature* **457** (7231), 910–914.

Stott, S. L., Lee, R. J., Nagrath, S., Yu, M., et al., (2010). Isolation and characterization of circulating tumor cells from patients with localized and metastatic prostate cancer. *Sci. Trans. Med.* **2**(25), 25ra23.

Sun, C., Dobi, A., Mohamed, A., Li, H., et al., (2008). *TMPRSS2–ERG* fusion, a common genomic alteration in prostate cancer activates C-*MYC* and abrogates prostate epithelial differentiation. *Oncogene* **27**(40), 5348–5353.

Sun, Q. -P., Li, L. -Y., Chen, Z., Pang, J., et al., (2010). Detection of *TMPRSS2–ETS* fusions by a multiprobe fluorescence *in situ* hybridization assay for the early diagnosis of prostate cancer: a pilot study. *J. Mol. Diagnostics* **12**(5), 718–724.

Sun, Y., and Gao, W. -Q. (2011). Human prostate cancer stem cells: new features unveiled. *Asian J. Androl.* **13**(3), 355–356.

Tang, D. G., Patrawala, L., Calhoun, T., Bhatia, B., et al., (2007). Prostate cancer stem/progenitor cells: identification, characterization, and implications. *Mol. Carcinogen.* **46** (1), 1–14.

Taylor, B. S., Schultz, N., Hieronymus, H., Gopalan, A., et al., (2010). Integrative genomic profiling of human prostate cancer. *Cancer Cell* **18**(1), 11–22.

Terry, S., Yang, X., Chen, M. -W., Vacherot, F., and Buttyan, R. (2006). Multifaceted interaction between the androgen and WNT signaling pathways and the implication for prostate cancer. *J. Cellular Biochem.* **99**(2), 402–410.

The American Cancer Society, 2011 "http://www.cancer.org/Research/CancerFactsFigures/index.".

Tomlins, S. A., Laxman, B., Dhanasekaran, S. M., Helgeson, B. E., Cao, X., Morris, D. S., Menon, A., Jing, X., Cao, Q., Han, B., Yu, J., Wang, L., Montie, J. E., Rubin, M. A., Pienta, K. J., Roulston, D., Shah, R. B., Varambally, S., Mehra, R., and Chinnaiyan, A. M. (2007). Distinct classes of chromosomal rearrangements create oncogenic ETS gene fusions in prostate cancer. *Nature* **448**(7153), 595–599.

Tomlins, S. A., Laxman, B., Varambally, S., Cao, X., et al., (2008). Role of the TMPRSS2–ERG gene fusion in prostate cancer. *Neoplasia* **10**(2), 177–188.

Tomlins, S. A., Rhodes, D. R., Perner, S., Dhanasekaran, S. M., et al., (2005). Recurrent fusion of TMPRSS2 and ETS transcription factor genes in prostate cancer. *Science* **310** (5748), 644–648.

Trojanowska, M. (2000). ETS factors and regulation of the extracellular matrix. *Oncogene* **19**(55), 6464–6471.

Vaarala, M. H., Porvari, K., Kyllönen, A., Lukkarinen, O., and Vihko, P. (2001). The *TMPRSS2* gene encoding transmembrane serine protease is overexpressed in a majority of prostate cancer patients: detection of mutated *TMPRSS2* form in a case of aggressive disease. *Int. J. Cancer* **94**(5), 705–710.

Verger, A., Buisine, E., Carrère, S., Wintjens, R., et al., (2001). Identification of amino acid residues in the *ETS* transcription factor Erg that mediate Erg-Jun/Fos-DNA ternary complex formation. *J. Bio. Chem.* **276**(20), 17181–17189.

Verhagen, P. C. M. S., Duijn van, P. W., Hermans, K. G. L., Looijenga, L. H. J., et al., (2006). The *PTEN* gene in locally progressive prostate cancer is preferentially inactivated by bi-allelic gene deletion. *J. Pathol.* **208**(5), 699–707.

Verhagen, A. P., Ramaekers, F. C., Aalders, T. W., Schaafsma, H. E., et al., (1992). Colocalization of basal and luminal cell-type cytokeratins in human prostate cancer. *Cancer Res.* **52**(22), 6182–6187.

Vis, A. N., and Schröder, F. H. (2009). Key targets of hormonal treatment of prostate cancer. Part 1: the androgen receptor and steroidogenic pathways. *BJU Int.* **104**(4), 438–448.

Vukovic, B., Park, P. C., Al-Maghrabi, J., Beheshti, B., et al., (2003). Evidence of multifocality of telomere erosion in high-grade prostatic intraepithelial neoplasia (HPIN) and concurrent carcinoma. *Oncogene* **22**(13), 1978–1987.

Wang, J., Cai, Y., Ren, C., and Ittmann, M. (2006). Expression of variant *TMPRSS2/ERG* fusion messenger RNAs is associated with aggressive prostate cancer. *Cancer Res.* **66**(17), 8347–8351.

Wang, J., Cai, Y., Yu, W., Ren, C., et al., (2008). Pleiotropic biological activities of alternatively spliced *TMPRSS2/ERG* fusion gene transcripts. *Cancer Res.* **68**(20), 8516–8524.

Wang, X., Kruithof-de Julio, M., Economides, K. D., Walker, D., et al., (2009). A luminal epithelial stem cell that is a cell of origin for prostate cancer. *Nature* **461**(7263), 495–500.

Wang, Y., Romigh, T., He, X., Orloff, M. S., et al., (2010). Resveratrol regulates the *PTEN/AKT* pathway through androgen receptor-dependent and -independent mechanisms in prostate cancer cell lines. *Human Mol. Gen.* **19**(22), 4319–4329.

Watson, S. K., Woolcock, B. W., Fee, J. N., Bainbridge, T. C., et al., (2009). Minimum altered regions in early prostate cancer progression identified by high resolution whole genome tiling path BAC array comparative hybridization. *Prostate* **69**(9), 961–975.

Wilson, S., Greer, B., Hooper, J., Zijlstra, A., et al., (2005). The membrane-anchored serine protease, *TMPRSS2*, activates PAR-2 in prostate cancer cells. *Biochem. J.* **388**(Pt 3), 967–972.

Wise, A. M., Stamey, T. A., McNeal, J. E., and Clayton, J. L. (2002). Morphologic and clinical significance of multifocal prostate cancers in radical prostatectomy specimens. *Urology* **60**(2), 264–269.

Yoshimoto, M., Cunha, I. W., Coudry, R. A., Fonseca, F. P., et al., (2007). FISH analysis of 107 prostate cancers shows that *PTEN* genomic deletion is associated with poor clinical outcome. *Br. J. Cancer* **97**(5), 678–685.

Yoshimoto, M., Cutz, J. -C., Nuin, P. A. S., Joshua, A. M., et al., (2006). Interphase FISH analysis of *PTEN* in histologic sections shows genomic deletions in 68% of primary prostate cancer and 23% of high-grade prostatic intra-epithelial neoplasias. *Cancer Genetics Cytogen.* **169**(2), 128–137.

Yoshimoto, M., Joshua, A. M., Chilton-Macneill, S., Bayani, J., et al., (2006a). Three-color FISH analysis of TMPRSS2/ERG fusions in prostate cancer indicates that genomic microdeletion of chromosome 21 is associated with rearrangement. *Neoplasia* **8**(6), 465–469.

Yoshimoto, M., Joshua, A. M., Cunha, I. W., Coudry, R. A., et al., (2008). Absence of *TMPRSS2:ERG* fusions and *PTEN* losses in prostate cancer is associated with a favorable outcome. *Mod. Pathol.* **21**(12), 1451–1460.

Yoshimoto, M., Ludkovski, O., Bayani, J., Graham, C., et al., (2007a). Microdeletion and concurrent translocation associated with a complex TMPRSS2:ERG prostate cancer gene fusion. *Genes Chromosomes and Cancer* **46**(9), 861–863.

Yu, Y. P., Landsittel, D., Jing, L., Nelson, J., et al., (2004). Gene expression alterations in prostate cancer predicting tumor aggression and preceding development of malignancy. *J. Clin. Oncol.* **22**(14), 2790–2799.

Zha, S., Gage, W. R., Sauvageot, J., Saria, E. A., et al., (2001). Cyclooxygenase-2 is up-regulated in proliferative inflammatory atrophy of the prostate, but not in prostate carcinoma. *Cancer Res.* **61**(24), 8617–8623.

Zhang, S., Pavlovitz, B., Tull, J., Wang, Y., et al., (2010). Detection of *TMPRSS2* gene deletions and translocations in carcinoma, intraepithelial neoplasia, and normal epithelium of the prostate by direct fluorescence in situ hybridization. *Diagn. Mol. Pathol.* **19**(3), 151–156.

Zhang, S., and Yu, D. (2010). PI(3)King apart *PTEN*'s role in cancer. *Clin. Cancer Res.* **16** (17), 4325–4330.

Zvereva, M. I., Shcherbakova, D. M., and Dontsova, O. A. (2010). Telomerase: structure, functions, and activity regulation. *Biochem. Biokhimiia* **75**(13), 1563–1583.

Evolutionary Mechanisms and Diversity in Cancer

Henry H.Q. Heng[*,†,‡], Joshua B. Stevens[*], Steven W. Bremer[*], Guo Liu[*], Batoul Y. Abdallah[*] and Christine J. Ye[¶]

[*] *Center for Molecular Medicine and Genetics, Wayne State University School of Medicine, MI, USA*
[†] *Karmanos Cancer Institute, Wayne State University School of Medicine, MI, USA*
[‡] *Department of Pathology, Wayne State University School of Medicine, MI, USA*
[¶] *Department of Internal Medicine, Wayne State University School of Medicine, MI, USA*

The recently introduced genome theory of cancer evolution provides a new framework for evolutionary studies on cancer. In particular, the established relationship between the large number of individual molecular mechanisms and the general evolutionary mechanism of cancer calls upon a change in our strategies that have been based on the characterization of common cancer gene mutations and their defined pathways. To further explain the significance of the genome theory of cancer evolution, a brief review will be presented describing the various attempts to

Advances in CANCER RESEARCH, Volume 112
Copyright 2011, Elsevier Inc. All right reserved.
0065-230X/10 $35.00
DOI: 10.1016/B978-0-12-387688-1.00008-9

illustrate the evolutionary mechanism of cancer, followed by further analysis of some key components of somatic cell evolution, including the diversity of biological systems, the multiple levels of information systems and control systems, the two phases (the punctuated or discontinuous phase and gradual Darwinian stepwise phase) and dynamic patterns of somatic cell evolution where genome replacement is the driving force. By linking various individual molecular mechanisms to the level of genome population diversity and tumorigenicity, the general mechanism of cancer has been identified as the evolutionary mechanism of cancer, which can be summarized by the following three steps including stress-induced genome instability, population diversity or heterogeneity, and genome-mediated macroevolution. Interestingly, the evolutionary mechanism is equal to the collective aggregate of all individual molecular mechanisms. This relationship explains why most of the known molecular mechanisms can contribute to cancer yet there is no single dominant mechanism for the majority of clinical cases. Despite the fact that each molecular mechanism can serve as a system stress and initiate the evolutionary process, to achieve cancer, multiple cycles of genome-mediated macroevolution are required and are a stochastically determined event. Finally, the potential clinical implications of the evolutionary mechanism of cancer are briefly reviewed.
© 2011 Elsevier Inc.

I. INTRODUCTION

Following decades of cancer research that extensively focused on the identification and characterization of individual cancer genes and their associated molecular mechanisms, the overwhelming genetic/epigenetic heterogeneity and system complexity now being identified forcefully challenge the way we conduct cancer research (Berger *et al.*, 2011; Esteller, 2008; Feinberg and Irizarry, 2010; Heng, 2007; Heng *et al.*, 2004, 2009, 2010; Jones *et al.*, 2008; Parsons *et al.*, 2008; Wood *et al.*, 2007). Despite the fact that there is an increasing support for the use of evolutionary principles as an approach that would unify the field (Anderson and Quaranta, 2008; Axelrod *et al.*, 2006; Breivik, 2005; Cahill *et al.*, 1999; Crespi and Summers, 2006; Gatenby, 2009; Glazier, 2006; Greaves, 2007; Maley *et al.*, 2006; Merlo *et al.*, 2006; Nowell, 1976; Schwab and Pienta, 1996; Shibata and Tavaré, 2006; Tsao *et al.*, 2000; Vincent and Brown, 2005), many important issues have not been addressed. For example, what is the general pattern of somatic cell evolution? Is it a slow stepwise microevolutionary process, a fast-punctuated process, or a combination thereof? What is the main driving force of cancer progression and drug resistance, gene mutations, genome alterations, or nongenetic alterations including epigenetic dis-regulation? Can genetic diversity be effectively measured using the degree of gene mutation, copy number variation, overall genome alteration, or a combination of all of these? Is the evolutionary process best monitored by tracing trees of methylation patterns, gene mutation trees, karyotype trees, or both gene and karyotype trees? What is the relationship between individual molecular mechanisms and evolutionary mechanisms of cancer (is there a general mechanism to unify all cancer cases)? To address these questions, a new conceptual framework is needed that is based on four important transitions: (1) emphasizing the previously ignored genome level rather than well-researched cancer genes; (2) increasing attention to stochastic genetic events during cancer progression rather than only studying commonly shared patterns; (3) switching from a stepwise linear model to a stochastic punctuated model of evolution; (4) and establishing a general evolutionary mechanism of cancer rather than continuing to complete the fast growing list of individual molecular mechanisms. By searching for such a new concept, the genome theory of cancer evolution has emerged and states that the driving force of somatic cell evolution is genome-mediated system replacement, and genome evolution can be described as multiple cycles of punctuated and stepwise evolution where stochastic genome-level alterations are the key (Heng, 2007, 2009; Heng *et al.*, 2006a–c, 2011; Ye *et al.*, 2007, 2009;).

II. A BRIEF REVIEW OF CANCER EVOLUTIONARY STUDIES

Cancer is a story of somatic cell evolution. During cancer initiation, progression, long-distance invasion, and drug resistance, each new advance is achieved by an evolutionary process that is coupled with the dominance of a new genome defined system. Unfortunately, for decades, evolutionary studies on cancer have only comprised a tiny portion of the effort to understand the mechanism of this terrifying disease.

In the field of cancer research, traditional evolutionary analyses can be summarized mainly into two types of study: the somatic selection of cancer genes and cells within individuals and natural selection within populations of genes that predispose or help to prevent cancer (Crespi and Summers, 2006). Even among those who study cancer evolution, research has focused on identification of stepwise patterns of genetic alterations (specific cancer genes or common chromosomal alterations) during cancer progression.

A. Linear Evolutionary Models

Evolutionary studies of bacteria mutations (Brash and Cairns, 2009; Cairns, 1975), hematological malignancies (Nowell, 1976; Rowley, 1988), and the conventional ideas of evolution that the accumulation of small alterations leads to big changes, have influenced the generally accepted idea that the evolutionary process of tumorigenesis occurs through multiple linear steps (Hanahan and Weinberg, 2000; Huang *et al.*, 2009; Nowak *et al.*, 2006; Nowell, 1976; Vogelstein and Kinzler, 1993). In such a linear model, the evolutionary process of cancer has been thought to be mainly based on the accumulation of gene mutations with chromosome change playing a primarily ancillary role. In many types of hematological cancers where commonly shared marker chromosomes can be identified, focus is placed on the unique pattern of the abnormal chromosome itself, assuming that the rest of the genome is unaffected, and other changes are insignificant as they are more or less random. According to the prevailing theory, a stepwise pattern based on gene trees can be identified during cancer evolution and key gene mutations driving cancer evolution can be identified and targeted for medical intervention. There are some very exciting studies of solid tumors that demonstrate the power of evolutionary analysis in this manner (Maley *et al.*, 2006). Interestingly, most successful stories are linked to slowly progressing types of cancer. In most cancers it is difficult to illustrate the main pathway of evolution, as there are too many genome-level alterations and large numbers of gene mutations that make building a gene tree difficult. The fact that every tumor would have a different tree limits the impact of this type of study. Increasing data show that the dynamics of cancer evolution is much higher than that of natural evolution (Heng, 2007b; Heng *et al.*, 2010a), as many conserved genetic elements are shared among different species, yet are drastically different among cancer samples (Calin *et al.*, 2007).

Mitelman and colleagues have systematically analyzed karyotype evolution in various types of cancers with a focus on patterns of clonal evolution. Through this they have established relationships between cytogenetic subgroups and karyotypic pathways. However, of the 50,000 tumor cases that have been reported, there are over 40,000 karyotypic aberrations suggesting that no commonly shared, recurrent patterns exist in the majority of cases, particularly in solid tumors (Mitelman, 2006). Even though clonal evolution can be observed in some cases during different stages of primary tumors and metastatic cancers, and diverse karyotypes can be grouped according to their statistical behavior (Höglund *et al.*, 2005), the presence of karyotypic heterogeneity is underreported and overwhelming. Chromosomal instability has been blamed for this high heterogeneity (Gisselsson, 2003; Heng *et al.*, 2004). Understanding the consequences of karyotypic alterations to the overall system and the meaning of the long ignored nonclonal chromosomal aberrations (NCCAs) is an additional challenge, as now we know that population heterogeneity can change the genetic landscape, affecting evolution.

Another important topic is the mathematical and computational modeling of somatic cell evolution in tumorigenesis. Various factors have been studied by Gatenby's group including the competition between normal and cancer tissue, the contributions of microenvironments, and the dynamics of the evolutionary landscape (Gatenby, 1996; Gatenby *et al.*, 2010; Vincent and Gatenby, 2008). In addition to using agent-based computational models to illustrate the promising strategies of benign cell boosters and chemosensitive boosters (Maley *et al.*, 2004), a number of key features of cellular alteration (self-sufficiency in growth signals; insensitivity to antigrowth signals; evasion of apoptosis; limitless replicative potential; sustained angiogenesis and tissue invasion and genetic instability) or hallmarks of cancer have been analyzed in a three-dimensional stochastic multistep model (Spencer *et al.*, 2006). The predictions regarding genetic instability are particularly interesting. By illustrating that early-onset and late-onset tumors take different mutational paths to cancer, and that genetic instability is often the cause of mutations in earlier onset tumors, this type of analysis demonstrates the mechanism of evolutionary dynamics. Similarly, simulations have been used to study tumor growth using the hybrid discrete continuum model (Anderson *et al.*, 2006). The colorectal tumor evolution process (waiting time and transition from benign, invasive and metastatic phase) has been analyzed based on common mutations (Beerenwinkel *et al.*, 2007; Jones *et al.*, 2008). However, most of these models are based on the gene theory while the key contributions of the genome have been ignored. For example, many of these approaches have considered carcinogenesis as a micro-evolutionary process where multiple genetic and epigenetic changes accumulate in a single cell (Foo *et al.*, 2011; Tarafa *et al.*, 2008).

B. Genome-Based Stochastic Evolutionary Models

In the past decade, we have witnessed the beginning of major shifts in modus operandi of cancer evolution. First, a shift from cancer gene-focused research to genome-focused research has begun to occur, as a large number of cancer genes have been identified. In particular, recent cancer whole genome sequencing has revealed that there are tens of thousands of gene mutations in each patient, most with low penetrance, that are not shared between patients. It is clear that a new synthesis is needed to make sense of the large amounts of diverse and often conflicting information. Second, there is a realization that previously ignored genetic "noise" such as seemingly random genome alterations do in fact play an important role in cancer evolution. Most cancer samples do not share chromosomal aberrations but display a high level of stochastic chromosomal change that represents the overall tumor heterogeneity (Atkin and Baker, 1990; Bartos *et al.*, 2004; Bayani *et al.*, 2007; Foulds, 1954; Heppner, 1984; Heppner and Miller, 1998; Klein *et al.*, 2002; Stoecklein and Klein, 2010). Due to the overwhelming presence of this heterogeneity, the specific alterations biologists have studied for so long (specific markers such as commonly shared genetic alterations in cancer) are not the best indicators to monitor system instability (Heng *et al.*, 2004, 2006b). These important shifts are driven by the technological advances used to profile genes and the genome. For example, multiple color fluorescence *in situ* hybridization techniques such as spectral karyotyping (SKY) have drastically improved our capability to classify karyoytypic alterations in cancer cells (Beatty and Heng, 2007; Heng *et al.*, 1997, 2001, 2003; Schröck *et al.*, 1996; Ye *et al.*, 2001, 2006, 2009). Quite the opposite as initially hoped for, however, SKY did not reveal many commonly shared patterns of chromosomal aberrations, but further delineated the complexity of karyotype changes and the presence of overwhelming genome heterogeneity. Clearly, a new concept is required to deal with this reality that the large amount of seemingly random genome alterations represents a typical feature of cancer rather than an exception. By applying a system approach, we realized that the elevated levels of random chromosome aberrations do not represent insignificant genetic noise, but are an indicator of system instability, and that the specific alterations biologists have studied for so long (such as commonly shared genetic alterations in cancer) are not the best indicators to monitor system instability (Heng *et al.*, 2004, 2006b).

If a nonspecific alteration is a key indicator of system dynamics, reexamination and reinterpretation of stochastic genetic alterations represents a novel and important approach. Reapplication of the previous focus on the gene level reveals a link between alterations at many levels and system instability to elements that contribute to cancer evolution. For example, an array of experimental manipulations, including inactivation of genes that are essential for genome stability, over expression of oncoproteins, treatment with carcinogens, and increasing system stresses, share the common feature of elevated frequencies of random chromosome aberrations. This reveals that among multiple levels of a genetic system, the genome is the highest level of genetic organization, and biological evolution must involve heritable elements, thus there must be more focus on chromosomal changes.

To firmly establish the biological function of these nonclonal chromosome aberrations (NCCAs) in cancer progression, in addition to analyzing clinical samples (the end products of cancer evolution), various *in vitro* models have been used to watch evolution in action. By tracing the progression of the immortalization process, for example, individual cells as well as cell populations have been studied using karyotype analysis. By comparing the karyotypes of cells from the same passage and across a long period of time, an evolutionary tree of karyotypes can be developed and analyzed. The rationale of this tedious examination is straightforward: the contribution of NCCAs to cancer evolution is a fundamental concern, as is the determination of the relationship between NCCAs and different clonal chromosome aberrations or CCAs, and the frequency of evolutionary convergence and divergence.

This analysis has delivered many surprises (Heng *et al.*, 2006a). In addition to confirming the importance of NCCAs (providing crucial material for somatic cell evolution and serving as an experimental index for measuring genome instability), the following conclusions have been reached: first, there are two phases of karyotypic evolution, a discontinuous or punctuated phase where no dominant karyotypes are shared among cells or across generations, and a stepwise evolution where common karyotypes are shared between cells and across generations (Heng *et al.*, 2006a–c). The observation of two phases co-existing in one process is highly significant for evolutionary studies as it illustrates the possible relationship between macro- and microevolution. Second, these two phases are changeable, depending on system instability. The highest level of NCCAs is observable within the punctuated phase while shared CCAs occur during the stepwise phase. Introduction of stress into the stepwise phase will cause a switch into the punctuated phase. Conversely, under certain conditions, the punctuated phase can switch to the stepwise phase. The punctuated phase is often shorter; however, it also can last for years (unpublished data). Third, when performing multiple runs of evolutionary experiments using cell populations derived from the same initial cells, the end results are not the same and thus are not predictable. Though many runs of evolution will stop at various dead ends, some runs are successful and establish key status changes such as immortalization. Each successful "winner," however, has a distinct karyotype, illustrating the stochastic nature of *in vitro* evolution (Fig. 1). Similar patterns have been observed in different systems, including human and mouse models of transformation and drug resistance.

The information generated from our evolutionary analyses explains many puzzling clinical phenomena, for example the inconsistency between animal models with artificially created dominant pathways and actual clinical situations where stochastic evolution is predominant, the drastically different responses between patients, the gap between tumor response and prolonged life (initial treatments can reduce the size of tumors by eliminating clonal cells, but the treatment can also accelerate evolution causing a switching to the nonclonal phase and a new run of evolution). Our evolutionary analysis also offers new insight into evolutionary principles and their implication to research. For example, genes and the genome represent two distinct levels of evolution and neither their

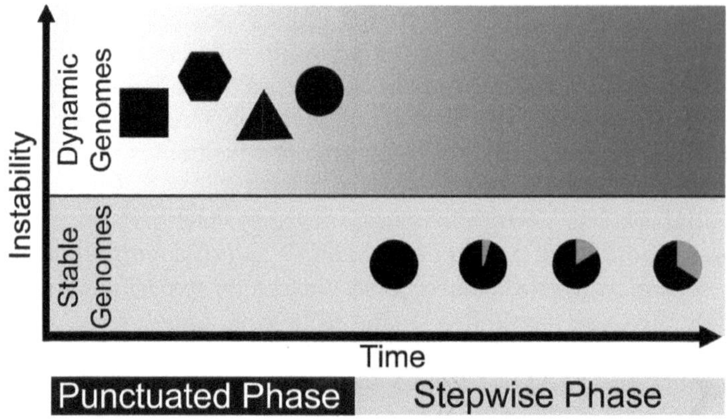

Fig. 1 Stochastic model of cancer evolution. Cancer evolution can be divided into two distinct evolutionary phases, the punctuated stochastic phase and the stepwise gradual phase. Punctuated phases are marked by extreme heterogeneity and rapid genome changes, represented by changes in the shapes, with each shape representing a unique genome. The punctuated phase is caused by system instability-mediated macroevolution. Following selection pressure a unique genome system is selected (circles). Over time this new genome system remains relatively stable, although it does acquire low-level change (expanding pie pieces) such as gene mutations, epigenetic alterations, as well as small traceable genome-level alterations that aide in adaptation. Thus the stepwise phase is mainly associated with system stability and microevolution (Heng *et al.*, 2006a, 2006b, 2006c, 2010a, 2010b). Only one run of NCCA/CCA cycle is presented.

dynamics nor their functions should be confused (Heng, 2007b, 2009). When genomes are constantly changing, it is difficult to trace evolutionary trees at the genome level, and hence building a gene tree is even less useful. Cancer cells within many aggressive solid tumors, such as high-grade carcinomas, display different karyotypes and therefore represent different systems, thus molecular profiles based on averages are flawed. We have no idea what the average is that we are profiling (in contrast, the karyotype diversity in leukemia's and gene-fusion driven sarcomas is much lower. For these cases, profile averaging might be workable). In addition, when all genomes are different, the function of the same genes may be very different. This viewpoint strongly challenges the current idea of tracing genes in evolutionary studies without incorporating information on the altered genome. Analyzing genes is effective for microevolutionary studies, but cancer involves macroevolution reflected as a series of genome replacements. Gene expression studies have shown that when karyotypes are significantly different, the gene expression profiles are different (Heng *et al.*, 2008; Pavelka *et al.*, 2010).

Tracing gene changes will not reveal the pattern of cancer evolution, and therefore, is it workable to focus on genome-level heterogeneity to study the potential of cancer progression? We systematically analyzed five cancer models with known molecular mechanisms (Ye *et al.,* 2009). Each model had diverse molecular mechanisms identified as drivers of cancer progression, in the end the only common link to tumorigenicity in all models was elevated NCCAs. The link between diverse molecular mechanisms and common tumorigenicity has promoted the establishment of the simple relationship between the evolutionary mechanism and the large number of molecular mechanisms and all of their potential interactions. That is, the general mechanism of evolution is equal to the combination of all individual molecular mechanisms (Heng *et al.*, 2010a). This relationship explains why there are so many different molecular mechanisms linked to cancer, yet, the clinical utility of each is low because most of the individual mechanisms are not highly penetrant across cancer patients due to the stochastic feature of cancer evolution.

III. GENETIC ORGANIZATION AND DIVERSITY

One of the key components of cancer evolution is the source of genetic diversity that has traditionally been viewed primarily as gene mutational diversity. There has been heated debate about how many mutations and how many waves of clonal expansion are required for cancer to form (Bodmer *et al.*, 2008; Loeb *et al.*, 2008), how driver mutations and by-stander mutations are distinguished, and finally how can the massive quantities of nonclonal gene mutations detected in individual tumors be reconciled. In recent years, increasing analyses of epigenetic contributions in cancer have generated a great deal of excitement, particularly when gene mutation studies fail to unify the field (Feinberg *et al.*, 2006; Jones and Baylin, 2007). However, there are challenges and limitations with epigenetic studies (Heng *et al.*, 2009). First, the contribution of epigenetic dis-regulation to cancer is mainly related to known cancer genes. If the cancer gene theory cannot explain the mystery of cancer, related studies will not do the job either. Second, the key to epigenetic regulation is its "prefixed" nature. Prototypic gene mutations result in high dynamics and low predictability, rendering use of this information for medical intervention extremely difficult and even dangerous (Heng *et al.*, 2010b). Third, it is difficult to estimate the epigenetic contribution to cancer within a background of high levels of genetic alteration.

Similar to the problem of gene-based research, when the entire system is altered (defined by the newly formed genome), comparing only the epigenetic status has limited utility. It is thus problematic to alter only the epigenetic status in order to reverse cancer, as when the entire system is changed, epigenetic regulation alone will not revert a genome change back to normal. Furthermore, epigenetic change is heterogeneous even between healthy tissues. In cancer there may be a global pattern of epigenetic change, such as hypomethylation, but that change varies from cell to cell. True, epigenetic changes are essential for triggering micro-evolution, linked to genome instability, but they differ from genome variations.

We will argue here that genome alterations are the primary and most important level of genetic diversity in cancer evolution. This is based on the following facts and reasoning: (1) most genome-level change impacts large numbers of genes. Aneuploidy changes the whole genome network, and chromosomal translocations not only impact directly on the gene at the breakpoint but also change the relationship of overall genome topology. (2) The genome defines the functional boundary of genes and epigenetic effects. (3) The biological function of genes can be changed when the genome is altered. (4) The vast majority of cancer cases display genome alterations and similar genome alterations are linked to prognosis.

Despite the fact that there is an interesting relationship among epigenetic–gene–genome alterations (they all can impact on each other, and most of the system changes can be reflected at these three levels), we should not confuse our priorities in research. According to our recent studies, the epigenetic level might be the most sensitive to stress, followed by the gene level and then the genome level. When the majority of a genome is altered, it reaches a point of no return as a new system becomes dominant. There is an urgent need to develop approaches that can incorporate both the micro- and macro evolutionary dynamics of cancer and to determine which level of genetic or epigenetic variation should be the focus of future cancer research (Heng *et al.*, 2009; Ye *et al.*, 2007).

IV. MULTIPLE LEVELS OF CONTROL SYSTEMS ARE ESSENTIAL FOR BIOLOGICAL FUNCTIONS

A common gene centric view is that genes are the most important biological unit in biology and that defective cancer genes are the direct cause of cancer. Those who also consider the genome to be important tend to think that the genome is secondary to the gene; hence, the more the genes are understood, the more the genome will be understood, as the accumulation of knowledge about genes will automatically extrapolate to how the genome works. After all, the genome is just the sum of all genes. To illustrate the foremost importance of the genome and its unique emergent properties that are not shared by genes, it is necessary to separate the main function of genes and the function of the genome in genetic systems both from a genetic information standpoint and from a system control point of view, and then illustrate their fundamental distinctive roles both in organismal and somatic cell evolution.

A. The Gene and the Genome Represent Different Genetic Coding Systems

To appreciate the distinct difference between genes and the genome, the key genetic difference between them must be understood. According to the gene theory, the information coded within the DNA sequence is all the genetic information needed to understand the mystery of life. The information exchange from DNA to RNA and to proteins has laid the foundation of modern molecular biology, and the majority of the molecular biological research. In contrast with the genome theory, DNA is not equal to the heritance (Gorelick and Laubichler, 2008), DNA and the genome context are totally different genetic coding systems, and the key genetic information regarding assembly of the parts most likely is not stored within stretches of DNA. Instead, DNA only codes the parts and some tools of the system, while the complete interactive plan is achieved by genome context (genomic topology) mediated self-organization (Heng *et al.*, 2010a, 2010b, 2011; Ye *et al.*, 2007). In other words, DNA codes the proteins and RNAs, while the interactive relationship among all genes and regulationary elements are achieved by 3D genome topology, which serves as the platform where the self-organization principle works to form a genetic network. The same karyotype ensures the same genome topology and sexual reproduction maintains the same genome (karyotype). Therefore, the genome context ensured by sexual reproduction codes the genetic network (Heng, 2009; Heng *et al.*, 2011; Ye *et al.*, 2007). No wonder most organisms display different genomes but often share a large portion of genes. In mammals, for example, most species have similar genes but these are organized within different genomes with variable genome context (Ferguson-Smith and Trifonov, 2007; Heng, 2009; Kohn *et al.*, 2006; Murphy *et al.*, 2005; Ye *et al.*, 2007). Most cancer systems undergo genome shattering and reshuffling as well and such genome reorganization can form new systems required for both organismal and cancer evolution. However, in contrast to the somatic evolution of cancer, it has been proposed that sexual reproduction ensures the genome context thus ensuring faithful dissemination of the genetic codes above the DNA level to future generations (Heng, 2007b).

B. The Function of Genes and the Genome is Mainly Responsible for Different System Controls

Importantly, different genetic coding includes not just where genetic information is stored, but is responsible for different levels of system control. It is well known that genes are responsible for many features of a system, such as skin color and the eye color of individuals. Yet the genome, despite not directly coding for proteins defines and controls the boundary of how the system can work when so many proteins interact with each other under various environmental challenges, which defines a system. Based on the information theory, when considering the system control of multiple levels of a system, the information of the lower level is easier to access, but unfortunately is often not as useful when it comes to system control. This explains why we have had limited success on the War on Cancer by focusing only on decoding gene mutations. We have surely sequenced large enough numbers of cancer genes and understood numerous molecular mechanisms of cancer. Many drugs have been designed accordingly, but we still do not understand how cancer is controlled. If the information needed to control cancer is not stored within cancer genes but lies within the genome as suggested by the genome theory, then we have to refocus our efforts to study a totally different level of system control. As cancer is a systems disease rather than an individual gene mutation disease, the 1D sequence of individual genes will not have the information of 3D interactions that occur within the genetic network (Heng *et al.*, 2011).

C. Conflict Between Genes and the Genome

The above reasoning needs more supporting evidence. One interesting consideration is the key biological conflict between the function of genes and the genome. Previously influenced by the gene-centric concept, the main function of the genome has been thought to serve as a vehicle to pass genes, and therefore if gene change can be traced during the evolutionary process and patterns can be identified, we should be able to dissect the cancer process. The gene-centric viewpoint completely ignores the crucial function of the genome organizing and controlling the system, as if in spite of what happens to the genome, the same gene has the same function. Furthermore, it is well known that genome alteration is a key feature of the vast majority of cancer cases and it is very puzzling why such important clinical facts have been ignored by many molecular researchers.

One way to illustrate the gene/genome conflict is to link genome- and gene-level changes to different evolutionary patterns. By synthesizing diverse information, genome alterations and gene mutations have been linked to macro-and microevolution respectively, and genome alteration serves as a driving force for cancer evolution. Interestingly, as the punctuated pattern of karyotypic evolution mimics the evolutionary pattern of asexual species, we have hypothesized that asexual species should display high levels of genetic diversity (Heng, 2007b). The corroboration of studies on this hypothesis has led us to propose that the main function of sex is to preserve the genome identity and to filter out altered genomes. Various studies now support this conclusion (Glansdorff *et al.*, 2009; Gorelick and Carpinone, 2009; Wilkins and Holiday, 2009). Sexual reproduction serves as a main mechanism maintaining genome stability, as all sexual species display genome-level stasis despite high levels of gene dynamics, which explains the paradoxical observation of active evolutionary adaptation in the short term yet stasis during long-term evolution (Futuyma, 2010; Gorelick and Heng, 2011). The gene and genome display different and conflicting functions in evolution: the gene provides system dynamics and short-term adaptation, while the genome maintains the system stability responsible for long-term stasis. Clearly, such a relationship in organismal evolution underscores the importance of genome instability in cancer evolution.

V. THE PATTERN OF EVOLUTIONARY DYNAMICS OF SOMATIC CELL EVOLUTION

As most of the gene-based evolutionary studies cannot satisfactorily explain the pattern of somatic cell evolution (Heng, 2008a, 2009, 2010), it is necessary to investigate the essential contribution of the genome. Over the past decade, by "watching evolution in action" and tracing both individual cells and cell populations in various experiments, high frequencies of genome dynamics have been unexpectedly documented. Analysis of these phenomena has revealed the most exciting insight into genome-based cancer evolution.

A. NCCA/CCA Cycles

Realizing the importance of NCCAs was the first step to understand karyotype dynamics during cancer evolution, which is the story of the significance of the insignificant. Prior to our analyses, these seemingly random chromosomal alterations were thought to be insignificant genetic noise (Albertson *et al.*, 2003; Mitelman, 2000). By screening large samples of tumors and cancer cell lines, we have realized that the frequency of NCCAs represents true genome instability in that NCCAs are constantly changing, whereas CCAs represent established, albeit changed from normal, systems. When karyotypic evolution was traced using a cellular immortalization model, NCCAs were found to provide the material for karyotypic evolution to occur (Heng *et al.*, 2006a–c, 2008; Ye *et al.*, 2007). Dominance of NCCAs or specific CCAs represents two distinctive phases of evolution, namely punctuated (or discontinuous) and gradual, stepwise evolution. In the punctuated phase, where karyotypical evolution is not traceable, large numbers of nonclonal chromosome aberrations are present, with some displaying massive genome reorganization. The key feature of this phase is the appearance of large amounts of instability (heterogeneity) and the overall slow growth, possibly due to a high rate of cell death. In the gradual stepwise phase, in contrast, a clonal population stands out immediately following selection pressure such as the crisis stage of immortalization and stepwise karyotypic changes can be traced as NCCA levels are reduced. This stage lasts longer, though it can eventually switch into a punctuated phase. These two phases repeat throughout the life of a tumor. As the stochastic discontinuous changes of the karyotypes mimic the pattern of punctuated evolution (Gould and Eldredge, 1993), we applied this term to describe the evolutionary dynamics of both somatic cell and organismal evolution (Heng, 2007b). Interestingly, the two evolutionary phases reflected by the NCCA/CCA interaction also provided needed information for understanding punctuated evolution (Heng, 2007, Heng, 2009; Heng *et al.*, 2006b,c, 2009, 2010). The NCCA/CCA cycle explains many puzzling phenomena including the rapid establishment of drug resistance, stochastic evolutionary patterns, and the high level of inconsistence among many molecular studies that are based on cell populations with high levels of NCCAs and even completely altered CCAs. Further analyses illustrates that the NCCA and CCA cycle represents both survival and growth advantages respectively (Ye *et al.*, 2007).

Interestingly, the NCCA/CCA cycle challenges the definition of clonal populations. From a gene perspective, if a cell population shares the same marker gene, this cell population is called clonal, as they are derived from the same progenitor cell with that gene marker. Although, as evidenced by single allele sequencing, the vast majority of gene mutations are nonclonal in tumors. However, divergent karyotypes, despite sharing the same gene marker, are clearly different systems, and should no longer be referred to as a clonal population. We do not consider cancer cells to be clones of normal cells, despite the fact that all are generated from one fertilized cell and share many DNA markers. This also emphasizes that the classification of clonal populations should focus on the genome level rather than the gene level, as most of the tumors are not clonal particularly since they have different karyotypes.

The stochastic relationship between NCCAs and CCAs also questions the generality of many molecular findings that are based on one run of highly historically contingent evolution. We have performed a number of parallel experiments to illustrate this point. When the same mouse ovarian surface epithelial cells (MOSEC) were divided into different dishes and grown continuously in culture, these primary cells went through a series of changes and some became malignant. We compared the karyotypes and gene expression profiles of the transformed cells from independent cultures, and surprisingly they were drastically different and did not share any common patterns (Lawrenson *et al.*, unpublished data). Similarly, when the human ovarian cancer cells were treated with drug cisplatin, multiple drug-resistant clones were selected. Again, all selected clones with stable drug-resistant phenotypes displayed drastically different karyotypes and protein profiles indicating that they frequently take different molecular paths (Heng *et al.*, unpublished observations). It is clear that the different molecular mechanisms we have detected only represent one of the many properties of a newly formed genome. The question is, when the genome context is constantly altered as evidenced by changed genomes in different cancer samples and for different stages of the same sample, what is the true value of any molecular analysis that is based on individual genes but ignores the genome context? The reality of high-level genome dynamics and heterogeneity also disproves the key assumption of many mathematical models that illustrate the accumulation of multiple genetic and epigenetic changes in a single cell.

In addition, the punctuated/stepwise cycle of evolution should also apply to the different system levels as it represents a general evolutionary pattern under stable and dynamic conditions. We have predicted that such cycles should be detected at the sub-chromosomal (including copy number variations) and gene mutation levels (Ye *et al.*, 2007). Interestingly, using single-cell sequencing, it has recently been illustrated that at the DNA level tumors grow by "punctuated clonal expansion with few persistent intermediates" (Navin *et al.*, 2011). It should be pointed out that even if the stepwise phase can be detected at the DNA level, it still can be punctuated at the genome level. The key to monitoring cancer evolution is at the genome level rather than the gene level.

There is an important difference between organismal evolution and cancer evolution in terms of the concept of population heterogeneity. Most of the organismal evolutionary studies refer to the population of a given species that is defined by the same genome context, while cancer cell populations are mixtures of cells with different genomes and mutations and the same genome with different gene mutations. Different methods of analysis must be applied accordingly.

The relationship of punctuated NCCAs/CCAs discovered during cancer evolution also favors the idea of treating cancer cells with altered genomes as different systems or even different species. Based on the analysis of karyotype differences among natural species, and the punctuated progression pattern of cancer genomes, our group has further suggested that the altered cancer genome might represent different genome defined species (Ye *et al.*, 2007). To consider cancer as a different species is an unconventional but not a new idea. This was suggested prior to the establishment of molecular biology and it has been further stated that cancer might represent a different species (Duesberg and Rasnick, 2000; Van Valen and Maiorana, 1991). For example, based on the ecological species concept, Van Valen suggested that Hela cells be defined as a new species, which was named *Helacyton gartleri*, not knowing that Hela cells are constantly changing their karyotypes as reflected by NCCA/CCA cycles. Similarly, defining all cancers as one species (by claiming cancer is a species of its own) is incorrect as judged by the presence of diverse genomes. In a sense, each different karyotype should be considered a distinct species.

To date, there is only limited evidence to support this idea. In the case of facial cancer in the Tasmanian devil, the chromosomes are less like those in the animal's normal cells but more like those in tumors from across different animals, suggesting that this cancer represents a new organism (e.g., a parasite, e.g.) (Murchison, 2008; Welsh, 2011). The NCCA/CCA cycles further illustrate the macroevolutionary process, and demonstrate that the most successful cancer might display a different genome context and therefore represent a different species, as speciation is associated with macroevolution. Despite the fact that it would be a hard sell to the research community, a recent analysis has restated this point in a mainstream evolutionary journal (Vincent, 2010), which should attract people's attention to this issue.

Clearly, the genome theory will help the research community accept the concept of considering various cancer cells as different species (Heng, 2009; Ye *et al.*, 2007). First, cancer is a separate biological system within an animal's body. One argument against considering cancer as a species is based on differences with natural species, primarily that they seem to be able to evolve indefinitely on earth (but who says that the life on earth can last forever?), while cancer cells will be eliminated by the death of the host which is the end of evolution). Different individuals have many unique bacterial populations that are not shared, and these bacterial populations will die with a specific individual. Therefore one could link these bacteria to cancer cells as they share the same fate with the demise of the host. Obviously, bacteria are a biological species as are cancers/tumors. Second, genomes rather than genes define a species (Heng, 2007b, 2009; Ye *et al.*, 2007). When the genomes differ among tumor cells and cancer progression is achieved through multiple cycles of the macro/micro-evolutionary process (reflected as various NCCA/CCA cycles), it is logical to link the altered genome to different biological systems or species. Third, the reinterpretation of the function of sex has demonstrated the key difference between sexual and asexual genomes, the conservation of and persistence of the genome context. As sex serves as a filter to maintain the identity of a genome, there is a defined genome context of a given species (Gorlick and Heng, 2011; Heng, 2007b; Ye *et al.*, 2007). In contrast, somatic cell evolution is a typical asexual process and there is no effective filter to eliminate altered genomes, making cancer evolution a highly dynamic and less predictable process. Each cell with a different genome represents a different species, even though most of them will not survive. Now, we understand the phenomenon that we observed during the punctuated phase, where parents and daughter cells display different genomes and are in fact different species.

To consider different tumors as different species is of practical importance. Such a drastic idea brings needed attention to the fact that tumors are unique and should be treated differently as could different cell lines arising from a tumor. For example, despite the fact that many cell lines can be generated from the same tumor, the varying genome within cells of a tumor distinguishes them into different systems or "species." Such an appreciation will lead to the systematic and standardized examination of karyotypes for cancer samples and cell lines, as currently, only gene level profiles have been systematically analyzed. Based on the analysis of genome dynamics, we have suggested that caution is needed when performing molecular analysis when the genome context is altered, as these changes often change the identity of species (Ye *et al.*, 2007).

B. Genome Chaos: One Effective Way to Achieve Genome Diversity Which is Essential for Cancer Evolution

Genome chaos is a phenomenon where major genome reorganization can be achieved within a short period of time. Genome chaos (karyotypic chaos, or chromosome chaos) has been used to describe a massively altered genome (both from a structural and numerical point of view) (Duesberg, 2007; Heng, 2007a; Heng *et al.*, 2006a–c). As illustrated by our *in vitro* immortalization model and drug resistance models, genome chaos is driven by both internal and external conditions, such as by telomere shortening and a drug challenge respectively. Figure 2 is a model of how drug induced genome chaos may occur, where the vast majority of the chromosomes display multiple, complex translocations. Massive genome shattering and quick rejoining occurs within a very short time window following drug treatment. Follow up of these chaotic genomes

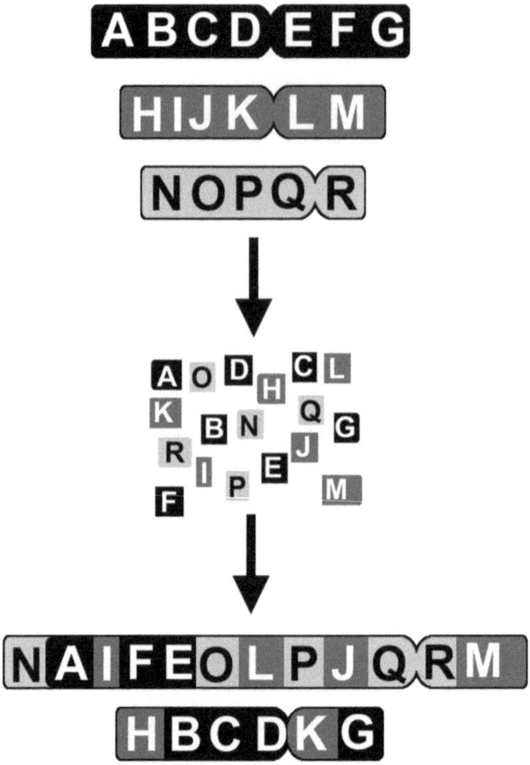

Fig. 2 Schematic demonstrating how various forms of genome chaos may occur. Normal chromosomes are shown at the top. Each letter within the chromosomes represents a distinct region. Following exposure to sufficient levels of various stressors, the genome is partially cut apart by processes such as chromosome fragmentation (Stevens *et al.*, 2007, 2011) (middle). Following partial degradation of the genome, regions are recombined and rejoined resulting in the genome chaos demonstrated at the bottom of the figure. The newly formed chimeric chromosomes can be a mixture of various chromosomal origins, or occasionally from a single chromosome. The process of genome chaos represents an evolutionary process of genome shattering that reorganizes the genetic network (Heng, 2009).

demonstrates that these unstable genomes will go through a series of alterations until some of them reach a more stable state, evidenced by a stable, karyotypically clonal population. Different clones are selected from these chaotic genomes under various conditions, demonstrating that chaotic genomes provide the needed diversity for cancer to evolve. Our experiments also linked genome chaos to a new mechanism of drug resistance (Heng *et al.*, 2011, Stevens *et al.*, unpublished data).

The detection of some highly complicated karyotypes with massive alterations in cancer cells is nothing new, particularly in tumors with high levels of genetic instability such as that induced by defective telomeres (Gisselsson *et al.*, 2000; Heng *et al.*, 2006a). However, the significance of genome chaos in cancer evolution has not been fully appreciated. Genome chaos occurs within the punctuated phase of evolution, underscoring its importance in producing new genomes to achieve macroevolution, the main force driving cancer evolution. Equally important, by following the evolutionary dynamics of these chaotic genomes, our analyses demonstrate that, these highly dynamic genomes are essential for, and can lead to successful evolution despite inducing massive instability-mediated death. In addition, genome chaos is induced under both experimental and pathological conditions through a general mechanism resultant from stressful conditions that are linked to chromosome fragmentation, a newly identified form of mitotic cell death (Stevens *et al.*, 2007, 2010, 2011). As chromosome fragmentation has been linked to chromosome shattering and genome reorganization (Meyerson and Pellman, 2011), it is likely that stress-induced chromosome fragmentation plays an important role in the creation of genome chaos. Recently, one form of genome chaos was reported in ALL patients. This type of highly localized genome chaos has been named chromothripsis (Stephens *et al.*, 2011). Even though the mechanism of generating massive genome reorganization is unclear, it has been suggested that the occurrence of fragmentation of the entire chromosome followed by random rejoining of these fragments is the way this happens (Fig. 2). It is interesting to point out that any genome-level alterations are likely coupled with lower level changes. For instance, translocated chromosomes identified by cytogenetics, as illustrated often contain hundreds of incidences of reorganization detectable at the DNA level (Kitada *et al.*, 2011).

To date, cancer research has focused on identification of specific gene mutations, and the mechanisms by which these arise. Other ideas have been explored within the mutation field including the concept of mutator phenotypes, which have been suggested to be linked to DNA repair deficiency (Prindle *et al.*, 2010). By comparison, genome chaos can achieve much greater and more drastic changes including massive changes at the DNA level that are more powerful than repair deficiency alone would cause. Increasing sequencing data support the involvement of genome chaos during cancer evolution, as a high level of genome rearrangements are detected from a large percentage of the tumors sequenced, and limited numbers of chromosomal translocations are coupled with hundreds of rearrangements at the DNA level (Kitada *et al.*, 2011; Stevens *et al.*, 2011). Current sequencing technologies mainly analyze relatively dominant clonal cell populations and the vast majority of cancer cells display nonclonal mutations or genome alterations that are therefore either not detected or are under-detected (Bielas *et al.*, 2006). High-level genome rearrangements should be a universal phenomenon, and this further suggests the importance of genome reorganization in cancer evolution.

C. Macro and Micro Somatic Cell Evolution and the Multiple Level Landscape Model

In organismal evolution, macroevolution refers to evolution that occurs at or above the level of species while micro-evolution refers to changes at or below the species level. Traditionally, it is thought that the accumulation of micro-evolution leads to macroevolution but direct evidence of this is lacking. Our experiments challenge this idea by illustrating the relationship between the fast punctuated phase, where drastic karyotype alterations dominate, and the gradual stepwise phase, where most cells share the same karyotypes indicating that most changes occur below the genome level. The alterations below and above the genome seem to involve different processes. It is likely that the punctuated phase is driven by genome-mediated evolution, and the stepwise phase occurs mainly by gene-mediated evolution in addition to a low rate of genome-level alteration. The vast majority of species display unique karyotypes (King, 1993; White, 1978; Ye *et al.*, 2007), and thus it is reasonable to consider karyotype changes as a key feature of macroevolution (Heng, 2009). When we applied this concept to somatic cell evolution, it became clear why tracing genes across the entire evolutionary process is less than meaningful, as there are multiple macro and micro-evolution processes occurring in cancer as multiple cycles of NCCAs/CCAs occur, and the unpredictability of cancer is mainly instilled by the punctuated phase of macroevolution.

Separating the macro and microevolution phases of cancer evolution is highly significant. It finally reveals the pattern and dynamics of somatic cell evolution and calls for a new approach to monitor the macroevolutionary phase that has been long ignored in both organismal and somatic cell evolution. In particular, it explains the limitations of current efforts that are contingent on a gene centric system viewpoint to understand cancer evolution, such as the regulation of complex networks with a genetic and epigenetic landscape (Ao, 2009; Huang *et al.*, 2009). When discussing cancer evolution, system biologists tend to not account for high levels of genome alterations, but focus on how gene mutations and epigenetic regulation rewires the network. Based on this assumption, there has been elegant quantitative work to model how different factors can alter the landscape and how cancer attractors can become dominant without the force of selection. However, the vast majority of cancers display drastically altered genomes, and according to the genome theory, the rewiring of the network and the reshuffling of the genome are distinct processes. Genome reshuffling leads to bulk network rewiring, but network rewiring generally does not cause genome reshuffling. Altered genomes define altered systems, the issue of how genome reorganization (macrochanges) occurs is fundamentally more important than the same system rewiring itself (micro-changes).

To solve this issue, we establish the concept of multiple landscapes and in particular to distinguish and integrate the different mechanisms of global and local landscapes. Here, the landscape concept is similar to Waddington's, except that there are two fundamental levels within the landscape. These levels include the Waddington model which fits into the local landscape and includes gene level change, and bigger "mountains" to describe the global, genome based landscapes (Fig. 3). Such an arrangement can clearly illustrate the scale difference between genes and the genome. We also borrowed various terms from the network dynamics of the epigenetic landscape and attractors (Huang *et al.*, 2009). In a sense, the energy potential (also described as elevation state stability) and stress condition-mediated genetic heterogeneity separates local and global landscapes. In regard to cancer, the cellular system heterogeneity occurring in cancer evolution can be divided into three categories: nongenetic heterogeneity, including epigenetic influences not involving genetic changes; gene mutation-mediated heterogeneity; and genome alteration-mediated heterogeneity. Each of these levels has its own landscape. Nongenetic heterogeneity affects the smallest local-level landscape, followed by the gene mutation level. Genome-mediated heterogeneity applies to the highest, global level landscape. As illustrated in Fig. 3, the local and global landscapes are very different in terms of the level of energy barriers, and degree of stress/heterogeneity. Global landscapes are the level of major mountains, which are high enough to not permit local attractor transition. In contrast, crossing small local hills requires low energy potential, which can be achieved by the same genome with different gene or epigenetic states. In addition to the 3D landscape, there is the possibility of involvement of time. In order to understand the local landscape, understanding how gene mutations and epigenetic regulation impact the gene regulatory network is required. To understand how to cross big "mountains," one has to focus on formation of new systems created by drastic genome-reshuffling (such as genome chaos events or genome shattering (Heng, 2009). The power of rewiring an existing network is limited by the genome defined survivable boundary. When stress is within the tolerable degree of such a boundary, rewiring might work. However, when stress exceeds the tolerance of the boundary, the system is no longer viable. That is the reason why many cells are eliminated under stress despite the capability of network rewiring (Heng, unpublished data). This argument further maintains that the genome provides a species constraint to ensure long-term stasis, despite the large number of gene mutations that can provide short-term adaptation advantages (Heng, in preparation).

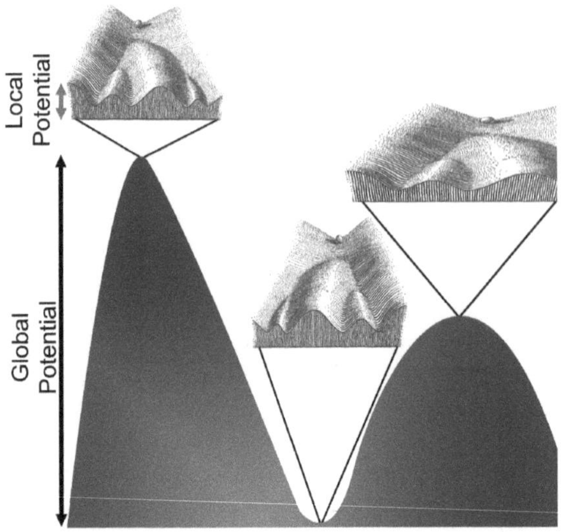

Fig. 3 The genetic landscape of cancer can be broken down to two levels of evolutionary potential. Local potential refers to adaptation potential provided primarily by gene-level or nongenetic changes (Brock *et al.*, 2009). While important for many biological processes such as development, local adaptive landscapes do not typically drive the evolutionary process of cancer. The global potential of the evolutionary landscape of cancer is derived primarily by genome-level change that drives macroevolution. It should be noted that local landscapes are found throughout the global landscape. The image of local landscape was adapted from Waddington.

We predict that the benefits of using the current adaptive landscape concept in studying cancer are extremely limited without integrating the key genome level, given the fact that genome-mediated macroevolution is the driving force of somatic cell evolution. Interestingly, in successful cases of application of an adaptive landscape to study biological issues including genetic switching analyses, protein folding, ecological studies and Wright's initial analysis of adaptive landscapes; (Dill and Chan, 1997; Wright, 1988; Zhu *et al.*, 2004), each case involved one distinct level of the biological system rather than cross level studies. A similar case is found in the original concept of Waddington's developmental landscape (Waddington, 1957), as during normal development, most of the genome is not altered, thus it focuses on the local landscape only.

The multiple levels of the landscape model also explain other phenomena. For example, how do high levels of overall system stability tolerate instability at the lower levels? Local adaptation is different from global adaptation, no matter how successful local adaptation is, it cannot overcome the global limitations. To achieve a global-level change, the cells need to jump onto another huge peak. As the energy potential required is too great to be accomplished by the gene level, only a new genome-defined system can accomplish this. It takes a change at the level of the genome (system) to change the global landscape. In addition, this explains why there are seemingly unlimited molecular pathways in cancer evolution, as globally there are unlimited roads from a mountain to the valleys. Locally, there are unlimited hills and valleys that can be considered as combinations of gene mutations and epigenetic regulation.

The concept of multiple levels of landscape should also apply to organismal evolutionary studies. The adaptation of the global landscape can only be achieved by macroevolution through genome reshuffling to form macroattractors (new species). The fact that a vast majority of species display different karyotypes supports this analysis. In contrast, microevolution through genome rewiring forms micro-attractors that are useful for local adaptation.

Like other landscape models, the mechanism of change for adaptive peaks is largely at the metaphor level. It is possible that the interaction between levels of system instability and system status plays an important role. Knowing that there is likely multiple cycles of the two phases of evolution in a given cancer, and time is also a crucial factor in cancer evolution, it is therefore important to integrate both time and stress factors into the picture as well.

VI. THE EVOLUTIONARY MECHANISM OF CANCER AND ITS MEDICAL IMPLICATIONS

Natural selection represents one of the main mechanisms of organismal evolution. As the accumulation of small genetic alterations favoring reproductive success is believed to be a key to organismal evolution, cancer evolutionary studies have been based on the same concept that has led many to focus on the tracing of changes in gene sequence during cancer progression. With the illustration of multiple cycles of the two phases of karyotype evolution, for most solid tumors, it has become clear that macroevolution drives cancer genome replacement and is the dominant feature. Cancer evolution via genome change is significant in that each change results in the formation of a new genome system. The function of a specific gene is defined by the genome system, questioning the traditional approach of studying cancer evolution by focusing only on cancer genes and linear models of cancer evolution. The stochastic nature of genome evolution also offers an explanation as to why every tumor is different, a much more complex reality than results from model systems indicate. In model systems the initial conditions (such as the over expression of specific oncogenes) form a dominant pathway through artificial selection that favors cancer progression. The initial conditions of cancer progression in patients however are so varied and can result in the selection of any of a massive number of stochastic genome alterations that can be linked to a specific pathway.

To study the dynamics of genome alteration-mediated macroevolution, experimental monitoring systems are needed that measure diversity at the genome level. The frequency of NCCAs, or seemingly random genome alterations, can now be used to measure population diversity as well as system instability. As described in the previous session, elevated NCCA frequency has been linked to many different types of stressful conditions, from defective genes responsible for genome integrity, to various drug and carcinogen treatments and different experimental manipulations. Furthermore, switching between NCCAs and CCAs represents the establishment of a macroevolution selected clone. Finally, a comparison of genome-level changes in a number of known cancer models with well characterized "causative" molecular mechanisms, including the activation and down regulation of various genes, micro-satellite instability, and centrosome amplification, reveals that each "mechanism" is also associated with elevated NCCA frequency. All of the above molecular mechanisms are different, but is there a common mechanism unifying these different mechanisms? Not surprisingly, the common feature these models share is elevated genome heterogeneity. We thus have linked the degree of genome heterogeneity (measured by the frequency of NCCAs) to tumorigenicity. Regardless of which molecular mechanisms are involved, as long as the genome system is unstable, evolution will favor cancer progression (Ye *et al.*, 2009). Similarly, we also studied the general mechanism of tumor response to treatment. Our results demonstrate that many tumors that display treatment resistance also display high levels of NCCAs and genome heterogeneity (Heng *et al.*, 2010a; unpublished data). Together, all of these studies led to the realization that cancer is truly an evolutionary story where genome heterogeneity is the key factor. Connecting the dots elucidated the general mechanism of cancer, which is summarized below.

Evolutionary mechanism of cancer = $\sum$ all molecular mechanisms of cancer

The evolutionary mechanism of cancer is stress-induced system instability resulting in population diversity (mainly at the genome level) and genome-mediated (not gene mutation driven) macroevolution. Interestingly, the complicated process can be explained by three simple components including stress, population diversity and genome evolution. However, the challenge is that these three components can be linked to unlimited molecular mechanisms, as there are many sources of system stress and we anticipate many factors can contribute to cancer evolution (Fig. 4). In addition, as each genome alteration can affect large numbers of genes, high frequencies of genome alteration during evolution will result in many potential molecular pathways rendering molecular characterization of cancer extremely challenging.

This relationship explains why such large numbers of cancer genes have been identified (and many more will be identified), which stands in stark contrast to the original prediction that only a handful of cancer genes

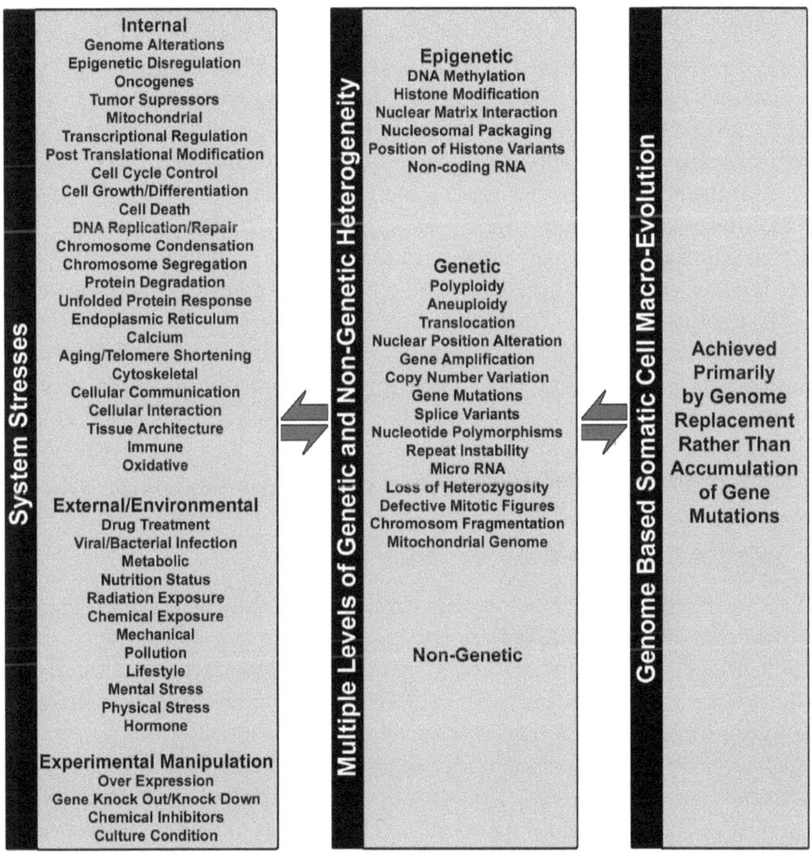

Fig. 4 System stress results in multiple levels of heterogeneity, which in turn drive somatic evolution. Various types of cellular stress include internal stresses, external stresses, and experimental manipulations. Response to these stresses results in heterogeneity of the cellular system at multiple levels including epigenetic, genetic and nongenetic. Due to the abundance of potential interactions present, the use of clinically relevant predictions based on an individual molecular pathway is extremely challenging (Heng *et al.*, 2009, 2010b).

are involved in all cancers (Weinberg, 1982). As cancer is a typical evolutionary process, where complexity and nonlinear progression is the key, any one of many molecular mechanisms could contribute to system instability ultimately promoting cancer evolution, though in reality, multiple mechanisms are often involved and contribute to form "the perfect storm." In theory, any gene that can impact on a cancer pathway could be linked to cancer under experimental conditions, and likely be "validated" among some patients. However, the vast majority of the gene mutations identified will not be shared. This combined with the large number of genes restricts the usefulness of individual molecular mechanisms within the clinic. Furthermore, the evolutionary mechanism of cancer has helped us to introduce the genome theory of somatic cell evolution, which defines genome-mediated evolution as macroevolution and gene, epigenetic regulation-mediated evolution as micro-evolution. For any given cancer formation, a number of cycles of macro/micro-evolution processes are required. By and large, cancer evolution is driven by macroevolution where genome replacement is the key. Of course, all gene mutations and epigenetic alterations contribute to this process, particularly those that can impact on system stability or increase the probability of evolution, can contribute to system dynamics and are a precondition for cancer evolution.

This conclusion needs to be carefully considered by the research community as it is not enough to just test different hypotheses under artificial conditions, but we need to develop a means of testing that has real medical implications. Clearly, moving away from gene oriented evolutionary research is the first crucial step.

A. Implications to Cancer Research

Previous cancer research has been limited to tracing individual molecular mechanisms, instead of focusing on the three key components of the evolutionary mechanism of cancer. Analyzing these factors will certainly provide new understanding and opportunities for research. For instance, the evolutionary mechanism of cancer solves the puzzle of why "cancer genes" can be demonstrated under experimental conditions yet they paradoxically seem to not be that important in clinical situations. Many different molecular mechanisms contribute to the evolutionary process of cancer by triggering genome-mediated evolution. Most cancer genes can be linked to cancer particularly under experimental conditions where one specific molecular mechanism is often manipulated into a supernatural dominant form. In genetically manipulated mouse models, all the cells of the animal share the same genetic defects, which is not the case in most sporadic cancers. Furthermore, in these models the manipulations artificially promote cancer evolution much more strongly than what occurs in nature. Most of the animal models drastically reduce genome heterogeneity, which helps investigators illustrate the importance of favored genes. However, in the clinical world, due to the low penetrance of these gene alleles or mutations within a patient population and the high degree of heterogeneous genetic and environmental effects behind most cancer cases, these well-studied gene mutations are only rarely evident in patients. In particular, due to the stochastic evolutionary process where the initial contributing factors might be very different from late stage genetic alterations, it is hard to predict the clinical outcomes based on the initial genetic alterations. As there are so many possible roads (mimicking the unlimited combinations of gene mutations) that can lead to cancer, each individual molecular mechanism becomes less clinically important.

This new reality demands new approaches. For example, if the degree of genetic heterogeneity is the driving force that changes the genetic landscape during cancer progression, genome chaos detected in cancer evolution is more important to somatic cell evolution compared to various gene mutations, as only genome-level alteration can change the global landscape, while gene- or epigenetic-level changes can reshape the local landscape. Microattractors are defined and constrained by the boundary of a genome or the domain of a system. At the global level, there are unlimited roads leading from peaks to the valleys, each of them representing a different network configuration or specific pathway. When an attraction status is formed, they form a CCA. At the local level, there are many roads leading from small hills to lower valleys each of them a different gene mutation or epigenetic alteration.

Similarly, many computational models need to be re-examined using the genome theory, as most of them have been established solely based on the gene mutation concept of cancer. Knowing that the key feature of cancer evolution is genome replacement-mediated macroevolution, the limitation of these models is obvious. There is now an urgent need to have new models that combine and simulate the two phases of cancer evolution. In addition, most models have different contributing factors that are separately monitored and which might not truly replicate the real dynamics of cancer evolution. Monitoring the dynamics at the genome level in contrast does not overly focus on one specific factor but focuses on the evolution of the overall system.

From an experimental point of view, the key to monitoring cellular population diversity or genome heterogeneity is measuring the frequencies of NCCAs. Given the fact that there are multiple levels of diversity including a nongenetic level, a gene level, an epigenetic level, a copy number level, and a genome level, it is necessary to study which level is most useful to reflect evolutionary potential. If all levels contribute significantly, there is a need to establish integrated models, which will be an extremely challenging task as the combination of the factors could reach an unmanageable number. If monitoring the higher-level results in more significant benefits, an effort is needed to make this transition. Clearly, the establishment of the genome theory of cancer evolution is an initial and crucial step for such a transition.

Another immediate implication is that there needs to be an adjustment in the expectation of the highly publicized cancer genome sequencing projects, for example the Cancer Genome Atlas (Collins and Barker, 2007; Stratton *et al.*, 2009). The genome is defined by the genome context rather than the gene content, and as the vast majority of cancers display altered genomes, there is therefore no common cancer genome. Thus, the Cancer Genome Atlas is very different from the original Human Genome Project that sequenced the normal human genome (Heng, 2007a). The Cancer Genome Atlas will sequence a large number of samples that are not really related to each other except their initial tissue origins. Following the stochastic evolutionary process, most of the solid tumors represent different systems where the same gene might not have the same biological function. Thus, it is possible that a certain pattern will be documented for patient populations after sequencing large numbers of cancer samples, but the clinical utility of these patterns is limited as the highly diverse gene mutation profiles have little predictive power for individual patients. Based on evolutionary analysis and the genome-centric concept, we have predicted that the expensive Cancer Genome Atlas project will fail to identify the expected commonly shared key genetic patterns among patients (Heng, 2007a). The publication of this analysis faced great resistance (taking over 2 years and reviews by a number of journals until initial data became available that were congruent with our analysis) as many reviewers believed that the commonly shared patterns would become evident following additional sequencing. Even today in the face of massive heterogeneous data, some still claim that the patterns have not been found yet and will be identified by just sequencing more samples. Given the increasing caution toward the clinical impact of the original Human Genome Project (Evans *et al.*, 2011), a realistic evaluation and revised expectation for the Cancer Genome Atlas project needs to be considered from the perspective of the genome theory of cancer evolution (Heng *et al.*, 2011).

B. Potential Clinical Implications

The evolutionary mechanism of cancer also offers many clinical implications. Based on the commonly accepted idea that sequential accumulation of gene mutations cause cancer, it would be ideal to detect early gene mutations even prior to tumor formation. New detection systems are being developed to detect single cells from billions of normal cells. However, knowing the evolutionary mechanisms of cancer and its related complexity (Coffey, 1998; Heng, 2008b), it might be problematic to push early detection to this extreme, as the existence of a limited number of abnormal cells only suggests the potential of becoming cancer and does not guarantee disease progression. Identification of limited numbers of precancer cells could cause stress to healthy individuals (such stress can indeed promote system instability), and some of the unnecessary diagnostics and treatments could paradoxically accelerate cancer evolution in addition to the problem of morbidity and mortality from such interventions (see discussion below). Additional studies are needed to establish the quantitative relationship between numbers and types of abnormal cells and their cancer potential. In the end, a more balanced approach is needed to address the issue of how early is too early. There are some encouraging ideas to apply biomarkers in cancer risk management by considering the evolutionary potential (Li *et al.*, 2011).

With regard to cancer prevention, it is obvious that reduced system stress is a key. According to our studies, an array of factors can contribute to system stress, triggering the evolution of cancer. These factors include errors of internal mechanisms including heritable genomic instability, defective genes, and varying metabolic types, and the influence of external/environmental stresses such as bacterial and viral infection, injury causing inflammation, chemical/radiation exposure, and drastically altering habitats (Greaves, 2007; Heng *et al.*, 2010b). Optimal healthy lifestyles should reduce system stress as well. One interesting topic would be to study the mechanism of chemoprevention and how treatment introduced stress might slow down or eliminate cancer progression. Questions such as "is there a specific change that follows chemoprevention in terms of the dynamics of cancer evolution?" are also important.

Regarding cancer treatment, more can be done to address the problems created by cancer evolution. The principle of cancer evolution has largely been ignored by those who also fail to realize that many of the popular approaches that are specifically designed to fight cancer actually facilitate cancer evolution. Thus, we need to change the mindset of the War on Cancer, which has been heavily influenced by success in treating bacterial infections, by not trying to hit cancer hard and attempt to destroy all cancer cells by any means. The fact that cancer and normal cells are closely related requires a different approach than fighting bacterial infections. Based on the relationship between NCCAs and CCAs, the evolutionary collaboration principle needs to be applied to influence the transition of NCCA/CCA dynamics and the speed of evolution. It is highly likely that promoting slow tumor growth with less diversity and physically removing slow growing tumors may be a more realistic strategy to manage cancer.

One important precaution is needed with regard to chemotherapy. It has been suggested that chemotherapy treatment can function as an evolutionary force to promote selection. For example, chemo-treatments can paradoxically speed up cancer evolution by selecting genetic variations (Maley *et al.*, 2004). Our recent data also illustrate that drug treatments can induce genome chaos leading to drastic new genomes formed by cellular macroevolution (Heng *et al.*, 2008, 2011). In this case, drug-mediated stress can speed up evolution by both selecting genetic variations and generating novel variations through genome chaos (a massive genome reorganization among chromosomes). If harsh chemotherapy eliminated all altered genomes while sparing normal genomes, this would be the method of choice. Unfortunately in the real world, the chaotic genome often finds a way to come back, as the ability to reorganize the genome system is so powerful. Regrettably, the treatment itself often promotes and accelerates cancer evolution by drastically and rapidly changing genome systems. A possible reason that this happens might be that the treatment removes system constraints on local adaptation. In particular, when treatment leads to genome alterations, it may lead to a new system being favored by cancer evolution. The clinical reality is that in many cancers with diverse and unstable genomes chemotherapy fails to prolong patient's lives despite effective initial tumor response supports our analysis (Heng, 2008b; Heng *et al.*, 2010b). It should be pointed out that standard chemotherapy works better in tumors with little diversity at the cytogenetic (genome) level. There seems to be a big difference between the common adult cancers with complex karyotypes and remarkable genome heterogeneity and the cancers in younger people with balanced translocations/gene fusions and little cytogenetic diversity. This latter category often responds to chemotherapy much better. However, among all cancers, these less heterogeneous types represent rarer tumors.

Systematic analysis is needed to re-evaluate the overall benefits of each specific treatment, using the standard of improving the quality of life or prolonging the overall survival rate, rather than focusing solely on the initial tumor response. In patient populations of the same cancer, evolutionary profiles are needed to predict the suitability of harsh treatment. For example, in patients with high genome instability, any additional stress from treatments might tip the balance and trigger genome chaos. Therefore, even a good treatment might be harmful to those who have unstable genomes. New methods to monitor overall genome instability are needed and extreme caution is needed with these patients.

Similarly, in the majority of patients, building a stronger system is as important as targeting the tumor directly. A stronger system (supporting the immunological system and homeostasis of the individual) can effectively constrain the dynamics of the lower-level system (Heng, 2007, 2008b). Using drastic methods to knock out cancer cells that are highly stressful to patients, from a genome theory point of view, is very harmful as it creates the opportunity for cancer to evolve. Clearly, we have to face these difficult questions and find the answers, as so many people's lives depend on the correct understanding of these strategies.

ACKNOWLEDGMENTS

This manuscript is part of a series of studies entitled, "The mechanisms of somatic cell and organismal evolution". We would like to thank Drs. Gloria Heppner; Gary Stein and Donald Coffey for their continuous support and interest in this project. Thanks also go to Drs. Markku Kurkinen and Leonard Lipovich for discussions. This work was partially supported by grants from SeeDNA Biotech Inc., the Susan G. Komen Breast Cancer Foundation and the DOD (GW093028).

REFERENCES

Albertson, D. G., Collins, C., McCormick, F., and Gray, J. W. (2003). Chromosome aberrations in solid tumors. *Nat. Genet.* **34**, 369–376.

Anderson, A. R., and Quaranta, V. (2008). Integrative mathematical oncology. *Nat. Rev. Cancer* 8, 227–234.

Anderson, A. R., Weaver, A. M., Cummings, P. T., and Quarant, V. (2006). Tumor morphology and phenotypic evolution driven by selective pressure from the microenvironment. *Cell* **127**, 905–915.

Ao, P. (2009). Global view of bionetwork dynamics: adaptive landscape. *J. Genet. Genomics* **36**, 63–73.

Atkin, N. B., and Baker, M. C. (1990). Are human cancers ever diploid–or often trisomic? Conflicting evidence from direct preparations and cultures. *Cytogenet. Cell Genet.* **53**, 58–60.

Axelrod, R., Axelrod, D. E., and Pienta, K. J. (2006). Evolution of cooperation among tumor cells. *Proc. Nat. Acad. Sci. U.S.A.* **103**, 13474–13479.

Bartos, J. D., Stoler, D. L., Matsui, S., Swede, H., Willmott, L. J., Sait, S. N., Petrelli, N. J., and Anderson, G. R. (2004). Genomic heterogeneity and instability in colorectal cancer: spectral karyotyping, glutathione transferase-Ml and ras. *Mutat Res.* **568**, 283–292.

Bayani, J., Selvarajah, S., Maire, G., Vukovic, B., Al-Romaih, K., Zielenska, M., and Squire, J. A. (2007). Genomic mechanisms and measurement of structural and numerical instability in cancer cells. *Semin. Cancer Biol.* **17**, 5–18.

Beatty, B., and Heng, H. H. Q. (2007). Gene mapping by fluorescence in situ hybridization. R. Meyer ed.), **Vol. 2**, pp. 713–748. Wiley-VCH, Weinheim.

Beerenwinkel, N., Antal, T., Dingli, D., Traulsen, A., Kinzler, K. W., Velculescu, V. E., Vogelstein, B., and Nowak, M. A. (2007). Genetic progression and the waiting time to cancer. *PLoS Comput. Biol.* **3**, e225.

Berger, M. F., Lawrence, M. S., Demichelis, F., Drier, Y., Cibulskis, K., Sivachenko, A. Y., Sboner, A., Esgueva, R., Pflueger, D., Sougnez, C., Onofrio, R., Carter, S. L., Park, K., Habegger, L., Ambrogio, L., Fennell, T., Parkin, M., Saksena, G., Voet, D., Ramos, A. H., Pugh, T. J., Wilkinson, J., Fisher, S., Winckler, W., Mahan, S., Ardlie, K., Baldwin, J., Simons, J. W., Kitabayashi, N., MacDonald, T. Y., Kantoff, P. W., Chin, L., Gabriel, S. B., Gerstein, M. B., Golub, T. R., Meyerson, M., Tewari, A., Lander, E. S., Getz, G., Rubin, M. A., and Garraway, L. A. (2011). The genomic complexity of primary human prostate cancer. *Nature* **470**, 214–220.

Bielas, J. H., Loeb, K. R., Rubin, B. P., True, L. D., and Loeb, L. A. (2006). Human cancers. express a mutator phenotype. *Proc. Nat. Acad. Sci. U.S.A.* **103**, 18238–18242.

Bodmer, W., Bielas, J. H., and Beckman, R. A. (2008). Genetic instability is not a requirement for tumor development. *Cancer Res* **68**, 3558–3560.

Brash, D., and Cairns, J. (2009). The mysterious steps in carcinogenesis. *Br. J. Cancer* **101**, 379–380.

Breivik, J. (2005). The evolutionary origin of genetic instability in cancer development. *Sem. Cancer Biol.* **15**, 51–60.

Brock, A., Chang, H., and Huang, S. (2009). Non-genetic heterogeneity—a mutation independent driving force for the somatic evolution of tumours. *Nat. Rev. Genet.* **10**, 336–342.

Cahill, D. P., Kinzler, K. W., Vogelstein, B., and Lengauer, C. (1999). Genetic instability. and Darwinian selection in tumours. *Trends Cell. Biol.* **9**, M57–M60.

Cairns, J. (1975). Mutation selection and the natural history of cancer. *Nature* **255**, 197–200.

Calin, G. A., Liu, C. G., Ferracin, M., Hyslop, T., Spizzo, R., Sevignani, C., Fabbri, M., Cimmino, A., Lee, E. J., Wojcik, S. E., Shimizu, M., Tili, E., Rossi, S., Taccioli, C., Pichiorri, F., Liu, X., Zupo, S., Herlea, V., Gramantieri, L., Lanza, G., Alder, H., Rassenti, L., Volinia, S., Schmittgen, T. D., Kipps, T. J., Negrini, M., and Croce, C. M. (2007). Ultraconserved regions encoding ncRNAs are altered in human leukemias and carcinomas. *Cancer Cell* **12**, 215–229.

Coffey, D. S. (1998). Self-organization, complexity and chaos: the new biology for medicine. *Nat. Med.* **4**, 882–885.

Collins, F. S., and Barker, A. D. (2007). Mapping the cancer genome. Pinpointing the genes involved in cancer will help chart a new course across the complex landscape of human malignancies. *Sci. Am.* **296**, 50–57.

Crespi, B., and Summers, K. (2005). Evolutionary biology of cancer. *Trends Ecol. Evol.* **20**, 545–552.

Dill, K. A., and Chan, H. S. (1997). From Levinthal to pathways to funnels. *Nat. Struct. Biol.* **4**, 10–19.

Duesberg, P. (2007). Chromosomal chaos and cancer. *Sci. Am.* **296**, 52–59.

Duesberg, P., and Rasnick, D. (2000). Aneuploidy, the somatic mutation that makes cancer a species of its own. *Cell Motil. Cytoskeleton* **47**, 81–107.

Esteller, M. (2008). Epigenetics in cancer. *N Engl. J. Med.* **358**, 1148–1159.

Evans, J. P., Meslin, E. M., Marteau, T. M., and Caulfield, T. (2011). Deflating the genomic bubble. *Science.* **331**, 861–862.

Feinberg, A. P., and Irizarry, R. A. (2010). Evolution in health and medicine Sackler colloquium: stochastic epigenetic variation as a driving force of development, evolutionary adaptation, and disease. *Proc. Nat. Acad. Sci. U.S.A. 107, Suppl* **1**, 1757–1764.

Feinberg, A. P., Ohlsson, R., and Henikoff, S. (2006). The epigenetic progenitor origin of human cancer. *Nat Rev Genet.* **7**, 21–33.

Ferguson-Smith, M. A., and Trifonov, V. (2007). Mammalian karyotype evolution. *Nat. Rev. Genet.* **8**, 950–962.

Foo, J., Leder, K., and Michor, F. (2011). Stochastic dynamics of cancer initiation. *Phys. Biol.* **8**, 015002.

Foulds, L. (1954). The experimental study of tumor progression: a review. *Can. Res.* **14**, 327–399.

Futuyma, D. J. (2010). Evolutionary constraint and ecological consequences. *Evolution* **64**, 1865–1884.

Gatenby, R. A. (1996). Application of competition theory to tumour growth: implications for tumour biology and treatment. *Eur. J. Cancer* **32A**, 722–726.

Gatenby, R. A. (2009). A change of strategy in the war on cancer. *Nature* **459**, 508–509.

Gatenby, R. A., Gillies, R. J., and Brown, J. S. (2010). Evolutionary dynamics of cancer prevention. *Nat. Rev. Cancer* **10**, 526–527.

Gisselsson, D. (2003). Chromosome instability in cancer: how, when, and why? *Adv. Cancer Res.* **87**, 1–29.

Gisselsson, D., Pettersso, L., Höglund, M., Heidenblad, M., Gorunov, a. L., Wiegant, J., Mertens, F., Dal Cin, P., Mitelman, F., and Mandahl, N. (2000). Chromosomal breakage-fusion-bridge events cause genetic intratumor heterogeneity. *Proc. Nat. Acad. Sci. U.S.A.* **97**, 5357–5362.

Glansdorff, N., Xu, Y., and Labedan, B. (2009). The conflict between horizontal gene transfer and the safeguard of identity: origin of meiotic sexuality. *J Mol Evol.* **69**, 470–480.

Glazier, A. (2006). Cure: Scientific, Social and Organizational Requirements for the Specific Cure of Cancer. Newton. MA.

Gorelick, R., and Carpinone, J. (2009). Origin and maintenance of sex: the evolutionary joys of self sex. *Biol. J. Linnean Soc.* **98**, 707–728.

Gorelick, R., and Heng, H. H. (2011). Sex reduces genetic variation: a multidisciplinary review. *Evolution* **65**, 1088–1098.

Gorelick, R., and Laubichler, M. (2008). Genetic = Heritable (Genetic ≠ DNA). *Biol. Theory* **3**, 79–84.

Gould, S. J., and Eldredge, N. (1993). Punctuated equilibrium comes of age. *Nature* **366**, 223–227.

Greaves, M. (2007). Darwinian medicine: a case for cancer. *Nat. Rev. Cancer* **7**, 213–221.

Hanahan, D., and Weinberg, R. A. (2000). The hallmarks of cancer. *Cell* **100**, 57–70.

Heng, H. H. (2007a). Cancer genome sequencing: the challenges ahead. *BioEssays* **29**, 783–794.

Heng, H. H. (2007b). Elimination of altered karyotypes by sexual reproduction preserves species identity. *Genome* **50**, 517–524.

Heng, H. H. (2008a). The gene-centric concept: a new liability? *BioEssays* **30**, 196–197.

Heng, H. H. (2008b). The conflict between complex system and reductionism. *J. Am. Med. Assoc* **300**, 1580–1581.

Heng, H. H. (2009). The genome-centric concept: re-synthesis of evolutionary theory. *BioEssays* **31**, 512–525.

Heng, H. H. (2010). Missing heritability and stochastic genome alterations. *Nat. Rev. Genet.* **11**, 813.

Heng, H. H., Bremer, S. W., Stevens, J., Ye, K. J., Miller, F., Liu, G., and Ye, C. J. (2006a). Cancer progression by non-clonal chromosome aberrations. *J. Cell Biochem* **98**, 1424–1435.

Heng, H. H., Bremer, S. W., Stevens, J. B., Ye, K. J., Liu, G., and Ye, C. J. (2009). Genetic and epigenetic heterogeneity in cancer: agenome-centric perspective. *J. Cell. Physiol.* **220**, 538–547.

Heng, H. H., Liu, G., Bremer, S., Ye, K. J., Stevens, J., and Ye, C. J. (2006b). Clonal and nonclonal chromosome aberrations and genome variation and aberration. *Genome* **49**, 195–204.

Heng, H. H., Liu, G., Lu, W., Bremer, S., Ye, C. J., Hughes, M., and Moens, P. (2001). Spectral karyotyping (SKY) of mouse meiotic chromosomes. *Genome* **4**, 293–298.

Heng, H. H., Liu, G., Stevens, J. B., Bremer, S. W., Ye, K. J., Abdallah, B. Y., Horne, S. D., and Ye, C. J. (2011). Decoding the genome beyond sequencing: the next phase of genomic research. *Genomics.* doi:101016/j.ygeno.2011.05.008.

Heng, H. H., Liu, G., Stevens, J. B., Bremer, S. W., Ye, K. J., and Ye, C. J. (2010a). Genetic and epigenetic heterogeneity in cancer: the ultimate challenge for drug therapy. *Curr. Drug Targets* **11**, 1304–1316.

Heng, H. H., Spyropoulos, B., and Moens, P. B. (1997). FISH technology in chromosome and genome research. *BioEssays* **19**, 75–84.

Heng, H. H., Stevens, J. B., Bremer, S. W., Ye, K. J., Liu, G., and Ye, C. J. (2010b). The evolutionary mechanism of cancer. *J. Cell Biochem* **220**, 538–547.

Heng, H. H., Stevens, J. B., Lawrenson, L., Liu, G., Ye, K. J., Bremer, S. W., and Ye, C. J. (2008). Patterns of genome dynamics and cancer evolution. *Cell Oncol.* **30**, 513–514.

Heng, H. H., Stevens, J. B., Liu, G., Bremer, S. W., and Ye, C. J. (2004). Imaging genome. abnormalities in cancer research. *Cell Chromosome* **3**, 1.

Heng, H. H., Stevens, J. B., Liu, G., Bremer, S. W., Ye, K. J., Reddy, P. V., Wu, G. S., Wang, Y. A., Tainsky, M. A., and Ye, C. J. (2006c). Stochastic cancer progression driven by nonclonal chromosome aberrations. *J. Cell Physiol* **208**, 461–472.

Heng, H. H., Ye, C. J., Yang, F., Ebrahim, S., Liu, G., Bremer, S. W., Thomas, C. M., Ye, J., Chen, T. J., Tuck-Muller, C., Yu, J. W., Krawetz, S. A., and Johnson, A. (2003). Analysis of marker or complex chromosomal rearrangements present in pre- and post-natal karyotypes utilizing a combination of G-banding, spectral karyotyping and fluorescence in situ hybridization. *Clin. Genet.* **6**, 358–367.

Heppner, G. H. (1984). Tumor heterogeneity. *Cancer Res.* **44**, 2259–2265.

Heppner, G. H., and Miller, F. R. (1998). The cellular basis of tumor progression. *Int. Rev. Cytol.* **177**, 1–56.

Höglund, M., Frigyesi, A., Säll, T., Gisselsson, D., and Mitelman, F. (2005). Statistical behavior of complex cancer karyotypes. *Genes Chromosomes Cancer.* **42**, 327–341.

Huang, S., Ernberg, I., and Kauffman, S. (2009). Cancer attractors: a systems view of tumors from a gene network dynamics and developmental perspective. *Semin. Cell Dev. Biol.* **20**, 869–876.

Jones, P. A., and Baylin, S. B. (2007). The epigenomics of cancer. *Cell.* **128**, 683–692.

Jones, S., Chen, W. D., Parmigiani, G., Diehl, F., Beerenwinkel, N., Antal, T., Traulsen, A., Nowak, M. A., Siegel, C., Velculescu, V. E., Kinzler, K. W., Vogelstein, B., Willis, J., and Markowitz, S. D. (2008a). Comparative lesion sequencing provides insights into tumor evolution. *Proc. Natl Acad. Sci. U.S.A.* **105**, 4283–4288.

Jones, S., Zhang, X., Parsons, D. W., Lin, J. C., Leary, R. J., Angenendt, P., Mankoo, P., Carter, H., Kamiyama, H., Jimeno, A., Hong, S. M., Fu, B., Lin, M. T., Calhoun, E. S., Kamiyama, M., Walter, K., Nikolskaya, T., Nikolsky, Y., Hartigan, J., Smith, D. R., Hidalgo, M., Leach, S. D., Klein, A. P., Jaffee, E. M., Goggins, M., Maitr, a. A., Iacobuzio-Donahue, C., Eshleman, J. R., Kern, S. E., Hruban, R. H., Karchin, R., Papadopoulos, N., Parmigiani, G., Vogelstein, B., Velculescu, V. E., and Kinzler, K. W. (2008b). Core. signaling pathways in human pancreatic cancers revealed by global genomic analyses. *Science* **321**, 1801–1806.

King, M. (1993). Species evolution. The role of chromosome change. Cambridge University Press, Cambridge, UK.

Kitada, K., Taima, A., Ogasawara, K., Metsugi, S., and Aikawa, S. (2011). Chromosome-specific segmentation revealed by structural analysis of individually isolated chromosomes. *Genes Chromosomes Cancer* 50, 217–227.

Klein, C. A., Blankenstein, T. J., Schmidt-Kittler, O., Petronio, M., Polzer, B., Stoecklein, N. H., and Riethmüller, G. (2002). Genetic heterogeneity of single disseminated tumour cells in minimal residual cancer. *Lancet.* **360**, 683–689.

Kohn, M., Hogel, J., Vogel, W., Minich, P., Kehrer-Sawatzki, H., Graves, J. A., and Hameister, H. (2006). Reconstruction of a 450-My-old ancestral vertebrate protokaryotype. *Trends Genet.* **22**, 203–210.

Li, X., Blount, P. L., Vaughan, T. L., and Reid, B. J. (2011). Application of biomarkers in cancer risk management: evaluation from stochastic clonal evolutionary and dynamic system optimization points of view. *PLoS. Comput. Biol.* **7**, e1001087.

Loeb, L. A., Bielas, J. H., and Beckman, R. A. (2008). Cancers exhibit a mutator phenotype: clinical implications. *Cancer Res.* **68**, 3551–3557.

Maley, C. C., Galipeau, P. C., Finley, J. C., Wongsurawat, V. J., Li, X., Sanchez, C. A., Paulson, T. G., Blount, P. L., Risques, R. A., Rabinovitch, P. S., and Reid, B. J. (2006). Genetic clonal diversity predicts progression to esophageal adenocarcinoma. *Nat. Genet.* **38**, 468–473.

Maley, C. C., Reid, B. J., and Forrest, S. (2004). Cancer prevention strategies that address the evolutionary dynamics of neoplastic cells: simulating benign cell boosters and selection for chemosensitivity. *Cancer Epidemiol. Biomarkers Prev.* **13**, 1375–1384.

Merlo, L. M., Pepper, J. W., Reid, B. J., and Maley, C. C. (2006). Cancer as an evolutionary and ecological process. *Nat. Rev. Cancer* **6**, 924–935.

Meyerson, M., and Pellman, D. (2011). Cancer genomes evolve by pulverizing single chromosomes. *Cell* **144**, 9–10.

Mitelman, F. (2000). Recurrent chromosome aberrations in cancer. *Mutat Res.* **462**, 247–453.

Mitelman, F. (2006). 50,000 tumors, 40,000 aberrations, and 300 fusion genes: how much remains? 50 years of 46 chromosomes: progress in cytogenetics. *NCI., NIH. USA.*

Murchison, E. P. (2008). Clonally transmissible cancers in dogs and Tasmanian devils. *Oncogene* **2**, S19–30.

Murphy, W. J., Larkin, D. M., Everts-van der Wind, A., Bourque, G., Tesler, G., Auvil, L., Beever, J. E., Chowdhary, B. P., Galibert, F., Gatzke, L., Hitte, C., Meyers, S. N., Milan, D., Ostrander, E. A., Pape, G., Parker, H. G., Raudsepp, T., Rogatcheva, M. B., Schook, L. B., Skow, L. C., Welge, M., Womack, J. E., O'brien, S. J., Pevzner, P. A., and Lewin, H. A. (2005). Dynamics of mammalian chromosome evolution inferred from multispecies comparative maps. *Science* **309**, 613–617.

Navin, N., Kendall, J., Troge, J., Andrews, P., Rodgers, L., McIndoo, J., Cook, K., Stepansky, A., Levy, D., Esposito, D., Muthuswamy, L., Krasnitz, A., McCombie, W. R., Hicks, J., and Wigler, M. (2011). Tumour evolution inferred by single-cell sequencing. *Nature* **472**, 90–94.

Nowak, M. A., Michor, F., and Iwasa, Y. (2006). Genetic instability and clonal expansion. *J. Theor. Biol.* **241**, 26–32.

Nowell, P. C. (1976). The clonal evolution of tumor cell populations. *Science* **194**, 23–28.

Parsons, D. W., Jones, S., Zhang, X., Lin, J. C., Leary, R. J., Angenend, P., Mankoo, P., Carter, H., Siu, I. M., Gallia, G. L., Olivi, A., McLendon, R., Rasheed, B. A., Keir, S., Nikolskaya, T., Nikolsky, Y., Busam, D. A., Tekleab, H., Diaz Jr., L. A., Hartigan, J., Smith, D. R., Strausberg, R. L., Marie, S. K., Shinjo, S. M., Yan, H., Riggins, G. J., Bigner, D. D., Karchin, R., Papadopoulos, N., Parmigiani, G., Vogelstein, B., Velculescu, V. E., and Kinzler, K. W. (2008). An integrated genomic analysis of human glioblastoma multiforme. *Science* **321**, 1807–1812.

Pavelka, N., Rancati, G., Zhu, J., Bradford, W. D., Saraf, A., Florens, L., Sanderson, B. W., Hattem, G. L., and Li, R. (2010). Aneuploidy confers quantitative proteome changes and phenotypic variation in budding yeast. *Nature* **468**, 321–325.

Prindle, M. J., Fox, E. J., and Loeb, L. A. (2010). The mutator phenotype in cancer: molecular mechanisms and targeting strategies. *Curr. Drug Targets* **11**, 1296–1303.

Rowley, J. D. (1988). The critical role of chromosome translocations inhuman leukemias. *Annu. Rev. Genet.* **32**, 495–519.

Schröck, E., du Manoir, S., Veldman, T., Schoell, B., Wienberg, J., Ferguson-Smith, M. A., Ning, Y., Ledbetter, D. H., Bar-Am, I., Soenksen, D., Garini, Y., and Ried, T. (1996). *Science* **273**, 494–497.

Schwab, E. D., and Pienta, K. J. (1996). Cancer as a complex adaptive system. *Med. Hypotheses* **47**, 235–241.

Shibata, D., and Tavaré, S. (2006). Counting divisions in a human somatic cell tree: how, what and why? *Cell Cycle* **5**, 610–614.

Spencer, S. L., Gerety, R. A., Pienta, K. J., and Forrest, S. (2006). Modeling somatic evolution in tumorigenesis. *PLoS. Comput. Biol.* **2**, e108.

Stephens, P. J., Greenman, C. D., Fu, B., Yang, F., Bignell, G. R., Mudie, L. J., Pleasance, E. D., Lau, K. W., Beare, D., Stebbings, L. A., McLaren, S., Lin, M. L., McBride, D. J., Varela, I., Nik-Zainal, S., Leroy, C., Jia, M., Menzies, A., Butler, A. P., Teague, J. W., Quail, M. A., Burton, J., Swerdlow, H., Carter, N. P., Morsberger, L. A., Iacobuzio-Donahue, C., Follows, G. A., Green, A. R., Flanagan, A. M., Stratton, M. R., Futreal, P. A., and Campbell, P. J. (2011). Massive genomic rearrangement acquired in a single catastrophic event during cancer development. *Cell* **144**, 27–40.

Stevens, J. B., Abdallah, B. Y., Regan, S. M., Liu, G., Bremer, S. W., Ye, C. J., and Heng, H. H. (2010). Comparison of mitotic cell death by chromosome fragmentation to premature chromosome condensation. *Mol. Cytogenet.* **3**, 20.

Stevens, J. B., Liu, G., Bremer, S. W., Ye, K. J., Xu, W., Xu, J., Sun, Y., Wu, G. S., Savasan, S., Krawetz, S. A., Ye, C. J., and Heng, H. H. (2007). Mitotic cell death by chromosome fragmentation. *Cancer Res.* **67**, 7686–7694.

Stevens, J. B., Abdallah, B. Y., Liu, G., Ye, C. J., Horne, S. D., Wang, G., Savasan, S., Shekhar, M., Krawetz, S. A., Hüttemann, M., Tainsky, M. A., Wu, G. S., Xie, Y., Zhang, K., and Heng, H. H. (2011). Diverse system stresses: common mechanisms of chromosome fragmentation. *Cell Death Disease* **2**, e178.

Stoecklein, N. H., and Klein, C. A. (2010). Genetic disparity between primary tumours, disseminated tumour cells, and manifest metastasis. *Int. J. Cancer* **126**, 589–598.

Stratton, M. R., Campbell, P. J., and Futreal, P. A. (2009). The cancer genome. *Nature.* **458**, 719–724.

Tarafa, G., Tuck, D., Ladner, D., Topazian, M., Brand, R., Deters, C., Moreno, V., Capella, G., Lynch, H., Lizardi, P., and Costa, J. (2008). Mutational load distribution analysis yields metrics reflecting genetic instability during pancreatic carcinogenesis. *Proc. Nat. Acad. Sci. U.S.A.* **105**, 4306–4311.

Tsao, J. L., Yatabe, Y., Salovaara, R., Järvinen, H. J., Mecklin, J. P., Aaltonen, L. A., Tavaré, S., and Shibata, D. (2000). Genetic reconstruction of individual colorectal tumor. histories. *Proc. Nat. Acad. Sci. U.S.A.* **9**, 1236–1241.

Van Valen, L. M., and Maiorana, V. C. (1991). HeLa, a new microbial species. *Evolutionary Theory* **10**, 71–74.

Vincent, M. D. (2010). The animal within: carcinogenesis and the clonal evolution of cancer cells are speciation events sensu stricto. *Evolution* **64**, 1173–1183.

Vincent, T. L., and Brown, J. S. (2005). Evolutionary Game Theory, Natural Selection, and Darwinian Dynamics. Cambridge University Press, Cambridge, UK.

Vincent, T. L., and Gatenby, R. A. (2008). An evolutionary model for initiation, promotion, and progression in carcinogenesis. *Int. J .Oncol.* **32**, 729–737.

Vogelstein, B., and Kinzler, K. W. (2004). Cancer genes and the pathways they control. *Nat. Med.* **10**, 789–799.

Waddington, C. H. (1957). The Strategy of the Genes: A Discussion of Some Aspect of Theoretical Biology. MacMillan Company, New York.

Weinberg, R. A. (1982). Fewer and fewer oncogenes. *Cell* **30**, 3–4 Review.

Welsh, J. S. (2011). Contagious cancer. *Oncologist* **16**, 1–4.

White, M. J. D. (1978). Modes of Speciation. W.H. Freman and Co., San Franciisco. .

Wilkins, A. S., and Holliday, R. (2009). The evolution of meiosis from mitosis. *Genetics* **181**, 3–12.

Wood, L. D., Parsons, D. W., Jones, S., Lin, J., Sjöblom, T., Leary, R. J., Shen, D., Boca, S. M., Barber, T., Ptak, J., Silliman, N., Szabo, S., Dezso, Z., Ustyanksky, V., Nikolskaya, T., Nikolsky, Y., Karchin, R., Wilson, P. A., Kaminker, J. S., Zhang, Z., Croshaw, R., Willis, J., Dawson, D., Shipitsin, M., Willson, J. K., Sukumar, S., Polyak, K., Park, B. H., Pethiyagod, C. L., Pant, P. V., Ballinger, D. G., Sparks, A. B., Hartigan, J., Smith, D. R., Suh, E., Papadopoulos, N., Buckhaults, P., Markowitz, S. D., Parmigiani, G., Kinzler, K. W., Velculescu, V. E., and Vogelstein, B. (2007). The genomic landscapes of human breast and colorectal cancers. *Science* **318**, 1108–1113.

Wright, S. (1988). Surfaces of selective value revisited. *Am. Nat.* **131**, 115–123.

Ye, C. J., Lawrenson, L., Liu, G., Stevens, J. B., Ye, K. J., Bremer, S. W., and Heng, H. H. (2009a). Chapter 19: Simultaneous fluorescence immunostaining and FISH. *In* "Fluorescence In situ Hybridization (FISH)—Application" (T. Liehr ed.), pp. 193–216. Springer, Berlin.

Ye, C. J., Liu, G., Bremer, S. W., and Heng, H. H. (2007). The dynamics of cancer chromosomes and genomes. *Cytogenet. Genome Res.* **118**, 237–246.

Ye, C. J., Lu, W., Liu, G., Bremer, S. W., Wang, Y. A., Moens, P., Hughes, M., Krawetz, S. A., and Heng, H. H. (2001). The combination of SKY and specific loci detection with FISH or immunostaining. *Cytogenet. Cell Genet.* **93**, 195–202.

Ye, C. J., Stevens, J. B., Liu, G., Bremer, S. W., Jaiswal, A. S., Ye, K. J., Lin, M. F., Lawrenson, L., Lancaster, W. D., Kurkinen, M., Liao, J. D., Gairola, C. G., Shekhar, M. P., Narayan, S., Miller, F. R., and Heng, H. H. (2009b). Genome based cell population heterogeneity promotes tumorigenicity: the evolutionary mechanism of cancer. *J. Cell Physiol.* **219**, 288–300.

Ye, C. J., Stevens, J. B., Liu, G., Ye, K. J., Yang, F., Bremer, S. W., and Heng, H. H. (2006). Combined multicolor-FISH and immunostaining. *Cytogenet. Genome Res.* **114**, 227–234.

Zhu, X. M., Yin, L., Hood, L., and Ao, P. (2004). Robustness, stability and efficiency of phage lambda genetic switch: dynamical structure analysis. *J. Bioinf. Comput. Biol.* **2**, 785–817.

Cancer Stem Cells in Tumor Heterogeneity

Alexander Pietras

Center for Molecular Pathology, CREATE Health, Department of Laboratory Medicine, Lund University, Skåne University Hospital, Malmö, Sweden

Cancer cells within a given tumor were long regarded as a largely homogeneous group of cells originating from a common progenitor cell. However, it is increasingly appreciated that there is a considerable heterogeneity within tumors also on the tumor cell level. This heterogeneity extends to virtually all measurable properties of cancer cells, ranging from differentiation state, proliferation rate, migratory and invasive capacity to size, and therapeutic response. Such heterogeneity likely represents a major therapeutic hurdle, but the mechanisms underlying its emergence remain poorly understood and a controversial topic. The cancer stem cell model of tumor progression has gained increasing support during the past several years. In this review, I will discuss some major implications of the cancer stem cell hypothesis on the origins of tumor heterogeneity, focusing both on heterogeneity within the tumor cells proper and on potential transdifferentiation of cancer stem cells into stromal and endothelial lineages, as well as on heterogeneity of the therapeutic response. Evidence for and against a direct and causal role of cancer stem cells in the emergence of tumor heterogeneity will be weighed and alternative explanations for apparently contradictory observations discussed. Finally, I will discuss the potential origins of cancer stem cells and the various implications of origin to the contribution to tumor heterogeneity, and outline some future directions. © 2011 Elsevier Inc.

Advances in CANCER RESEARCH, Volume 112
Copyright 2011, Elsevier Inc. All right reserved.
0065-230X/10 $35.00
DOI: 10.1016/B978-0-12-387688-1.00009-0

I. INTRODUCTION

Cancers are generally characterized by a remarkable phenotypical and functional inter- and intratumoral heterogeneity. Intratumoral heterogeneity results in part from the co-existence of tumor cells and a large number of varying stromal cells within tumors, but importantly, even within the tumor cell compartment there are plentiful diverse geno- and phenotypes. Such heterogeneity is likely a major therapeutic hurdle, yet the mechanisms underlying the development of tumor heterogeneity remain poorly understood. In this review, I will discuss the potential implications of the cancer stem cell hypothesis for the emergence of intratumoral tumor cell heterogeneity. Even though intratumoral heterogeneity is likely evident in virtually all measurable properties of tumor cells, like proliferation rate, invasiveness, migratory potential, or therapeutic responses (Heppner, 1984), a lot of recent attention has been given heterogeneity with respect to tumor cell differentiation status (Pietras et al., 2008; Pietras, Johnsson, and Påhlman, 2010; Shipitsin *et al.*, 2007).

Although there is little doubt that clonal evolution of tumor cells within any tumor inevitably occurs and drives development of distinct tumor cell subclones, the extent to which such clonal evolution within any cell type contributes to the massive heterogeneity within any individual cancer remains unclear. A major controversy in contemporary cancer research relates to how tumors are thought to be maintained and progress to metastatic disease. The classical Darwinistic view that cancers advance through a process of clonal evolution has recently been challenged by the increasingly supported cancer stem cell hypothesis. The cancer stem cell hypothesis, if true, could have major implications for how we think of the origins of cellular heterogeneity in tumors and the role of clonal evolution.

II. A BRIEF INTRODUCTION TO MODELS OF TUMOR DEVELOPMENT

A. The Clonal Evolution Model

As first proposed by Nowell (1976), the clonal evolution model for tumor progression suggests that within a population of tumor cells, a natural selection occurs that favors cells that have acquired (e.g., through additional mutations) the most aggressive phenotype. The model suggests that cancers arise when a sufficient number of mutations have occurred in any given cell. Genetic instability within the tumor cell pool then inevitably leads to the accumulation of additional mutations within single cells;

mutations that confer a growth advantage (e.g., by making cells less sensitive to cell death cues or more prone to proliferate) to the cell will survive as the cells give rise to new offspring. Thus, heterogeneity within tumors according to the clonal evolution hypothesis first and foremost arises from different genetic hits affecting different cells that originate from the same initial cell of origin. Importantly, it is also stipulated that such evolution within the tumor is what eventually gives rise to tumor cells with the capacities of invasion into neighboring tissues and metastasis. Furthermore, the same clonal evolution is believed to cause resistance to anti-cancer drugs and irradiation.

While direct evidence for any model of tumor progression is difficult to come by, the clonal evolution hypothesis is compatible with several characteristics of cancer including phenotypic heterogeneity (Marusyk and Polyak, 2009), genetic instability (Hanahan and Weinberg, 2000), and the frequent emergence of chemotherapy-resistant clones after treatment (Shah and Sawyers, 2003). The clonal evolution model further rests on the widespread notion that primary tumors and metastases frequently share the vast majority of genetic aberrations, but that metastases (that are thought to arise from cells generated by clonal evolution within the primary tumor) have acquired additional hits not present in the primary tumor. However, only few studies have addressed this matter by studying genetic aberrations in matched pairs of primary tumor and metastases. Intriguingly, although many reports at least partially support the prediction of the clonal evolution theory (Fujii *et al.*, 1996; Jones *et al.*, 2008; Kuukasjarvi *et al.*, 1997), several of the studies performed to date report remarkably diverging patterns of genetic alterations in primary tumors as compared to metastases from the same patient (Albanese *et al.*, 2004; Artale *et al.*, 2008; Kalikaki *et al.*, 2008; Katona *et al.*, 2007). These data open for the possibility that in some tumors, metastases may arise from early disseminated tumor cells, or conceivably, a genetically more stable founder cell common to primary tumor and metastasis, rather than cells that have acquired a number of additional genetic aberrations by clonal evolution within the primary tumor.

Essentially, the clonal evolution hypothesis assumes that all cells within a tumor hold an equal potential to maintain and advance the tumor to metastasis, pending the acquirement of the necessary capacities.

B. The Cancer Stem Cell Model

Although Julius Cohnheim first put an early version of the cancer stem cell hypothesis forward as early as 1875 (Cohnheim, 1875), it is not until recently that it has gained wide support in the cancer research community. Cohnheim suggested that cancers arise from "embryonic rests," remnants from development that were transformed into cancer, based on similarities between certain cancer cells and embryonic cells. Increased understanding of developmental and stem cell biology as well as technological advances has enabled researchers to investigate these ideas more thoroughly.

Unlike the clonal evolution hypothesis, the cancer stem cell model assumes that cancers are maintained and propagated by a presumably (but not necessarily) small subpopulation of cells within the tumor that have or have acquired properties of stem cells. Like most organs are believed to have a pool of organ-specific stem cells that have the capacity to self-renew as well as give rise to the various cell types within their respective tissue (Kim *et al.*, 2005; Osawa *et al.*, 1996; Reynolds and Weiss, 1996), so would cancer stem cells give rise to the multitude of cancer cell types within a tumor. Cancer stem cells themselves would have a near-unlimited proliferation capacity, whereas the more differentiated tumor cells it gives rise to would hold only a limited potential for proliferation. Thus, seeding of metastases and maintenance of tumors would primarily and in the long run depend on the stem cell population. While many believe that organ-specific stem cells may indeed be the cells of origin for most cancers, the cancer stem cell hypothesis itself does not stipulate how stem-like properties were acquired. The properties of cancer stem cells may equally occur through transformation of normal stem cells as dedifferentiation of more mature transformed cells (Heddleston *et al.*, 2009; Jögi *et al.*, 2002; Mani *et al.*, 2008). Partly because of this potential confusion, cancer stem cells are frequently referred to as tumor-initiating cells (TICs), a term strictly based on the experimental evidence backing the hypothesis (see below). The cancer stem cell model suggests that metastatic spread occurs through dissemination of cancer stem cells, and that resistance to chemo- and radiotherapy is likely due to inherent resistance in the cancer stem cell pool rather than induced selection by clonal evolution.

Early experimental support for the cancer stem cell model comes from studies performed in the 1960s that would be (and should be) impossible to conduct today due to ethical considerations (Southam and Brunschwig, 1961). Transplanting human cancer cells back to patients subcutaneously post-surgery allowed Southam and Brunschwig to conclude that at least 1,000,000 injected tumor cells were required to seed a new tumor, suggesting that not all tumor cells had an equal capacity to induce tumor growth. The first successful isolation of a population of cells from a human cancer that appeared to have an exclusive ability to form new tumors was performed in John Dick's laboratory in 1994 (Lapidot *et al.*, 1994). By transplanting cells from human acute myeloid leukemias into SCID mice, they found that the only cells capable of engrafting a new leukemia were cells that were isolated as positive for the cell surface marker CD34, but negative for CD38 (CD34 + /CD38–). Xenotrasplantation of CD34 + /CD38+ or CD34– cells rendered no tumor growth in mice. Similar studies in a wide variety of leukemias and solid tumors (Al-Hajj *et al.*, 2003; Eramo *et al.*, 2008; Hermann *et al.*, 2007; O'Brien *et al.*, 2007; Patrawala *et al.*, 2006; Ricci-Vitiani *et al.*, 2007; Schatton *et al.*, 2008; Singh *et al.*, 2004; Taylor *et al.*, 2005) have followed to identify cell populations with an increased tumor-forming ability as compared to other populations from the same tumor. The interpretation of these and similar results has recently been somewhat complicated by the finding that the tumor formation efficiency of different cell populations can be greatly affected by the host animal, most notably by the level of immunodeficiency in recipient mice (Quintana *et al.*, 2008). Despite this insight, however, there are clearly qualitative differences in the tumor-forming ability of cancer stem cells versus nonstem-like cells in the most commonly used mouse models. Whether or not these differences represent an actual and important biological issue, however, remains an open question.

C. Cancer Stem Cells versus Cell of Origin: An Important Distinction

Some controversial issues surrounding the cancer stem cell hypothesis relates to the origin of cancer stem cells, and the origin of cancer stem cells versus the cell of origin of the tumor as a whole. Many of these discussions boil down to more or less entirely semantic issues instigated by the use of the term "stem cells' in referring to "cancer stem cells," and thus several other terms have been suggested to overcome these issues such as TICs, tumor-propagating cells, and cancer stem-like cells. Nevertheless, the issue of origin of the cancer stem cells is an interesting one and one of potential importance for the functional and practical implications of the cancer stem cell hypothesis for cancer biology and ultimately cancer therapeutics.

The first issue: does the cancer stem cell hypothesis imply that cancer stem cells are derived from normal tissue stem cells? Does the cancer stem cell hypothesis thereby imply that normal tissue stem cells act as cells of origin for most cancers? While normal tissue stem cells have been widely suggested as the most likely cells of origin for many cancers (Gerdes and Yuspa, 2005; Holland, 2001; Hubbard and Gargett, 2010; Sell, 2010; Shupe and Petersen, 2005; Visvader, 2011; Waters, Newman, and Levy, 2010), there is nothing in the cancer stem cell hypothesis that requires that cancer stem cells themselves are indeed the original tumor cell of the cancer. Instead, it is equally likely and equally compatible with the cancer stem cell model that cancers arise in a more differentiated cell and that cancer stem cells occur later in tumor progression. It is likely, however, that normal tissue stem cells require fewer alterations or hits in order to acquire cancer stem cell properties, due to their inherent stem cell abilities.

One likely origin of cancer stem cells in tumors arising from more differentiated cells is simply dedifferentiation of relatively differentiated tumor cells towards a stem cell-like state. Such dedifferentiation, although largely unheard of or at least still controversial during normal tissue development, has been demonstrated to occur in several tumor models and to be triggered by a number of extrinsic environmental stimuli such as tumor hypoxia (Helczynska *et al.*, 2003; Jögi *et al.*, 2002), or occur as a natural consequence of tumor progression (Delahunt, 1999), long before the cancer stem cell model regained mainstream interest in the research community.

In 2008, Weinberg and colleagues investigated the relationship between normal and neoplastic stem cells and cells that have undergone the process of epithelial-to-mesenchymal transition (for a general review on EMT in cancer, see Guarino, Rubino, and Ballabio, 2007) in a much-discussed paper (Mani *et al.*, 2008) that initiated a possible bridge between the cancer stem cell and stochastic clonal evolution models of tumor progression. Importantly, this study clearly makes the point that cells with properties of cancer stem cells need not be derived from actual tissue stem cells. The authors conclude that both normal and malignant human breast stem cells express high levels of molecular markers traditionally associated with epithelial-to-mesenchymal transition, such as N-cadherin and FOXC2. In line with that observation, induction of epithelial-to-mesenchymal transition in breast cancer cells lead to an upregulation of markers associated with the stem cell state (Mani *et al.*, 2008). Importantly, forced epithelial-to-mesenchymal transition resulted in cells with the functional properties of breast cancer stem cells. These cells formed more mammospheres and colonies in soft agar as compared to wild-type controls, and interestingly formed tumors when xenotransplanted in mice with a significantly higher efficiency suggesting that epithelial-to-mesenchymal transition indeed generated cells that fulfill most criteria used to define cancer stem cells (Mani *et al.*, 2008). Similar data have since been presented for several other epithelial cancers such as prostate (Giannoni *et al.*, 2010) and lung (DiMeo *et al.*, 2009) cancer and again for breast cancer (Santisteban *et al.*, 2009), suggesting that epithelial-to-mesenchymal transition in these common cancers may be one general source of cells with cancer stem cell properties.

These concepts may be of more than strictly academic interest as they can significantly impact what strategies can be successfully used clinically to target the cancer stem cell pool. In the original and most strict version of the cancer stem cell model, it is inferred that the cancer stem cell pool is more or less constant and static. Successful targeting of these cells should, in theory, eliminate the tumor with time because of the supposedly limited survival/proliferation potential of the more differentiated tumor bulk cells (Fig. 1A). However, this strategy rests heavily on the assumption that tumor bulk cells cannot reacquire the properties of stem cells and effectively become part of the cancer stem cell pool. Dedifferentiation and epithelial-to-mesenchymal transition as discussed above could clearly generate cancer stem cells from more differentiated bulk cells, and thus targeting the cancer stem cells in any given tumor does not permanently exclude the possible existence of cancer stem cells in that tumor (Fig. 1B). Does this apparent plasticity mean, then, that targeting the cancer stem cell pool is useless and that the cancer stem cell model adds no further insight as compared to the clonal evolution model? Not necessarily. The presence of tumor cells with stem cell properties may still be of crucial importance for long-term tumor maintenance

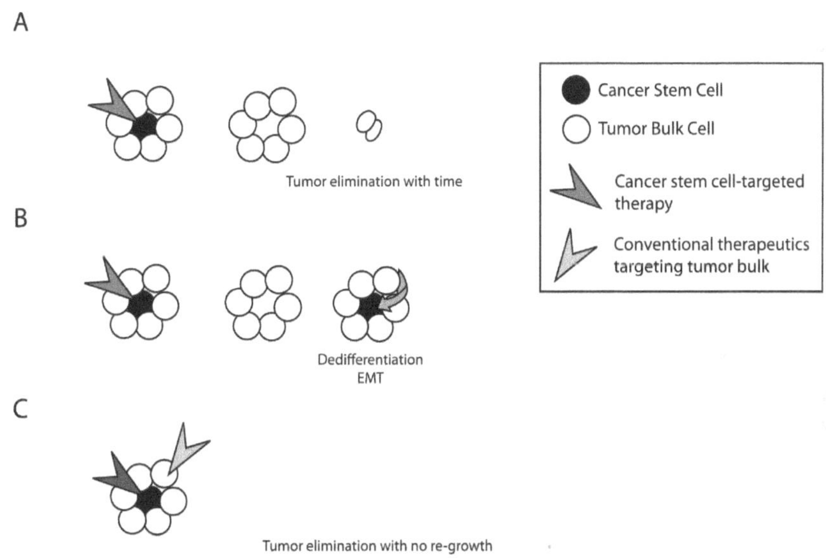

Fig. 1 Strategic implications for therapeutics of the cancer stem cell model. (A) In the most strict interpretation of the cancer stem cell model, therapeutic targeting of this subset of tumor cells will eliminate tumor growth in the long run, as the more differentiated tumor cells of the tumor bulk will have a limited proliferative capacity. (B) Dedifferentiation and epithelial-to-mesenchymal transition may render differentiated tumor bulk cells capable of acquiring cancer stem cell properties, and thus therapeutic targeting of the cancer stem cell pool only may not exclude the existence of tumor cells with cancer stem cell properties with time. (C) Targeting of the cancer stem cell compartment in parallel with conventional therapeutics that target the bulk of tumor cells should eliminate tumor growth.

and progression. It is generally assumed that current anti-cancer therapeutics such as irradiation and chemotherapy successfully targets the bulk of tumor cells (see below), but not the cancer stem cell pool. This opens for the opportunity to target in parallel the cancer stem cell pool with novel and targeted therapeutics, while also targeting the bulk of the tumor with more traditional means (Fig. 1C). This strategy, if successful, would leave the tumor without viable cells that can repopulate the tumor and reinitiate tumor growth.

III. EVIDENCE FOR A ROLE OF CANCER STEM CELLS IN PHENOTYPIC TUMOR CELL HETEROGENEITY

The mere existence of cancer stem cells, whatever their role in tumor biology in a broader sense may be, can be considered evidence of a role in promoting tumor heterogeneity. Cancer stem cells differ from their nonstem cancer cell counterparts in a wide variety of measurable properties, some of which include: proliferation rate (Moore and Lyle, 2011), migratory and invasive behavior (Wei *et al.*, 2010), metastatic potential (Liu *et al.*, 2010), and DNA repair activation/mechanisms (Bao *et al.*, 2006a).

A biologically more intriguing possibility, however, is that the whole range of phenotypically diverse tumor cells within any given tumor are all derived directly from the cancer stem cell pool. Compelling evidence in support of the cancer stem cell hypothesis demonstrate that putative cancer stem cells isolated from leukemias, breast cancers, as well as gliomas, when xenotransplanted into immunocompromised mice, give rise to cancers very similar to and containing all the different cell types of the original tumor (Al-Hajj *et al.*, 2003; Lapidot *et al.*, 1994; Lee *et al.*, 2006). In contrast, nonstem cancer cells give rise to either no tumors at all or tumors that lack many of the key characteristics of the cancer observed in the patient (Al-Hajj and Clarke, 2004; Al-Hajj *et al.*, 2003, 2004; Lapidot *et al.*, 1994; Lee *et al.*, 2006).

In their landmark contribution, Al-Hajj *et al.* (2003) were first to identify *bona fide* cancer stem cells in a solid tumor when they isolated CD44 + /CD24– lineage cells from human breast cancer patients. These cells illustrate an important point in the biology of cancer stem and nonstem cells; in addition to having the exclusive ability to propagate tumors in recipient mice, they also give rise to tumors containing nonstem cancer cells while nonstem cancer cells never give rise to cancer stem cells. Importantly, nonstem cancer cells generated by cancer stem cells too are unable to form new tumors upon transplantation.

In the paper by Howard Fine and colleagues, the authors describe the parallel isolation of cancer cells from human gliomas in two different ways; one culture was established traditionally in medium containing serum (similar to how most cancer cell lines are established and cultured), while the other culture from the same specimen was established in cancer stem cell media (notably without added serum). Within weeks, the traditional culture had altered its phenotype and developed into a homogeneous cell line with an epithelial-like morphology. Importantly, while cells cultured under stem cell conditions formed xenograft tumors in mice that recapitulated most features of the clinical gliomas, conventionally cultured cells formed tumors that were remarkably well circumscribed and did not show any tumor cell infiltration into surrounding tissues. SNP analysis and karyotyping revealed that serum-cultured cells had acquired several novel genetic aberrations not seen in the original tumor, while the cancer stem cell culture had remained true to the original genotype. Overall, data presented by Lee *et al.* emphasize that cancer stem cell cultures under certain circumstances may serve as more reliable models of human cancer biology than conventional cancer cell lines do, whether or not tumor progression follows the cancer stem cell model.

These data indicate that the heterogeneity seen in tumors may arise from differentiation of a cancer stem cell into different lineages rather than from clonal evolution of equal tumor cells. This is particularly compelling, perhaps, because the cancer stem cells isolated from human tumors typically represent only a minor fraction of all tumor cells. The inherent heterogeneity in the nonstem cell population at isolation should allow for fully heterogeneous tumors to form, if this was not dependent on the cancer stem cell compartment as a driver of tumor heterogeneity. That is, most of the heterogeneity seen in the clinical specimen should be represented instantly in the nonstem cell injection, while only a minor population would be represented in the cancer stem cell injection. Thus, formation of tumors with all cell lineages of the clinical specimen from the cancer stem cell but not from the nonstem cancer cell compartment strongly suggests that the cancer stem cells give rise to the various lineages also in the tumor.

A recent study of heterogeneity in human malignant melanoma cultures gives further support for the role of cancer stem cells in tumor heterogeneity. Melanomas are derived from the neural crest, and melanoma biology, in some ways, mimics normal neural crest development. Thus, in a malignant melanoma lesion, cells expressing neural crest stem cell markers such as SOX10 and CD271 are frequently present together with more differentiated tumor cells reminiscent of neural crest derivatives. Civenni *et al.* (2011) not only found that sorting of melanoma cells expressing the neural crest stem cell marker CD271 resulted in cells with an exclusive ability to form melanomas when transplanted into recipient mice, but further found that the resulting tumors contained all the various lineages expected from the original tumor. This study, however, also shed some further light on one of the major caveats of the predominant method to identify and define cancer stem cells by xenotransplantation (and as also discussed elsewhere in this review). When transplanting sorted cells into mice that were more severely immunocompromised, both CD271 positive and negative fractions gave rise to xenograft tumors. Intriguingly, in these mice none of the fractions gave rise to tumors with the full heterogeneity of the original melanoma represented, suggesting that cancer stem cells themselves are not sufficient for multi-lineage differentiation.

IV. CLUES FROM METASTATIC AND RECURRING LESIONS

Clues to how tumors are maintained and progress to metastatic disease may be found in comparisons of the genetic identity of the diagnostic clone of the primary tumor with that of recurring or metastatic lesions. In some interpretations of the cancer stem cell model, it is assumed that the bulk of tumor cells contains clones with additional genetic alterations as compared to the cancer stem cell compartment. As the recurring and/or metastatic tumors theoretically stem from the cancer stem cell compartment, these would likely be clonally related to, but significantly different from, the dominant clone in the primary tumor. In the more classical view of the clonal evolution model, on the other hand, metastases and recurring tumors, given similar selective pressures, are likely to be highly similar to the most advanced clone in the primary tumor. In the relatively few studies dealing with this issue directly to date, interestingly, most of the time the dominant clone at diagnosis differed significantly from the recurring or metastatic lesion's identity (Li *et al.*, 2003; Mullighan *et al.*, 2008; Zuna *et al.*, 2004). These data suggest that the recurring tumor did not evolve directly from the dominant clone at diagnosis; instead, the recurring tumor and the primary tumor likely shared a common ancestor further back in evolution.

Intriguingly, despite these data there is frequently a striking histological similarity between primary tumor and corresponding metastases (Ma *et al.*, 2003; Weigelt *et al.*, 2003). While this fact could conceivably be explained by any number of reasons, it is also consistent with the idea of a common progenitor cell giving rise to both tumors, that is the cancer stem cell. It is conceivable that a cancer stem cell, that according to the hypothesis remains only a minor population of the primary and metastatic lesion, has acquired a number of genetic hits, and then gives rise to the bulk of tumor cells that will in turn acquire additional aberrations. The metastatic or recurring tumor, then, would also be seeded directly by the cancer stem cell. The bulk of tumor cells within the metastatic or recurring lesion would then acquire its own set of additional aberrations, each likely different from the additional hits present in the bulk of the primary tumor. A comparison of the primary tumor and the metastatic or recurring tumor would then, as was frequently the case in the studies referenced above, suggest that indeed these lesions were clonally related, but from a common ancestor that lies further back evolutionary than the dominant clone in the bulk of each tumor type. Furthermore, it is likely that two individual tumors initiated and maintained by the same pool of cancer stem cells would obtain strikingly similar histological features, despite the fact that they each acquire a unique set of additional genetic aberrations.

V. CANCER STEM CELLS AND ENDOTHELIAL/STROMAL DIFFERENTIATION

Recent publications have hinted at further complexity in the contribution of cancer stem cells to intratumor heterogeneity beyond the diversity of tumor cells proper (Fig. 2). An emerging concept in cancer biology is the

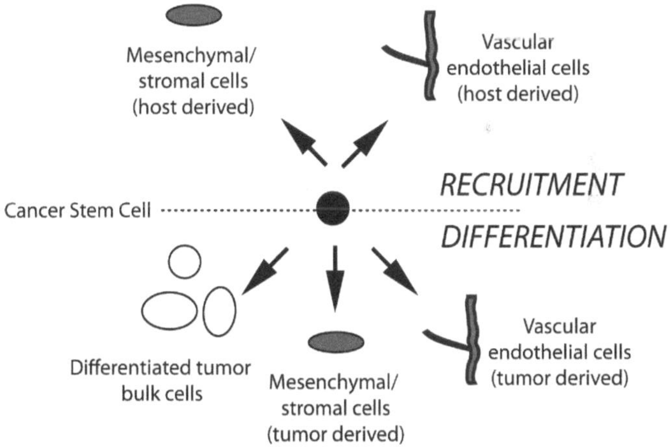

Fig. 2 Cancer stem cells in tumor heterogeneity. Cancer stem cells contribute to tumor heterogeneity by differentiating and transdifferentiating into various tumor lineages as well as mesenchymal/stromal and endothelial cells of tumor origin. In addition, cancer stem cells aid in the recruitment of host-derived endothelial and mesenchymal/stromal cells through secretion of growth factors.

transdifferentiation of actual tumor cells into various lineages that were not typically perceived as tumor-derived. In a landmark contribution, Maniotis *et al.* first described the process of vasculogenic mimicry of malignant melanoma cells over 10 years ago (Bissell, 1999; Maniotis *et al.*, 1999). While blood vessels in tumors are generally thought to form from the recruitment and differentiation of normal host endothelial cells (Phng and Gerhardt, 2009), the authors described the existence of apparently functional vascular channels in human malignant melanomas were lined not by classical endothelial cells, but instead, by tumor cells proper (Maniotis *et al.*, 1999). Similar results have since been obtained in other tumor forms including the childhood tumor neuroblastoma, adding to the notion that vascular support in tumors can sometimes come from the tumor itself (Pezzolo *et al.*, 2007). However, until recently, the mechanisms and the precise cellular players involved have been largely unknown.

Two recent reports have indicated that the ability of human glioblastoma multiforme tumors to form tumor-derived endothelial cells relies on the cancer stem cell compartment (Ricci-Vitiani *et al.*, 2010; Wang *et al.*, 2010). The authors of both papers identify apparently functional vessels within human glioblastomas that carry genetic alterations typical of the bulk of tumor cells within their respective tumors. Isolating and culturing cancer stem cells by use of cell surface markers, both studies go on to show that under *in vitro* conditions, at least a proportion of cancer stem cells, but not their nonstem cell counterparts, can form capillary-like networks in matrigel, much like normal cultured endothelial cells (Ricci-Vitiani *et al.*, 2010; Wang *et al.*, 2010). Furthermore, *in vivo* studies confirmed that cancer stem cell-derived xenograft tumors in mice contained vascular endothelial cells derived directly from the injected tumor cells (Wang *et al.*, 2010). Importantly, Ricci-Vitiani *et al.* demonstrated that specifically targeting cancer stem cell-derived endothelial cells in xenograft tumors significantly impaired tumor growth, suggesting a functional importance of endothelial differentiation of cancer stem cells (Ricci-Vitiani *et al.*, 2010).

The concept of tumor-derived endothelial cells carries major implications for anti-tumor therapeutics. A key advantage in therapeutic targeting of angiogenesis and tumor vasculature is, or is thought to be, the relative genetic stability and "normal" behavior of the tumor-associated vasculature as compared to the tumor cells proper. If tumor cells themselves, on the other hand, are capable of forming functional blood vessels within the tumor, classical strategies to target endothelial cells may be inefficient in targeting these plastic and genetically unstable tumor-derived cells with endothelial differentiation. If this heterogeneity within the endothelial cell pool in a tumor is a direct consequence of the multi-lineage differentiation potential of the cancer stem cells, therapeutic targeting of this subpopulation of cells may aid cancer therapeutics regardless of whether these cells are the only cells capable of propagating and maintaining a particular tumor.

Similarly, cancer stem cells of glioblastoma multiforme tumors have been shown to have the ability to differentiate into mesenchymal lineages (Ricci-Vitiani *et al.*, 2008). Ricci-Vitiani *et al.* demonstrated that both *in vivo* and *in vitro*, a subset of glioblastoma stem cells was able to give rise to both neuronal and osteo-chondrogenic lineage cells. These studies again highlight that targeting "normal" cells of the stromal compartment in tumors may be complicated by stromal differentiation of a small subset of bona fide tumor cells. Furthermore, they are suggestive of an even broader role for cancer stem cells in regulating intratumoral cellular heterogeneity (Fig. 2).

VI. CANCER STEM CELLS IN ANGIOGENESIS

Accumulating evidence additionally support a direct and crucial role for cancer stem cells in the recruitment, expansion, and differentiation of normal host-derived tumor vasculature, thereby contributing to another well-documented intratumoral heterogeneity (Fig. 2). Different regions of the same tumor may have dramatically different vascular density and tumor cells themselves play a key role in the recruitment of vessels. Early on, VEGF was identified as a key secreted factor produced by CD133+ human glioblastoma cancer stem cells (Bao *et al.*, 2006b), and Bao *et al.* demonstrated that VEGF secreted from cancer stem cells contributed to and promoted glioblastoma angiogenesis. Subsequent studies have revealed that cancer stem cells or putative cancer stem cells in a number of tumor forms exhibit higher than normal levels of the hypoxia-inducible transcription factors and specifically, HIF-2α, regardless of oxygen tension (Bar *et al.*, 2010; Kim *et al.*, 2009; Li *et al.*, 2009; Mendez *et al.*, 2010; Pietras *et al.*, 2008, 2010; Pistollato *et al.*, 2010a; Seidel *et al.*, 2010). The mechanisms underlying this pseudo-hypoxic phenotype of cancer stem cells remain poorly understood, but it is increasingly clear that both metabolic and angiogenic heterogeneity within tumors are affected by aberrant expression of key players of the hypoxic response.

It is increasingly evident that the same pathways that regulate maintenance of stem cell properties in both normal and malignant stem cells play important roles also in regulation of tumor angiogenesis. This is particularly true perhaps for the Notch signaling pathway, which has emerged as a popular target in tumor biology for both targeting stem cell phenotypes and aberrant tumor vascularization (Bar *et al.*, 2010; Charles *et al.*, 2010; Gustafsson *et al.*, 2005; Noguera-Troise *et al.*, 2006; Pistollato *et al.*, 2010b; Ridgway *et al.*, 2006; Siekmann and Lawson, 2007; Thurston *et al.*, 2007; Yustein *et al.*, 2010; Zeng *et al.*, 2005). Indeed, there is accumulating evidence that the pseudo-hypoxic phenotype of cancer stem cells itself will lead to increased Notch signaling or specific expression of elements of the Notch signaling pathway, subsequently leading to an increase in both tumor "stemness" and tumor vascularization (Chen *et al.*, 2010; Eliasz *et al.*, 2010; Gustafsson *et al.*, 2005; Jögi *et al.*, 2002; Pietras *et al.*, 2011; Sahlgren *et al.*, 2008).

VII. CANCER STEM CELLS IN HETEROGENEITY OF THE THERAPEUTIC RESPONSE

Data obtained from studies of putative cancer stem cells provide an alternative explanation for how resistance to chemo- and radiotherapy occurs in tumors. Several experiments have demonstrated in human and animal systems that when treated with these modalities, the cancer stem cell pool survives whereas the nonstem cell pool dies (Al-Hajj *et al.*, 2004; Bao *et al.*, 2006a; Bleau *et al.*, 2009; Chappell and Dalton, 2010; Creighton *et al.*, 2009; Hambardzumyan *et al.*, 2008a, 2008b; Singh and Settleman, 2010). The point that stem-like cells in tumors may be more resistant to radiation therapy has been best demonstrated in brain tumors. Bao *et al.* (2006a) sorted stem-like cells from gliomas based on their expression of CD133, and subjected them and their corresponding non stem-like CD133 negative cells to irradiation *in vitro* and *in vivo*. Specifically, the authors showed that the stem-like cells more efficiently activated the DNA damage checkpoint and thus were better at repairing radiation-induced DNA damage. While these data were mainly derived from isolates of stem-like and nonstem-like cells from human tumors, Holland and colleagues elegantly demonstrated similar findings in several physiologically relevant mouse models of the childhood brain tumor medulloblastoma (Hambardzumyan *et al.*, 2008b). The mouse models used, induced by Sonic Hedgehog in combination with N-Myc or AKT, show remarkable similarity to human brain tumors in general and medulloblastomas in particular, with a well-defined and relatively small stem cell compartment located in a perivascular niche, clearly separated from the bulk of tumor cells with more differentiated phenotypes. These mouse models can thus be used to study the biology of cancer stem cells without isolation of cell populations and disturbing the tumor microenvironment. Irradiating these medulloblastomas, Hambardzumyan *et al.* (2008b) demonstrated that indeed, while the bulk of tumor cells rapidly underwent apoptotic cell death upon irradiation, the stem-like cells of the perivascular niche simply went into cell cycle arrest for up to 72 h, and then begun to proliferate again. These data clearly demonstrated that in these models, tumor resistance to irradiation and most likely recurrence after treatment despite successful "debulking" of the tumor by radiotherapy were due to resistance of the cancer stem cell pool. Interestingly, the cancer stem cell compartment appeared to preferentially activate PI3K/AKT signaling in these tumors, and inhibiting these pathways during irradiation greatly increased the level of cell death in the perivascular compartment. These data promise a therapeutic benefit of specifically targeting pathways hyperactivated in cancer stem cells specifically in combination with more conventional anti-tumor therapeutics.

Similar mechanisms appear to be in place for resistance to chemotherapy in gliomas. In PDGF-induced murine gliomas, Bleau *et al.* first demonstrated that the side population assay (based on Hoechst 33342 exclusion via the drug efflux pump ABCG2, reviewed in Goodell *et al.*, 2005) can be used to isolate highly tumorigenic cells with properties of cancer stem cells (Bleau *et al.*, 2009). Strikingly, treating neurospheres *in vitro* with the standard chemotherapy for human gliomas—temozolomide—resulted in a greatly increased proportion of side population cells. Remarkably, this effect was seen despite that temozolomide is not a known substrate for the ABCG2 pump and, indeed, the increase in side population cells did not appear to be the result of selection for ABCG2 expressing cells as temozolomide was as efficient in killing ABCG2 positive and negative cells (Bleau *et al.*, 2009). Like in the medulloblastoma models, it was found that the PI3K/AKT pathways regulate the side population phenotype in stem-like cancer cells.

Taken together, these results suggest that resistance may occur not due to a selection of tumor cells that happen to have acquired properties of resistance (as generally assumed by the clonal evolution hypothesis), but rather due to a cancer stem cell pool that is intrinsically resistant to chemo- and radiotherapy.

VIII. UNRESOLVED ISSUES REGARDING THE CANCER STEM CELL MODEL IN TUMOR HETEROGENEITY

While the cancer stem cell model attractively answers several key questions about the origins of intratumoral phenotypic heterogeneity, a number of issues remain unresolved. Surprisingly, only a limited number of studies have convincingly addressed the clonal relationship between the cancer stem cell compartment and the bulk of tumor cells within any given tumor. Early studies from the Polyak laboratory on breast tumor heterogeneity were indeed consistent with the cancer stem cell model: isolating cancer stem cells in breast tumors based on the expression of CD44, the authors were able to show that CD44+ and CD24+ are clonally related, with the more differentiated nonstem cancer cell CD24+ pool having acquired additional genetic hits not present in the CD44+ cancer stem cell pool (Campbell and Polyak, 2007; Shipitsin *et al.*, 2007). However, more recently, another study from the same laboratory (Park *et al.*, 2010) paints a more complex picture. Within the same tumor, Park *et al.* found remarkable genetic heterogeneity within phenotypically similar tumor cell subgroups; that is, even within the cancer stem cell pool there were a multitude of genetic clones. The authors conclude that their data are inconsistent with the cancer stem cell model altogether and specifically with the concept that CD24+ more epithelial-like cells would be derived from the CD44+ cancer stem cell pool. However, these data are inconsistent only with the strictest definition of the cancer stem cell model. The cancer stem cell model per se arguably does not necessarily exclude the possibility that there is significant diversity within the cancer stem cell pool. In fact, genetic diversity within the cancer stem cell pool would be a prerequisite for long-term expansion of diverse subclones within a tumor. Clonal evolution will occur in any population of cancer cells; however, only those genetic aberrations acquired in the cancer stem cell pool would survive in the long run (Gupta *et al.*, 2009). Similar results were recently reported in a study of leukemia-initiating cells in *BCR-ABL1*-positive lymphoblastic leukemia from the Dick laboratory (Notta *et al.*, 2011). Notta *et al.* found that in samples from *BCR-ABL1*-related acute lymphoblastic leukemia patients, functionally defined TICs of diverse genetic clones were present, suggesting a multi-clonal evolution of acute lymphoblastic leukemias. Intriguing data showed that in mouse xenografts derived from patient samples, the dominant clone at diagnosis in the clinic was frequently not the clone that grew as a xenograft. Rather, a genetically related clone from an earlier evolutionary stage, as a few genetic aberrations besides the *BCR-ABL1* fusion were indeed often shared. Of note, the same patient sample could give rise to xenografts in mice arising from distinct clones that were all different from the dominant clone at diagnosis. These data are in line with the recent report indicating that ALL relapse tumors frequently arise from a clone that is related to but not identical to the predominant clone at diagnosis (Mullighan *et al.*, 2008). Together, these reports indicate that there may be a considerable heterogeneity and clonal evolution going on even within the cancer stem cell pool. The cancer stem cell hypothesis, thus, would take a giant leap toward finally merging with the stochastic clonal evolution model of tumor progression. It is noteworthy, however, when interpreting and extrapolating the results of Notta *et al.*, that although they have studied functionally defined TICs, *bona fide* cancer stem cells have, to my knowledge, not yet been isolated from acute lymphoblastic leukemias.

Further evidence pointing toward a high plasticity in the tumor-initiating cell pool comes from studies on melanoma-initiating cells in severely immunocompromised animals (Quintana *et al.*, 2010). In addition to concluding that, as previously demonstrated by the same group (Quintana *et al.*, 2008), melanomas contain a high proportion of cells capable of forming xenograft tumors if the recipient mouse is the right one, there were no consistent markers or no consistent phenotype that represented the TICs (Quintana *et al.*, 2010). Furthermore, the authors demonstrate that a large variety of tumor cell subpopulations were able to recapitulate the full heterogeneity of the parent tumor, again arguing strongly against a simple cancer stem cell model in malignant melanoma (Quintana *et al.*, 2010).

There is little direct evidence to prove that in fact cancer stem cells do give rise to the bulk of tumor cells. One intriguing possibility arises from the fact that many normal cell types with properties of stem cells are recruited into solid tumors. It is well established that mesenchymal stem cells are recruited from the bone marrow into tumors (Birnbaum *et al.*, 2007) and in brain tumors, neural stem cells exhibit great tropism toward the tumors (Aboody *et al.*, 2000). Indeed, such tropism has been widely suggested to potentially function as a way to specifically deliver anticancer therapeutics into tumor tissue (Bexell *et al.*, 2010). Although the topic remains controversial, mesenchymal stem cells have been suggested to have a more direct role in tumorigenesis than previously anticipated. In a mouse model of gastric cancer, using a GFP-tagged mesenchymal lineage, Wang and colleagues demonstrated that, in fact, resulting gastric cancers after inflammation consisted largely of epithelial bona fide tumor cells derived directly from the mesenchymal (i.e., GFP positive) cells (Houghton *et al.*, 2004). These data demonstrate clearly that at sites of injury, mesenchymal stem cells are prone to acquire alterations in behavior ultimately leading to tumor formation. The milieu within tumors typically contains many elements that may conceivably alter normal cell behavior toward malignancy. The abundance of growth factors produced by tumor cells themselves (Hanahan and Weinberg, 2000) will undoubtedly exert effects also on adjacent nontumor cells that are present within the tumor. Specifically in gliomas, Holland and co-workers demonstrated that platelet-derived growth factor exposure induces dedifferentiation of neural cells (Dai *et al.*, 2001). In line with these results, paracrine PDGF-PDGFR signaling from tumor cells to adjacent neural progenitor cells was demonstrated to result in seemingly unrestricted expansion of the neural progenitor cell compartment in a murine model of brain tumors (Assanah *et al.*, 2006). Similarly and as discussed above and at length elsewhere (Axelson *et al.*, 2005; Edsjo *et al.*, 2007; Pietras *et al.*, 2010), hypoxic environments have been shown dedifferentiate human tumor cells of various lineages toward stem cell-like phenotypes (Heddleston *et al.*, 2009; Jögi *et al.*, 2002, 2004). Furthermore, microenvironmental cues can lead to the induction of epithelial-to-mesenchymal transition (Lopez-Novoa and Nieto, 2009; Sullivan *et al.*, 2009), a process that in itself may render cells with stem cell-like properties (Mani *et al.*, 2008). Taken together, it is not unlikely that neural and mesenchymal stem- and progenitor cells that are recruited into sites of tumor growth are affected by the tumor microenvironment in ways that ultimately increase the risk of these cells turning tumorous themselves (Fomchenko and Holland, 2005). Given that these cells already possess many of the features associated with stem cells in general and cancer stem cells in particular, it is certainly possible that cancer stem cells in fact represent recruited stem- and progenitor cells unrelated to the bulk of tumor cells, or at least, unrelated to the TICs, that have turned cancerous within the tumor microenvironment. This speculation calls for a rigid separation of the cancer stem cell hypothesis from issues of cell of origin, as it implies that cancer stem cells in most cases by no means have a direct relationship to the tumor-founding cell that acquired the first tumorigenic genetic hits. However, this would not exclude the possibility that stem- and progenitor cells recruited to sites of injury in fact do represent both cell of origin and cancer stem cell in certain specific cases (Houghton *et al.*, 2004).

IX. FUTURE DIRECTIONS

Complete understanding of tumor biology and investigations into the true contribution of cancer stem cells to tumor heterogeneity have thus far largely been hampered by the lack of reliable and advanced experimental systems. As discussed elsewhere, the most commonly used *in vivo* assays to study the biology of cancer stem cells come with several serious caveats (Quintana *et al.*, 2008), and importantly, frequently require a total disruption of the heterogeneous tumor microenvironment in which putative cancer stem cells normally act. For instance, cancer stem cells like their nonmalignant counterparts are increasingly recognized to depend on a cancerous equivalent to the stem cell niche. Such dependence on micro-environmental elements such as tumor vasculature or specific oxygen tensions have been demonstrated for both leukemic and solid cancer stem cells. It is likely that most other cell types present in tumors, too, contribute to maintenance of cancer stem cells and/or differentiation of cancer stem cells into more differentiated progeny, as was recently demonstrated in transplantation models of melanoma into immunodeficient mice with varying levels of immunodeficiency (Civenni *et al.*, 2011). It is thus likely that tumor models in immunodeficient mice, where tumors essentially lack key components of the normal tumor microenvironment, may never fully mimic human tumor biology.

A major improvement would be studying mouse models of cancers that naturally contain stem-like cancer cells within a complete tumor microenvironment. Many genetically engineered mouse models, however, appear to not necessarily follow the cancer stem cell model, an issue that is likely derived from the way these cancers are initiated and may call for caution when interpreting and extrapolating from data on tumor progression generated by such models. Considering this, it would be of great interest to characterize the presence and possible function of cancer stem cells in the most commonly used mouse tumor models. While the vast majority of studies on cancer stem cells are presently performed on human tumor material with xenotransplantation into immunocompromised mice, some of the key studies mentioned in this review indeed investigated these phenomena in mouse models of cancer. For both of the brain tumors glioma and medulloblastoma, the RCAS/tva system (reviewed in Huse and Holland, 2009; Momota and Holland, 2009) can give rise to tumors that appear remarkably faithful to human brain tumor biology given that the right vectors and initiating cells are employed (Bleau *et al.*, 2009; Charles *et al.*, 2010; Hambardzumyan *et al.*, 2008b). First, as was described for a large variety of human brain (and other) tumors (Calabrese *et al.*, 2007; Pietras *et al.*, 2008), a cell compartment expressing high levels of stem cell-associated markers as compared to the bulk of tumor cells appears located preferentially in a perivascular niche (Bleau *et al.*, 2009; Hambardzumyan *et al.*, 2008b). More importantly, upon irradiation, these cells persist while the bulk of the tumor undergoes apoptosis (Hambardzumyan *et al.*, 2008b), again mimicking what has been widely proposed for human cancer stem cells. Clearly, the identification of and subsequent study of such models will have great advantages as compared to using transplantation models of FACS-sorted human tumor cells.

To resolve the remaining issues regarding the origin of and the contribution to tumor heterogeneity and therapy resistance of cancer stem cells, true lineage tracing models of cancer *in vivo* must be used to study these phenomena without disrupting tumors and their microenvironment by cell sorting and transplantation. Mapping the fate of individual tumor cells and their progeny during tumor progression and therapy should aid in defining both origins and functions of cancer stem cells and tumor bulk. Given the recent advances in lineage tracing models and the ability to functionally define the role of normal tissue stem cells *in vivo* (Snippert *et al.*, 2010; Weber *et al.*, 2011), there is hope that similar studies can inform us further on the functional properties of cancer stem cells.

ACKNOWLEDGMENTS

A.P. receives support from the Tegger foundation and the Swedish Research Council.

REFERENCES

Aboody, K. S., Brown, A., Rainov, N. G., Bower, K. A., Liu, S., Yang, W., Small, J. E., Herrlinger, U., Ourednik, V., Black, P. M., Breakefield, X. O., and Snyder, E. Y. (2000). Neural stem cells display extensive tropism for pathology in adult brain: evidence from intracranial gliomas. *Proc. Natl. Acad. Sci. U.S.A.* **97**, 12846–12851.

Albanese, I., Scibetta, A. G., Migliavacca, M., Russo, A., Bazan, V., Tomasino, R. M., Colomba, P., Tagliavia, M., and La Farina, M. (2004). Heterogeneity within and between primary colorectal carcinomas and matched metastases as revealed by analysis of Ki-ras and p53 mutations. *Biochem. Biophys. Res. Commun.* **325**, 784–791.

Al-Hajj, M., Becker, M. W., Wicha, M., Weissman, I., and Clarke, M. F. (2004). Therapeutic implications of cancer stem cells. *Curr. Opin. Genet. Dev.* **14**, 43–47.

Al-Hajj, M., and Clarke, M. F. (2004). Self-renewal and solid tumor stem cells. *Oncogene* **23**, 7274–7282.

Al-Hajj, M., Wicha, M. S., Benito-Hernandez, A., Morrison, S. J., and Clarke, M. F. (2003). Prospective identification of tumorigenic breast cancer cells. *Proc. Natl. Acad. Sci. U.S.A.* **100**, 3983–3988.

Artale, S., Sartore-Bianchi, A., Veronese, S. M., Gambi, V., Sarnataro, C. S., Gambacorta, M., Lauricella, C., and Siena, S. (2008). Mutations of KRAS and BRAF in primary and matched metastatic sites of colorectal cancer. *J. Clin. Oncol.* **26**, 4217–4219.

Assanah, M., Lochhead, R., Ogden, A., Bruce, J., Goldman, J., and Canoll, P. (2006). Glial progenitors in adult white matter are driven to form malignant gliomas by platelet-derived growth factor-expressing retroviruses. *J. Neurosci.* **26**, 6781–6790.

Axelson, H., Fredlund, E., Ovenberger, M., Landberg, G., and Påhlman, S. (2005). Hypoxia-induced dedifferentiation of tumor cells—a mechanism behind heterogeneity and aggressiveness of solid tumors. *Semin. Cell. Dev. Biol.* **16**, 554–563.

Bao, S., Wu, Q., McLendon, R. E., Hao, Y., Shi, Q., Hjelmeland, A. B., Dewhirst, M. W., Bigner, D. D., and Rich, J. N. (2006a). Glioma stem cells promote radioresistance by preferential activation of the DNA damage response. *Nature* **444**, 756–760.

Bao, S., Wu, Q., Sathornsumetee, S., Hao, Y., Li, Z., Hjelmeland, A. B., Shi, Q., McLendon, R. E., Bigner, D. D., and Rich, J. N. (2006b). Stem cell-like glioma cells promote tumor angiogenesis through vascular endothelial growth factor. *Cancer Res.* **66**, 7843–7848.

Bar, E. E., Lin, A., Mahairaki, V., Matsui, W., and Eberhart, C. G. (2010). Hypoxia increases the expression of stem-cell markers and promotes clonogenicity in glioblastoma neurospheres. *Am. J. Pathol.* **177**, 1491–1502.

Bexell, D., Scheding, S., and Bengzon, J. (2010). Toward brain tumor gene therapy using multipotent mesenchymal stromal cell vectors. *Mol. Ther.* **18**, 1067–1075.

Birnbaum, T., Roider, J., Schankin, C. J., Padovan, C. S., Schichor, C., Goldbrunner, R., and Straube, A. (2007). Malignant gliomas actively recruit bone marrow stromal cells by secreting angiogenic cytokines. *J. Neurooncol.* **83**, 241–247.

Bissell, M. J. (1999). Tumor plasticity allows vasculogenic mimicry, a novel form of angiogenic switch. A rose by any other name? *Am. J. Pathol.* **155**, 675–679.

Bleau, A. M., Hambardzumyan, D., Ozawa, T., Fomchenko, E. I., Huse, J. T., Brennan, C. W., and Holland, E. C. (2009). PTEN/PI3K/Akt pathway regulates the side population phenotype and ABCG2 activity in glioma tumor stem-like cells. *Cell Stem Cell.* **4**, 226–235.

Calabrese, C., Poppleton, H., Kocak, M., Hogg, T. L., Fuller, C., Hamner, B., Oh, E. Y., Gaber, M. W., Finklestein, D., Allen, M., Frank, A., Bayazitov, I. T., Zakharenko, S. S., Gajjar, A., Davidoff, A., and Gilbertson, R. J. (2007). A perivascular niche for brain tumor stem cells. *Cancer Cell* **11**, 69–82.

Campbell, L. L., and Polyak, K. (2007). Breast tumor heterogeneity: cancer stem cells or clonal evolution? *Cell Cycle* **6**, 2332–2338.

Chappell, J., and Dalton, S. (2010). Altered cell cycle regulation helps stem-like carcinoma cells resist apoptosis. *BMC Biol.* **8**, 63.

Charles, N., Ozawa, T., Squatrito, M., Bleau, A. M., Brennan, C. W., Hambardzumyan, D., and Holland, E. C. (2010). Perivascular nitric oxide activates notch signaling and promotes stem-like character in PDGF-induced glioma cells. *Cell Stem Cell.* **6**, 141–152.

Chen, J., Imanaka, N., and Griffin, J. D. (2010). Hypoxia potentiates Notch signaling in breast cancer leading to decreased E-cadherin expression and increased cell migration and invasion. *Br J. Cancer.* **102**, 351–360.

Civenni, G., Walter, A., Kobert, N., Mihic-Probst, D., Zipser, M., Belloni, B., Seifert, B., Moch, H., Dummer, R., van den Broek, M., and Sommer, L. (2011). Human CD271-positive melanoma stem cells associated with metastasis establish tumor heterogeneity and long-term growth. *Cancer Res.* **71**(8), 3098–3109.

Cohnheim, J. (1875). Congenitales, quergestreiftes Muskelsarkom der Nieren. *Virchows Archiv.* **65**, 64–69.

Creighton, C. J., Li, X., Landis, M., Dixon, J. M., Neumeister, V. M., Sjolund, A., Rimm, D. L., Wong, H., Rodriguez, A., Herschkowitz, J. I., Fan, C., Zhang, X., He, X., Pavlick, A., Gutierrez, M. C., Renshaw, L., Larionov, A. A., Faratian, D., Hilsenbeck, S. G., Perou, C. M., Lewis, M. T., Rosen, J. M., and Chang, J. C. (2009). Residual breast cancers after conventional therapy display mesenchymal as well as tumor-initiating features. *Proc. Natl. Acad. Sci. U.S.A.* **106**, 13820–13825.

Dai, C., Celestino, J. C., Okada, Y., Louis, D. N., Fuller, G. N., and Holland, E. C. (2001). PDGF autocrine stimulation dedifferentiates cultured astrocytes and induces oligodendrogliomas and oligoastrocytomas from neural progenitors and astrocytes *in vivo*. *Genes Dev.* **15**, 1913–1925.

Delahunt, B. (1999). Sarcomatoid renal carcinoma: the final common dedifferentiation pathway of renal epithelial malignancies. *Pathology* **31**, 185–190.

DiMeo, T. A., Anderson, K., Phadke, P., Fan, C., Perou, C. M., Naber, S., and Kuperwasser, C. (2009). A novel lung metastasis signature links Wnt signaling with cancer cell self-renewal and epithelial-mesenchymal transition in basal-like breast cancer. *Cancer Res.* **69**, 5364–5373.

Edsjo, A., Holmquist, L., and Påhlman, S. (2007). Neuroblastoma as an experimental model for neuronal differentiation and hypoxia-induced tumor cell dedifferentiation. *Semin. Cancer Biol.* **17**, 248–256.

Eliasz, S., Liang, S., Chen, Y., De Marco, M. A., Machek, O., Skucha, S., Miele, L., and Bocchetta, M. (2010). Notch-1 stimulates survival of lung adenocarcinoma cells during hypoxia by activating the IGF-1R pathway. *Oncogene* **29**, 2488–2498.

Eramo, A., Lotti, F., Sette, G., Pilozzi, E., Biffoni, M., Di Virgilio, A., Conticello, C., Ruco, L., Peschle, C., and De Maria, R. (2008). Identification and expansion of the tumorigenic lung cancer stem cell population. *Cell Death. Differ.* **15**, 504–514.

Fomchenko, E. I., and Holland, E. C. (2005). Stem cells and brain cancer. *Exp. Cell Res.* **306**, 323–329.

Fujii, H., Marsh, C., Cairns, P., Sidransky, D., and Gabrielson, E. (1996). Genetic divergence in the clonal evolution of breast cancer. *Cancer Res.* **56**, 1493–1497.

Gerdes, M. J., and Yuspa, S. H. (2005). The contribution of epidermal stem cells to skin cancer. *Stem Cell. Rev.* **1**, 225–231.

Giannoni, E., Bianchini, F., Masieri, L., Serni, S., Torre, E., Calorini, L., and Chiarugi, P. (2010). Reciprocal activation of prostate cancer cells and cancer-associated fibroblasts stimulates epithelial-mesenchymal transition and cancer stemness. *Cancer Res.* **70**, 6945–6956.

Goodell, M. A., McKinney-Freeman, S., and Camargo, F. D. (2005). Isolation and characterization of side population cells. *Methods Mol. Biol.* **290**, 343–352.

Guarino, M., Rubino, B., and Ballabio, G. (2007). The role of epithelial-mesenchymal transition in cancer pathology. *Pathology* **39**, 305–318.

Gupta, P. B., Chaffer, C. L., and Weinberg, R. A. (2009). Cancer stem cells: mirage or reality? *Nat. Med.* **15**, 1010–1012.

Gustafsson, M. V., Zheng, X., Pereira, T., Gradin, K., Jin, S., Lundkvist, J., Ruas, J. L., Poellinger, L., Lendahl, U., and Bondesson, M. (2005). Hypoxia requires notch signaling to maintain the undifferentiated cell state. *Dev. Cell.* **9**, 617–628.

Hambardzumyan, D., Becher, O. J., and Holland, E. C. (2008a). Cancer stem cells and survival pathways. *Cell Cycle* **7**, 1371–1378.

Hambardzumyan, D., Becher, O. J., Rosenblum, M. K., Pandolfi, P. P., Manova-Todorova, K., and Holland, E. C. (2008b). PI3K pathway regulates survival of cancer stem cells residing in the perivascular niche following radiation in medulloblastoma *in vivo*. *Genes Dev.* **22**, 436–448.

Hanahan, D., and Weinberg, R. A. (2000). The hallmarks of cancer. *Cell* **100**, 57–70.

Heddleston, J. M., Li, Z., McLendon, R. E., Hjelmeland, A. B., and Rich, J. N. (2009). The hypoxic microenvironment maintains glioblastoma stem cells and promotes reprogramming towards a cancer stem cell phenotype. *Cell Cycle* **8**, 3274–3284.

Helczynska, K., Kronblad, A., Jögi, A., Nilsson, E., Beckman, S., Landberg, G., and Påhlman, S. (2003). Hypoxia promotes a dedifferentiated phenotype in ductal breast carcinoma in situ. *Cancer Res.* **63**, 1441–1444.

Heppner, G. H. (1984). Tumor heterogeneity. *Cancer Res.* **44**, 2259–2265.

Hermann, P. C., Huber, S. L., Herrler, T., Aicher, A., Ellwart, J. W., Guba, M., Bruns, C. J., and Heeschen, C. (2007). Distinct populations of cancer stem cells determine tumor growth and metastatic activity in human pancreatic cancer. *Cell Stem Cell.* **1**, 313–323.

Holland, E. C. (2001). Progenitor cells and glioma formation. *Curr. Opin. Neurol.* **14**, 683–688.

Houghton, J., Stoicov, C., Nomura, S., Rogers, A. B., Carlson, J., Li, H., Cai, X., Fox, J. G., Goldenring, J. R., and Wang, T. C. (2004). Gastric cancer originating from bone marrow-derived cells. *Science* **306**, 1568–1571.

Hubbard, S. A., and Gargett, C. E. (2010). A cancer stem cell origin for human endometrial carcinoma? *Reproduction* **140**, 23–32.

Huse, J. T., and Holland, E. C. (2009). Genetically engineered mouse models of brain cancer and the promise of preclinical testing. *Brain Pathol.* **19**, 132–143.

Jögi, A., Ora, I., Nilsson, H., Lindeheim, A., Makino, Y., Poellinger, L., Axelson, H., and Påhlman, S. (2002). Hypoxia alters gene expression in human neuroblastoma cells toward an immature and neural crest-like phenotype. *Proc. Natl. Acad. Sci. U.S.A.* **99**, 7021–7026.

Jögi, A., Vallon-Christersson, J., Holmquist, L., Axelson, H., Borg, A., and Påhlman, S. (2004). Human neuroblastoma cells exposed to hypoxia: induction of genes associated with growth, survival, and aggressive behavior. *Exp. Cell Res.* **295**, 469–487.

Jones, S., Zhang, X., Parsons, D. W., Lin, J. C., Leary, R. J., Angenendt, P., Mankoo, P., Carter, H., Kamiyama, H., Jimeno, A., Hong, S. M., Fu, B., Lin, M. T., Calhoun, E. S., Kamiyama, M., Walter, K., Nikolskaya, T., Nikolsky, Y., Hartigan, J., Smith, D. R., Hidalgo, M., Leach, S. D., Klein, A. P., Jaffee, E. M., Goggins, M., Maitra, A., Iacobuzio-Donahue, C., Eshleman, J. R., Kern, S. E., Hruban, R. H., Karchin, R., Papadopoulos, N., Parmigiani, G., Vogelstein, B., Velculescu, V. E., and Kinzler, K. W. (2008). Core signaling pathways in human pancreatic cancers revealed by global genomic analyses. *Science* **321**, 1801–1816.

Kalikaki, A., Koutsopoulos, A., Trypaki, M., Souglakos, J., Stathopoulos, E., Georgoulias, V., Mavroudis, D., and Voutsina, A. (2008). Comparison of EGFR and K-RAS gene status between primary tumours and corresponding metastases in NSCLC. *Br. J. Cancer* **99**, 923–929.

Katona, T. M., Jones, T. D., Wang, M., Eble, J. N., Billings, S. D., and Cheng, L. (2007). Genetically heterogeneous and clonally unrelated metastases may arise in patients with cutaneous melanoma. *Am. J. Surg. Pathol.* **31**, 1029–1037.

Kim, C. F., Jackson, E. L., Woolfenden, A. E., Lawrence, S., Babar, I., Vogel, S., Crowley, D., Bronson, R. T., and Jacks, T. (2005). Identification of bronchioalveolar stem cells in normal lung and lung cancer. *Cell* **121**, 823–835.

Kim, Y., Lin, Q., Zelterman, D., and Yun, Z. (2009). Hypoxia-regulated delta-like 1 homologue enhances cancer cell stemness and tumorigenicity. *Cancer Res.* **69**, 9271–9280.

Kuukasjarvi, T., Karhu, R., Tanner, M., Kahkonen, M., Schaffer, A., Nupponen, N., Pennanen, S., Kallioniemi, A., Kallioniemi, O. P., and Isola, J. (1997). Genetic heterogeneity and clonal evolution underlying development of asynchronous metastasis in human breast cancer. *Cancer Res.* **57**, 1597–1604.

Lapidot, T., Sirard, C., Vormoor, J., Murdoch, B., Hoang, T., Caceres-Cortes, J., Minden, M., Paterson, B., Caligiuri, M. A., and Dick, J. E. (1994). A cell initiating human acute myeloid leukaemia after transplantation into SCID mice. *Nature* **367**, 645–648.

Lee, J., Kotliarova, S., Kotliarov, Y., Li, A., Su, Q., Donin, N. M., Pastorino, S., Purow, B. W., Christopher, N., Zhang, W., Park, J. K., and Fine, H. A. (2006). Tumor stem cells derived from glioblastomas cultured in bFGF and EGF more closely mirror the phenotype and genotype of primary tumors than do serum-cultured cell lines. *Cancer Cell* **9**, 391–403.

Li, A., Zhou, J., Zuckerman, D., Rue, M., Dalton, V., Lyons, C., Silverman, L. B., Sallan, S. E., and Gribben, J. G. (2003). Sequence analysis of clonal immunoglobulin and T-cell receptor gene rearrangements in children with acute lymphoblastic leukemia at diagnosis and at relapse: implications for pathogenesis and for the clinical utility of PCR-based methods of minimal residual disease detection. *Blood* **102**, 4520–4526.

Li, Z., Bao, S., Wu, Q., Wang, H., Eyler, C., Sathornsumetee, S., Shi, Q., Cao, Y., Lathia, J., McLendon, R. E., Hjelmeland, A. B., and Rich, J. N. (2009). Hypoxia-inducible factors regulate tumorigenic capacity of glioma stem cells. *Cancer Cell* **15**, 501–513.

Liu, H., Patel, M. R., Prescher, J. A., Patsialou, A., Qian, D., Lin, J., Wen, S., Chang, Y. F., Bachmann, M. H., Shimono, Y., Dalerba, P., Adorno, M., Lobo, N., Bueno, J., Dirbas, F. M., Goswami, S., Somlo, G., Condeelis, J., Contag, C. H., Gambhir, S. S., and Clarke, M. F. (2010). Cancer stem cells from human breast tumors are involved in spontaneous metastases in orthotopic mouse models. *Proc. Natl. Acad. Sci. U.S.A.* **107**, 18115–18120.

Lopez-Novoa, J. M., and Nieto, M. A. (2009). Inflammation and EMT: an alliance towards organ fibrosis and cancer progression. *EMBO Mol. Med.* **1**, 303–314.

Ma, X. J., Salunga, R., Tuggle, J. T., Gaudet, J., Enright, E., McQuary, P., Payette, T., Pistone, M., Stecker, K., Zhang, B. M., Zhou, Y. X., Varnholt, H., Smith, B., Gadd, M., Chatfield, E., Kessler, J., Baer, T. M., Erlander, M. G., and Sgroi, D. C. (2003). Gene expression profiles of human breast cancer progression. *Proc. Natl. Acad. Sci. U. S.A.* **100**, 5974–5979.

Mani, S. A., Guo, W., Liao, M. J., Eaton, E. N., Ayyanan, A., Zhou, A. Y., Brooks, M., Reinhard, F., Zhang, C. C., Shipitsin, M., Campbell, L. L., Polyak, K., Brisken, C., Yang, J., and Weinberg, R. A. (2008). The epithelial-mesenchymal transition generates cells with properties of stem cells. *Cell* **133**, 704–715.

Maniotis, A. J., Folberg, R., Hess, A., Seftor, E. A., Gardner, L. M., Pe'er, J., Trent, J. M., Meltzer, P. S., and Hendrix, M. J. (1999). Vascular channel formation by human melanoma cells *in vivo* and *in vitro*: vasculogenic mimicry. *Am. J. Pathol.* **155**, 739–752.

Marusyk, A., and Polyak, K. (2009). Tumor heterogeneity: Causes and consequences. *Biochim. Biophys. Acta.*

Mendez, O., Zavadil, J., Esencay, M., Lukyanov, Y., Santovasi, D., Wang, S. C., Newcomb, E. W., and Zagzag, D. (2010). Knock down of HIF-1alpha in glioma cells reduces migration *in vitro* and invasion *in vivo* and impairs their ability to form tumor spheres. *Mol. Cancer* **9**, 133.

Momota, H., and Holland, E. C. (2009). Mouse models of CNS embryonal tumors. *Brain Tumor Pathol.* **26**, 43–50.

Moore, N., and Lyle, S. (2011). Quiescent, slow-cycling stem cell populations in cancer: a review of the evidence and discussion of significance. *J. Oncol* **2011**, 396076.

Mullighan, C. G., Phillips, L. A., Su, X., Ma, J., Miller, C. B., Shurtleff, S. A., and Downing, J. R. (2008). Genomic analysis of the clonal origins of relapsed acute lymphoblastic leukemia. *Science* **322**, 1377–1380.

Noguera-Troise, I., Daly, C., Papadopoulos, N. J., Coetzee, S., Boland, P., Gale, N. W., Lin, H. C., Yancopoulos, G. D., and Thurston, G. (2006). Blockade of Dll4 inhibits tumour growth by promoting non-productive angiogenesis. *Nature* **444**, 1032–1037.

Notta, F., Mullighan, C. G., Wang, J. C., Poeppl, A., Doulatov, S., Phillips, L. A., Ma, J., Minden, M. D., Downing, J. R., and Dick, J. E. (2011). Evolution of human BCR-ABL1 lymphoblastic leukaemia-initiating cells. *Nature* **469**, 362–367.

Nowell, P. C. (1976). The clonal evolution of tumor cell populations. *Science* **194**, 23–28.

O'Brien, C. A., Pollett, A., Gallinger, S., and Dick, J. E. (2007). A human colon cancer cell capable of initiating tumour growth in immunodeficient mice. *Nature* **445**, 106–110.

Osawa, M., Hanada, K., Hamada, H., and Nakauchi, H. (1996). Long-term lymphohematopoietic reconstitution by a single CD34-low/negative hematopoietic stem cell. *Science* **273**, 242–245.

Park, S. Y., Gonen, M., Kim, H. J., Michor, F., and Polyak, K. (2010). Cellular and genetic diversity in the progression of in situ human breast carcinomas to an invasive phenotype. *J. Clin. Invest.* **120**, 636–644.

Patrawala, L., Calhoun, T., Schneider-Broussard, R., Li, H., Bhatia, B., Tang, S., Reilly, J. G., Chandra, D., Zhou, J., Claypool, K., Coghlan, L., and Tang, D. G. (2006). Highly purified CD44+ prostate cancer cells from xenograft human tumors are enriched in tumorigenic and metastatic progenitor cells. *Oncogene* **25**, 1696–1708.

Pezzolo, A., Parodi, F., Corrias, M. V., Cinti, R., Gambini, C., and Pistoia, V. (2007). Tumor origin of endothelial cells in human neuroblastoma. *J. Clin. Oncol.* **25**, 376–383.

Phng, L. K., and Gerhardt, H. (2009). Angiogenesis: a team effort coordinated by notch. *Dev. Cell.* **16**, 196–208.

Pietras, A., Gisselsson, D., Ora, I., Noguera, R., Beckman, S., Navarro, S., and Påhlman, S. (2008). High levels of HIF-2alpha highlight an immature neural crest-like neuroblastoma cell cohort located in a perivascular niche. *J. Pathol.* **214**, 482–488.

Pietras, A., Johnsson, A. S., and Påhlman, S. (2010). The HIF-2alpha-driven pseudo-hypoxic phenotype in tumor aggressiveness, differentiation, and vascularization. *Curr. Top Microbiol. Immunol.* **345**, 1–20.

Pietras, A., von Stedingk, K., Lindgren, D., Påhlman, S., and Axelson, H. (2011). JAG2 induction in hypoxic tumor cells alters Notch signaling and enhances endothelial cell tube formation. *Mol Cancer Res.* 10.1158/1541MCR-0508.

Pistollato, F., Abbadi, S., Rampazzo, E., Persano, L., Della Puppa, A., Frasson, C., Sarto, E., Scienza, R., D'Avella, D., and Basso, G. (2010a). Intratumoral hypoxic gradient drives stem cells distribution and MGMT expression in glioblastoma. *Stem Cells* **28**, 851–862.

Pistollato, F., Rampazzo, E., Persano, L., Abbadi, S., Frasson, C., Denaro, L., D'Avella, D., Panchision, D. M., Della Puppa, A., Scienza, R., and Basso, G. (2010 b). Interaction of HIF1alpha and notch signaling regulates medulloblastoma precursor proliferation and fate. *Stem Cells* **28**(11), 1918–1929.

Quintana, E., Shackleton, M., Foster, H. R., Fullen, D. R., Sabel, M. S., Johnson, T. M., and Morrison, S. J. (2010). Phenotypic heterogeneity among tumorigenic melanoma cells from patients that is reversible and not hierarchically organized. *Cancer Cell.* **18**, 510–523.

Quintana, E., Shackleton, M., Sabel, M. S., Fullen, D. R., Johnson, T. M., and Morrison, S. J. (2008). Efficient tumour formation by single human melanoma cells. *Nature* **456**, 593–598.

Reynolds, B. A., and Weiss, S. (1996). Clonal and population analyses demonstrate that an EGF-responsive mammalian embryonic CNS precursor is a stem cell. *Dev. Biol.* **175**, 1–13.

Ricci-Vitiani, L., Lombardi, D. G., Pilozzi, E., Biffoni, M., Todaro, M., Peschle, C., and De Maria, R. (2007). Identification and expansion of human colon-cancer-initiating cells. *Nature* **445**, 111–115.

Ricci-Vitiani, L., Pallini, R., Biffoni, M., Todaro, M., Invernici, G., Cenci, T., Maira, G., Parati, E. A., Stassi, G., Larocca, L. M., and De Maria, R. (2010). Tumour vascularization via endothelial differentiation of glioblastoma stem-like cells. *Nature* **468**, 824–828.

Ricci-Vitiani, L., Pallini, R., Larocca, L. M., Lombardi, D. G., Signore, M., Pierconti, F., Petrucci, G., Montano, N., Maira, G., and De Maria, R. (2008). Mesenchymal differentiation of glioblastoma stem cells. *Cell Death Differ.* **15**, 1491–1498.

Ridgway, J., Zhang, G., Wu, Y., Stawicki, S., Liang, W. C., Chanthery, Y., Kowalski, J., Watts, R. J., Callahan, C., Kasman, I., Singh, M., Chien, M., Tan, C., Hongo, J. A., de Sauvage, F., Plowman, G., and Yan, M. (2006). Inhibition of Dll4 signalling inhibits tumour growth by deregulating angiogenesis. *Nature* **444**, 1083–1087.

Sahlgren, C., Gustafsson, M. V., Jin, S., Poellinger, L., and Lendahl, U. (2008). Notch signaling mediates hypoxia-induced tumor cell migration and invasion. *Proc. Natl. Acad. Sci. U.S.A.* **105**, 6392–6397.

Santisteban, M., Reiman, J. M., Asiedu, M. K., Behrens, M. D., Nassar, A., Kalli, K. R., Haluska, P., Ingle, J. N., Hartmann, L. C., Manjili, M. H., Radisky, D. C., Ferrone, S., and Knutson, K. L. (2009). Immune-induced epithelial to mesenchymal transition *in vivo* generates breast cancer stem cells. *Cancer Res.* **69**, 2887–2895.

Schatton, T., Murphy, G. F., Frank, N. Y., Yamaura, K., Waaga-Gasser, A. M., Gasser, M., Zhan, Q., Jordan, S., Duncan, L. M., Weishaupt, C., Fuhlbrigge, R. C., Kupper, T. S., Sayegh, M. H., and Frank, M. H. (2008). Identification of cells initiating human melanomas. *Nature* **451**, 345–349.

Seidel, S., Garvalov, B. K., Wirta, V., von Stechow, L., Schanzer, A., Meletis, K., Wolter, M., Sommerlad, D., Henze, A. T., Nister, M., Reifenberger, G., Lundeberg, J., Frisen, J., and

Acker, T. (2010). A hypoxic niche regulates glioblastoma stem cells through hypoxia inducible factor 2 alpha. *Brain* **133**, 983–995.

Sell, S. (2010). On the stem cell origin of cancer. *Am. J. Pathol.* **176**, 2584–2594.

Shah, N. P., and Sawyers, C. L. (2003). Mechanisms of resistance to STI571 in Philadelphia chromosome-associated leukemias. *Oncogene* **22**, 7389–7395.

Shipitsin, M., Campbell, L. L., Argani, P., Weremowicz, S., Bloushtain-Qimron, N., Yao, J., Nikolskaya, T., Serebryiskaya, T., Beroukhim, R., Hu, M., Halushka, M. K., Sukumar, S., Parker, L. M., Anderson, K. S., Harris, L. N., Garber, J. E., Richardson, A. L., Schnitt, S. J., Nikolsky, Y., Gelman, R. S., and Polyak, K. (2007). Molecular definition of breast tumor heterogeneity. *Cancer Cell.* **11**, 259–273.

Shupe, T., and Petersen, B. E. (2005). Evidence regarding a stem cell origin of hepatocellular carcinoma. *Stem Cell Rev.* **1**, 261–264.

Siekmann, A. F., and Lawson, N. D. (2007). Notch signalling limits angiogenic cell behaviour in developing zebrafish arteries. *Nature* **445**, 781–784.

Singh, A., and Settleman, J. (2010). EMT, cancer stem cells and drug resistance: an emerging axis of evil in the war on cancer. *Oncogene* **29**, 4741–4751.

Singh, S. K., Hawkins, C., Clarke, I. D., Squire, J. A., Bayani, J., Hide, T., Henkelman, R. M., Cusimano, M. D., and Dirks, P. B. (2004). Identification of human brain tumour initiating cells. *Nature* **432**, 396–401.

Snippert, H. J., van der Flier, L. G., Sato, T., van Es, J. H., van den Born, M., Kroon-Veenboer, C., Barker, N., Klein, A. M., van Rheenen, J., Simons, B. D., and Clevers, H. (2010). Intestinal crypt homeostasis results from neutral competition between symmetrically dividing Lgr5 stem cells. *Cell* **143**, 134–144.

Southam, C., and Brunschwig, A. (1961). Quantitative studies of autotransplantation of human cancer. *Cancer* **14**, 461–463.

Sullivan, N. J., Sasser, A. K., Axel, A. E., Vesuna, F., Raman, V., Ramirez, N., Oberyszyn, T. M., and Hall, B. M. (2009). Interleukin-6 induces an epithelial-mesenchymal transition phenotype in human breast cancer cells. *Oncogene* **28**, 2940–2947.

Taylor, M. D., Poppleton, H., Fuller, C., Su, X., Liu, Y., Jensen, P., Magdaleno, S., Dalton, J., Calabrese, C., Board, J., Macdonald, T., Rutka, J., Guha, A., Gajjar, A., Curran, T., and Gilbertson, R. J. (2005). Radial glia cells are candidate stem cells of ependymoma. *Cancer Cell.* **8**, 323–335.

Thurston, G., Noguera-Troise, I., and Yancopoulos, G. D. (2007). The Delta paradox: DLL4 blockade leads to more tumour vessels but less tumour growth. *Nat. Rev. Cancer* **7**, 327–331.

Visvader, J. E. (2011). Cells of origin in cancer. *Nature* **469**, 314–322.

Wang, R., Chadalavada, K., Wilshire, J., Kowalik, U., Hovinga, K. E., Geber, A., Fligelman, B., Leversha, M., Brennan, C., and Tabar, V. (2010). Glioblastoma stem-like cells give rise to tumour endothelium. *Nature* **468**, 829–833.

Waters, D., Newman, B., and Levy, M. L. (2010). Stem cell origin of brain tumors. *Adv. Exp. Med. Biol.* **671**, 58–66.

Weber, T., Baier, V., Pauly, R., Sahay, A., Baur, M., Herrmann, E., Ciccolini, F., Hen, R., Kronenberg, G., and Bartsch, D. (2011). Inducible gene expression in GFAP+ progenitor cells of the SGZ and the dorsal wall of the SVZ-A novel tool to manipulate and trace adult neurogenesis. *Glia.* **59**, 615–626.

Wei, X., Dombkowski, D., Meirelles, K., Pieretti-Vanmarcke, R., Szotek, P. P., Chang, H. L., Preffer, F. I., Mueller, P. R., Teixeira, J., MacLaughlin, D. T., and Donahoe, P. K. (2010). Mullerian inhibiting substance preferentially inhibits stem/progenitors in human ovarian cancer cell lines compared with chemotherapeutics. *Proc. Natl. Acad. Sci. U.S.A.* **107**, 18874–18879.

Weigelt, B., Glas, A. M., Wessels, L. F., Witteveen, A. T., Peterse, J. L., and van't Veer, L. J. (2003). Gene expression profiles of primary breast tumors maintained in distant metastases. *Proc. Natl. Acad. Sci. U.S.A.* **100**, 15901–15905.

Yustein, J. T., Liu, Y. C., Gao, P., Jie, C., Le, A., Vuica-Ross, M., Chng, W. J., Eberhart, C. G., Bergsagel, P. L., and Dang, C. V. (2010). Induction of ectopic Myc target gene JAG2 augments hypoxic growth and tumorigenesis in a human B-cell model. *Proc. Natl. Acad. Sci. U.S.A.* **107**, 3534–3539.

Zeng, Q., Li, S., Chepeha, D. B., Giordano, T. J., Li, J., Zhang, H., Polverini, P. J., Nor, J., Kitajewski, J., and Wang, C. Y. (2005). Crosstalk between tumor and endothelial cells promotes tumor angiogenesis by MAPK activation of Notch signaling. *Cancer Cell.* **8**, 13–23.

Zuna, J., Ford, A. M., Peham, M., Patel, N., Saha, V., Eckert, C., Kochling, J., Panzer-Grumayer, R., Trka, J., and Greaves, M. (2004). TEL deletion analysis supports a novel view of relapse in childhood acute lymphoblastic leukemia. *Clin. Cancer Res.* **10**, 5355–5360.

Cancer: Beyond Speciation

Mark D. Vincent

Department of Medical Oncology, London Regional Cancer Centre, London Health Sciences Centre, Ontario, Canada

Advances in CANCER RESEARCH, Volume 112
Copyright 2011, Elsevier Inc. All right reserved.
0065-230X/10 $35.00
DOI: 10.1016/B978-0-12-387688-1.00010-7

A good account of the nature of cancer should provide not only a description of its consistent features, but also how they arise, how they are maintained, why conventional chemotherapy succeeds, and fails, and where to look for better targets. Cancer was once regarded as enigmatic and inexplicable; more recently, the "mutation theory," based on random alterations in a relatively small set of proto-oncogenes and tumor suppressor genes, has enjoyed widespread acceptance.

The "mutation theory," however, is noticeable for its failure to explain the basis of differential chemosensitivity, for providing a paucity of targets, especially druggable ones, and for justifying the development of targeted therapies with, in general, disappointingly abbreviated clinical benefit. Furthermore, this theory has mistakenly predicted a widespread commonality of consistent genetic abnormalities across the range of cancers, whereas the opposite, that is, roiling macrogenomic instability, is generally the rule. In contrast, concerning what actually is consistent, that is, the suite of metabolic derangements common to virtually all, especially aggressive, cancers, the "Mutation Theory" has nothing to say.

Other hypotheses merit serious consideration "aneuploidy theories" posit whole-genome instability and imbalance as causally responsible for the propagation of the tumor. Another approach, that is, "derepression atavism," suggests cancer results from the release of an ancient survival program, characterized by the emergence of remarkably primitive features such as unicellularity, fermentation, and immortality; existential goals are served by heuristic genomic instability coupled with host-to-tumor biomass interconversion, mediated by the Warburg effect, a major component of the program.

Carcinogenesis is here seen as a process of de-speciation; however, genomic nonrestabilization raises issues as to where on the tree of life cancers belong, as a genuinely alternative modus vivendi. Philosophical considerations aside, genomic instability offers the prospect of subtle new therapies based on loss of information rather than gain; and the consistent, specific, and broad-spectrum perfidy of the Warburg effect highlights a supplemental target of the highest priority. © 2011 Elsevier Inc.

I. INTRODUCTION

A. Limitations of Heterogeneity and the "Many Disease" Model

The "many disease" view of cancer has, over time, successfully negotiated the transition from opinion to truism. This journey, as with all such journeys, has been rewarded by the removal of any obligation to exercise critical scrutiny, and so the "many disease" view has passed into conventional wisdom.

For is it not obvious that cancers involving different organs, different demographics, even different types of animals, with a wide range of distinctive etiologies, often specific treatments, and radically different outcomes, must surely be different diseases? And isn't this, in turn, now well supported by a panoply of unique molecular aberrations?

Well, these are facts and cannot be disputed. We should acknowledge the heterogeneity so minutely described and documented; we are even prepared to extend the concept, for example, to a group of patients with the same "type" and stage of cancer (e.g., "advanced non-small-cell lung cancer") who seem, for no obvious reason, to behave very differently from one another, whether treated or not. We also do not hesitate to invoke "clonal heterogeneity" within a given patient's cancer, as the automatic explanation for any change in tumor behavior: a sudden growth spurt or the development of drug resistance, for example.

But to accept the *fact* of heterogeneity, whatever its boundless utility across the multiple levels of the nosological hierarchy, does not mandate that we necessarily accept the whole *conceptual package*. In particular, heterogeneity is more of a description of, than an explanation for, the spectrum of cancer. Heterogeneity, even though it might in some superficial way, "explain" differential drug sensitivity, is not in itself an explanation of cancer; rather, it is the heterogeneity itself that requires explanation. At the very least, we should try to interrogate what has now become the dominant paradigm in clinical oncology, not only to extract what is there that is genuinely useful, but also to appreciate the limits of the concept, and, in the hope that deeper understanding might lead to therapeutic solutions, to obtain direction as to where else we need to search for better treatments.

B. The Modern Cancer Research Enterprise: A "Ship of Fools?"

In the clinic today, even in well-resourced facilities, and with patients not restricted by financial considerations, it should be thoroughly understood how bad the state of affairs is, in most cases of advanced common cancers. For example, in many aerodigestive cancers, we are flailing around with technologies from the 1960s (cisplatin), 1950s (fluorouracil), 19th century (radiation) and early 20th century (surgery); more recent innovations (e.g., taxanes, biologicals) are a decade or more old, and epitomizing the law of diminishing returns, tend to add disproportionately more to the cost than the efficacy. These and similar cancers, once metastatic, remain stubbornly incurable and survival increments over best supportive care (BSC) alone tally scarcely a year or two, on average. These modest clinical achievements, although worthwhile and even sometimes impressive, are meant to be constantly improved by the flow of knowledge from basic and applied research; however, in practice, this happens infrequently, despite the deluge of information, and the enormous dedication of thousands of people.

That such a vast and costly enterprise as modern cancer research should generate such meager returns on investment is cause for concern, and should prompt a question: as a society, do we really understand what we are dealing with, and if not, do we understand that we do not understand?

This essay represents an attempt to come to grips with these issues.

II. HOW TO "UNDERSTAND" CANCER

A. Characteristics Shared by all Cancers

All cancers, whatever their differences, are characterized by a number of common phenotypic features, without which they could not be grouped into a single categoric type. These features include progressive biomass accumulation of clonally derived cells, which arise within a pre-existing multicellular organism, but from inception, constitute tissue superfluous to adaptive requirements. Other definitions are possible (Vincent, 1987) but broadly similar. Cancers are a subtype of neoplasm, which is what is strictly being defined here, with the added proviso that all cancers manifest "invasiveness" in addition to superfluous cellular proliferation. Hanahan and Weinberg (Hanahan and Weinberg, 2000) identified six core traits that characterize cancer, that is, self-sufficiency in growth signals, insensitivity to antigrowth signals, apoptosis evasion, tissue invasion and metastasis, sustained angiogenesis, and limitless replicative potential. It is possible to consider expanding this "core" list to include the Warburg effect (WE), immortality, immune tolerance induction, autonomy, and "stemness"; but one at least which should be mentioned as "missing in action" from the original Hanahan–Weinberg is genetic instability; the recent update by Hanahan and Weinberg acknowledges "genome instability and mutation" as an "enabling characteristic" (Hanahan and Weinberg, 2011).

From this line of thinking, which so far has been very mainstream, we can draw at least two conclusions. Firstly, if all cancers are characterized by a core of certain invariant features, then this is what we should be trying to understand, and so there is a sense in which the "many diseases" paradigm is misleading; and secondly, the constituent cells are clearly "transformed" in some heritable way, and therefore have to have evolved out of normality.

These two conclusions are neither contradictory, nor without mutual relevance. Taking them together, we can come to two further conclusions: parsimony would seem to suggest that a single or at least a limited set of molecular alterations underlies this common core of phenotypic characteristics, and might be common to all cancers; and furthermore, what is common to all cancers has to be understood as an evolutionary phenomenon.

Note that the "single or at least a limited set of molecular alterations" does not necessarily imply in a concrete way that these changes are all exactly the same in each cancer cell (although some may be), but that the command and control structure in every cancer cell is affected in a manner that results in the same sort of functional abnormality. In other words, there may or may not be identical molecular abnormalities, but there always has to be at least one conceptually similar type of abnormality in cellular control mechanisms. To date, this shared type of abnormality is generally accepted to exist, and to be represented by a broadly similar set of aberrations affecting control of the cell cycle; so far these have proven difficult to "drug." I will argue here that additional, higher order commonalities characterize all cancer cells, which might prove superior targets.

B. Therapeutic Implications

From a therapeutic perspective, it would of course be an enormous advantage for there to be a limited or identical set of molecular abnormalities at which to target cancer drugs, across a range of cancers. Drugs, however subtle or indirect their mechanisms of action, are obliged to interact in a very concrete way with molecular targets, and the prospect of every cancer having a uniquely different set of molecular targets is actually a formidable, even insurmountable challenge to the prospects of ever developing curative therapy. Yet this is what the "many diseases" model of cancer seems to allow, and even imply, as does any naïve interpretation of "heterogeneity."

The "many diseases" model of oncogenesis is popular today not least because it matches so well the vogue for "personalized medicine," which is sweeping the cancer community as well as the broader health care industry. Personalized cancer medicine involves not just the patient-specific features that always mandated individualized care (e.g., co-morbidities, performance status, age, patient's wishes, etc.) but also the opportunity to use drugs that work in some patients but not others. These agents, moreover, should preferably be selected on measureable parameters in the tumor, that is, biomarkers, and most especially those that actually "drive" the cancer such as hormone receptor over-expression, growth factor receptor amplification/mutation, or other accessible oncogenic changes; in other words, the use of the so-called biologically targeted drugs. Successful drugs of this type would include tamoxifen for estrogen-receptor positive breast cancer; trastuzumab for HER2-positive breast cancer; imatinib for *BCR-ABL* transformed chronic myeloid leukemia (CML); sorafenib in renal and hepatocellular cancer; and gefitinib/erlotinib, two epidermal growth factor receptor tyrosine kinase inhibitors (e.g., EGFR-TKIs) for EGFR-mutated adenocarcinoma of the lung.

C. Personalized? Hardly

These therapies are not truly personalized in an individual sense, but apply (or not) to quite large subgroups of patients. In lung adenocarcinoma, for example, about 60% of never or light ex-smokers in East Asia are EGFR-mutated, and respond well to an EGFR-TKI (Mok *et al.*, 2009). Nonetheless, their total lack of efficacy in the remaining 40% of marker-negative patients mandates a sorting process that can be enacted in every patient of this nonsmoking profile. Just as it is, today, unthinkable to manage any breast cancer patient without the knowledge of the ER, PR, and HER2 status, it will soon be unthinkable to manage advanced non-small-cell lung cancer (A-NSCLC) adenocarcinoma (the commonest subtype) without EGFR mutational analysis. Within a few years, most of the oncogenic drivers of never-smoking lung adenocarcinomas will be routinely measured and even druggable, for example *β-RAF,* HER2, ALK translocations, etc. In this march of progress, never-smoking lung cancer (actually 25% of all lung cancer) (Parkin *et al.*, 2005) may be seen as a leading paradigm for other cancers in whom the druggable drivers (if indeed there are any) currently remain obscure.

Therefore, these trends do not exemplify personalized cancer medicine in the strict sense of the term; rather, they represent "subgroup cancer medicine," a term that does not have quite the same cachet. Nonetheless, there is a central problem with most of these targeted drugs; in patients with advanced disease, their beneficial effects are temporary (Table I) and acquired resistance is almost inevitable. For example, in EGFR-mt+ A-NSCLC, EGFR-TKI are able to control the disease for around 9–10.5 months (gefitinib) or perhaps a bit longer with erlotinib (12–14

Table I Duration of Benefit

Some inhibitory treatments directed against oncogenic drivers in advanced cancers						
Drugs	**Genes**	**Disease**	**Median duration of benefit**	**Parameter**	**Cure?**	**References**
EGFR-TKI	EGFR-Mt+	Adenocarcinoma of the lung	9.5–14 m	PFS	No	Mok *et al.* ibid 2009; Rosell *et al.*, ibid 2009; Paz-Ares *et al.*, ibid 2010
Crizotinib	EML4-ALK	Adenocarcinoma of the lung	>6 m	PFS	No	Kwak *et al.*, 2010
Sorafenib	Down- stream of ras	Renal cell; hepatocellular carcinoma	5.5 m 2.6 m	PFS PFS	No No	Kim *et al.*, 2011; Arranz *et al.*, 2011
Imatinib	Bcr-abl	CML	>5 yr	PFS	Maybe	Druker *et al.*, 2006
Trastuzumab	Her-2	Breast cancer	±3.6 m	PFS	No	Vogel *et al.*, 2002
Bevacizumab	VEGF	E.g., colorectal cancer	4.7 m	Incremental OS	No	Hurwitz *et al.*, 2004
Imatinib	C-kit	GIST	18–20 m	PFS	No	Blanke *et al.*, 2008
ARQ-197	C-MET	Adenocarcinoma lung	4.3 m	PFS	No	Schiller *et al.*, 2010
Afatanib	T790M mt of EGFR	Adenocarcinoma lung	3.3 m	PFS	No	Miller *et al.*, ibid 2010
Tamoxifen	Estrogen Receptor	Breast cancer	5–7 m	PFS	No	Falkson and Falkson, 1996; Bonneterre *et al.*, 2001
PLX-4032	ß-Raf V600E	Melanoma	6–9 m (est)	PFS	No	Sullivan and Atkins, 2010

months) (Mok *et al.* ibid; Paz-Ares *et al.*, 2010; Rosell *et al.*, 2009; Zhou *et al.*, 2010b). This is typical also of tamoxifen and other SERMs in ER+ breast cancer, for hormonal blocking agents in that disease (e.g., anastrazole, letrazole), in prostate cancer (with anti-androgens), in colorectal cancer initially responsive to anti-EGFR monoclonal antibodies (cetuximab, panitumumab), and renal cancer with respect to drugs like sunitinib, sorafenib, or everolimus. The only exceptions to this might (controversially) be imatinib in chronic phase CML or tamoxifen (or trastuzumab) in the adjuvant (post curative-intent surgical) settings, in which some patients may be nudged into cure via elimination of a small population of microscopic disease that surgery failed to eliminate. But, by-and-large, in the advanced setting, these drugs work for only a relatively short time; a 1 year period of good disease control, while very welcome in, say, a 47 year old with metastatic EGFR mt+ lung cancer, goes nowhere near indemnifying her against the loss of her life-years, since she will likely die about a year after failing EGFR-TKI, in the prime of her adult life.

Furthermore, these drugs do not work initially in everyone who is marker-positive, for reasons that are gradually being uncovered and which relate to genomic plasticity in the cancer cell. They work poorly or not at all in marker-negative patients; for instance, the most common subgroup of A-NSCLC is (ex)smoking adenocarcinoma, about 92% of whom are EGFR-wt, and which is typically lethal in approximately one year (Scagliotti *et al.*, 2008). For these much more common patients who are not candidates for currently available targeted drugs, conventional chemotherapy is the default option and may provide some useful palliation and modest life extension that is typically measured in 6–18 months additional time. Conventional chemotherapy is, of course, commonly, but wrongly regarded as untargeted, but I shall return to this egregious misconception later.

Inevitably, drug resistance emerges sooner or later in nearly all patients with the advanced, common cancers, and most of the uncommon ones as well, whether they are candidates for highly targeted drugs or the so-called untargeted chemotherapy. It is also worth noting a paradox here that few if any advanced cancers have become curable as a direct result of modern biologically targeted drugs; and indeed the cures of aggressive, and otherwise lethal cancers wrought in the 1970s in respect of, say, lymphoma and germ cell cancers, were due to the actions of classical chemotherapeutic agents.

D. Resistance—You Can Count on It

Although there is persistent the hope that drug resistance might be reducible to a limited number of pharmacologically tractable mechanisms, this has rarely proven possible. Drugs that "deal" with the T790M exon 20 mutations and C-MET amplification in EGFR mt+ lung cancers have emerged and appear to be modestly active (Miller *et al.*, 2010; Schiller *et al.*, 2010;) but attempts to deal with the drug extrusion pumps like p-glycoprotein (ABC B1), described over 30 years age (Hall *et al.*, 2009), have proven almost uniformly unsuccessful (McHugh and Callaghan, 2008) despite *in vitro* work that appeared to guarantee that p-gp inhibition was not only possible, but should work in patients with this form of "pleiotropic" drug resistance, and which led to major investment in drug development and failed clinical trials.

Furthermore, some undoubted drivers (e.g., mutated *K-RAS*), after 3 decades of intensive research, remain stubbornly undruggable (Young *et al.*, 2009). *K-RAS* is certainly associated with, and may actually cause, relative or absolute drug resistance to the EGFR-TKIs in lung cancer (Mao *et al.*, 2010), and to the anti-EGFR moAbs in colorectal cancer (Linardou *et al.*, 2008). *K-RAS* mutations, furthermore, may be associated with lack of benefit to adjuvant chemotherapy with cisplatin/vinorelbine in resected NSCLC (Butts *et al.*, 2010); however, data from other trials in A-NSCLC with chemotherapy does not necessarily substantiate *K-RAS* as a factor particularly associated with chemotherapy resistance, and the author has personally documented a dramatic response to pemetrexed single agent in a proven case of *K-RAS* mutant lung adenocarcinoma in an ex-smoker. Nevertheless mutant *K-RAS* is a prominent oncogenic driver and yet remains undruggable (McCormick, 2010).

Likewise, multiple other resistance mechanisms have been well described, but remain largely beyond the reach of modern ingenuity. Drugs like bevacizumab seem to exercise modest benefits not necessarily by cutting off blood supply, but paradoxically, by improving blood flow into tumors and hence drug delivery (Cao, 2010; Jain, 2005; Willett *et al.*, 2007). Generally, the literature is full of optimistic early results, and dashed hopes thereafter; the recent relatively disappointing results in advanced breast cancer with a PARP-inhibitor may be a case in point (Sanofi Aventis, 2011).

In summary, drug resistance, especially "acquired" resistance after a period of responsiveness, is indeed further evidence of the fact of heterogeneity in cancer, and of the underlying genomic plasticity and clonal evolution. It is proof that cancer is not only an evolutionary phenomenon in its emergence (carcinogenesis), but that further evolution occurs after the initial transformation, and which is characterized by heritable variation ("descent with modification"), the grist to the mill of natural (and unnatural) selection. The specter of most concern however, is that both oncogenic drivers and drug-resistance mechanisms, at least those that might be druggable, could be highly variable patient-to-patient and even within the same patient's cancer. This nihilistic perception is driven by, or at least highly congruent with the standard evolutionary interpretation of cancer (*vide infra*), which relies heavily on serial random mutations.

However else cancer is understood (and there is no monopoly on this) it is inescapably true that a full understanding requires a clear appreciation of its evolutionary origins and character. This is not new, of course, but it is nevertheless appropriate to describe what the conventional wisdom has to say about this before advancing an alternative (or at least complementary) hypothesis, as an antidote to the naïvite, or the therapeutic nihilism that the conventional wisdom might seem to encourage. This complementary hypothesis is, it should be immediately stated, is one that is also supportive of, reliant upon and consistent with Darwinian principles.

III. "CONVENTIONAL WISDOM": THE DARWINIAN VIEW OF CANCER

A. Evolutionary Principles

Darwin, Wallace and Herbert Spencer, heavily influenced by Malthus, advanced several ideas to account for the development of biological diversity on earth (Table II). These ideas form the core of what became known as Darwinism.

Although initially controversial, the elegance of these ideas, their explanatory power, and the enormous and painstaking evidence collected by Darwin and others gradually led to their acceptance, not only in mainstream biology, but also in common parlance. However, the mechanism(s) by which evolution operated were entirely unknown, until the re-discovery of Mendelian genetics and its reconciliation with Darwinism in the decade 1936–1947: the "Modern (evolutionary) Synthesis' (brought about by Huxley (Huxley, 1942), Fisher (Fisher, 1930); Dhobzansky

Table II Key Ideas of "Darwinism"

1. Common descent
2. Variation within populations
3. Descent with modification
4. Overpopulation
5. The struggle for existence (competition)
6. Natural selection
7. Survival of the fittest

(Dhobzansky, 1937); Mayr (Mayr, 1942); Simpson (Simpson, 1944); Haldane (Haldane, 1924–1934), Wright (Larson, 2004), and others).

Recent developments in molecular biology have only served to confirm both Darwinian evolution and the modern synthesis. Dawkins is associated with the "selfish gene" concept, which implies that gene preservation and transmission is an evolutionary priority (Dawkins, 1976). Nonetheless, surprisingly large areas of controversy remain, such as the precise mechanisms of speciation; the extent to which saltationism ("large jump" evolution, as opposed to gradualism) has contributed to, and continues to contribute to evolution; the role of epigenetic imprinting; the reasons for the evolution of sexual as opposed to asexual reproduction; and the precise grounding of the *tree of life*. Some of these controversies remain relevant to the evolution of cancer; in particular, the disputes around the primacy of the gene, as well as gradualism versus saltationism, should be noted.

Saltationism involves evolutionary change by large jumps in which intermediate forms do not, and never did, exist. This sort of evolution was named the "hopeful monster" hypothesis by the geneticist Richard Goldschmidt, who supported it as a kind of instantaneous speciation involving ". . . a complete change of the primary pattern or reaction system into a new one" (Goldschmidt R, 1940). By contrast, Darwin himself favored gradualism, encapsulated in the quote "Natura non facit saltum," (i.e., nature does not make large jumps). Although it is currently difficult to obtain a true consensus statement about this, it seems that modern biologists generally favor gradualism while admitting occasional saltation-type exceptions, most notoriously the appearance of life itself, and the development of the eukaryotes. The "discovery" of sexual reproduction and the appearance of multicellular organisms are also candidate examples of saltationism. Furthermore, just how rapidly new species can evolve is also a matter of contention, although most seem to suggest that speciation usually occurs by reproductive isolation ("allopatric speciation"), coupled with a gradualistic series of small mutations and, perhaps, genetic drift. The prudent position for nonspecialists seems to be agnosticism on this issue.

The consistent message of the conventional wisdom is that however evolution proceeds, whether slowly or rapidly, it proceeds mainly, if not exclusively, by the action of serial mutations, which are themselves utterly random and stochastic. It is often said that "selection (in evolution) has no eyes for the future" (Williams, 1966), by which is meant that the accumulation of damage in DNA does not proceed according to any pattern, and that the phenotypic consequences are solely the result of the cumulative effect of these mutations interacting with each other and the unmutated DNA, resulting in a system that is complex, largely unpredictable and indeed truly chaotic in a mathematical sense.

B. Evolution of Cancer: The "Standard Model"

The standard model of carcinogenesis states that serial, random unrepaired mutations in regions of DNA, central to the command and control of the cell cycle, produce, in surviving cells, a condition of perpetual clonal expansion insensitive to growth-inhibitory signals. Further mutations accumulate randomly, disrupting not only cell number homeostasis but also positional homeostasis, such that invasion, a crucial hallmark of malignant as opposed to benign neoplasia, then occurs. Variants that are "more aggressive" are simply selected out because they have a "fitness" advantage, that is, an advantage in growth and proliferation, and succeed in not only expanding at the expense of the normal body, but also at the expense of less aggressive clones of erstwhile sibling cancer cells. This concept, incidentally, is pure Malthus, who believed that over time, population growth rate would come to approximate that of the most fertile group (Malthus, 2000).

In the version explicated by Greaves (Greaves, 2007), this paradigm is imposed upon normal stem cells, which are then unable to exit the cell cycle, via the usual pathways of apoptosis, senescence, post-mitotic differentiation, or return to quiescence, because each of these pathways has been disabled by a mutation. Thus the normally constrained program of clonal expansion is unleashed for no particular reason, except the occasional failure of DNA repair mechanisms, which then permit rare cells to serially and progressively escape the various bottlenecks imposed by multicellular existence to prevent just this exigency. In this view, serial, rare, but random hits on key DNA control machinery progressively free occasional cells to compete successfully in harsher and harsher environments, in which only the most robust and aggressive are able to survive and proliferate. In this cascade, the mechanisms underlying genetic stability are eventually themselves affected, leading to "genetic diversity" and, inevitably, drug resistance, on a purely random basis. Because these mutational events are undirected, and need to be multiple in order to fully transform the cell, this process usually takes decades. Notwithstanding this, cancer is made possible in the first place by the "lack of perfection" in evolutionary engineering, which is required to provide some plasticity in DNA to allow embryonic sculpting and indeed evolution itself.

In short, the conventional view might be summarized as saying that everything of importance in a cancer cell is due to a random mutation, and the variants so produced are simply selected based on relative proliferative advantage, itself a hereditary trait.

This description combines Darwinian mechanisms of random variation, differential fitness, and natural selection, with the action of initiating exogenous mutagens, working with various promoting factors, acting in the context of cellular hierarchies, and resulting in the unrestricted and inappropriate depression of normally well-controlled programs of cell renewal and migration. Critically, in this view, there is absolutely no purpose to this, and although all cancers are meant to exhibit causative lesions in the same pathways, the exact position of these lesions is random; furthermore, they exist with multiple other relevant and irrelevant mutations, which at a level of detail are unique to each cancer, and to some extent to each cancer cell. Therefore although a naive version of the standard model would have only the same, very few genes always implicated in every patients across the range of cancers, a more sophisticated understanding requires only the same *type* of genetic aberration, with much irrelevant variation in the detailed minutiae.

Some versions acknowledge higher levels of genetic instability than can adequately be accounted for purely by random mutations, and indicate that DNA stabilization mechanisms might themselves have suffered disablement via the same random mutations, and having failed to fix themselves up, then contribute to DNA hyperinstability, with the net result that variant form production is increased. Nonetheless, most of this "secondary" variation is perceived, in the standard model, to be "genetic noise" (Heng, 2010 ibid).

The question, of course, is how much of this variation is cause and how much is effect, in terms of driving the malignant phenotype and/or drug resistance—and exactly how much has to be "targeted" therapeutically to regress the tumor, given that causality-reversal is conventionally considered a therapeutic strategy worthy of pursuit. Irrespective, these observations, on the fluidity of the cancer genome, and which seem now to be multiply confirmed, pose the gravest problems for the current simplistic "targeted" paradigm of therapy, as it struggles to seek validation in common cancers like those of the smoker's lung, the breast, the pancreas, etc.

C. Problems with the Standard Model

As others have realized (Israël, 1996; Heng, 2010 ibid), the standard model has done little to help in the clinic, except to predict that drug-resistant forms will emerge as an adaptation to the particular stress of anticancer therapies, and may already have emerged ("by chance") even before the start of therapy. However, we knew this already from experience. The standard model does not tell us which targets exist, which to attack, or how to do it. These shortcomings, *per se*, do not mean that the standard model is necessarily all wrong; inevitably, however, we are encouraged to look elsewhere for a more helpful theoretical framework. Nonetheless, while one purpose of this chapter is to critique the standard model, it is also important to recognize, but not to overfocus on, the "straw man," or naive version of the standard model (Table III).

Table III Models of Carcinogenesis: Criteria

	(Standard model, basic) "straw man"	**(Standard model, advanced) "mutation theory"**	**Genomic or karyotypic theory (Heng–Duesberg causality)**	**Programatic: de-repression atavism ("the animal within")**
Mutation	Series of random mutations	Series of random mutations	Both cause and consequence, but more the latter	Carcinogens represent a survival threat via mutation
Pathogenesis	Dependency on a few consistent changes in a few consistent genes; causal consistency	Ongogene/antioncogene dependence, continuously; aneuploidy just "noise"; causal consistency	Oncogenes initiate but later redundant or even absent, and not required; unstable aneuploidy the driver, takes over; causal inconsistency	Mutations dysregulate cell cycle, and trigger an ancient survival program, the Warburg effect, as the enabler; agnostic as to causal consistency
Natural selection	Irrelevant	Operative	Operative	Operative
Common phenotypic traits	Determined by a small set of genetic mutations	Stem cell self-renewal program de-repressed; WE not implicated, under emphasized; other traits explained by convergent evolution	Not emphasized	Re-emergent animal phylogenation; reprimitivization; adaptive resilience; Warburg metabolism; genomic scrambling
Genetic instability	Not envisaged	Acknowledged; due to secondary mutations; not central, but allows further evolution in aggressivity and resistance	Causally fundamental, the real driver	Key component of survival program; purpose is discovery of better-adapted genome

Heterogeneity	De-emphasized; only due to maturation hierarchy, persistent tissue-of-origin features	Acknowledged; accepts role of serial but random mutations, usually causing resistance	Emphasized, due to whole genome instability; unavoidable	Emphasized as central purpose; due to endogenous mutagen and "tolerator" function of the program
Priority	Gene-centric	Gene-centric; genetic instability an emergent epiphenomenon, but not central	Genes less important, genomic mayhem more important	Gene propagation depriorized; priority now is "any cost cellular survival" via genomic novelty-seeking
Targets	A few, common genetic aberrations across a whole range of cancers	Constant continuing drivers exist, just need to be found and inhibited; confounded by secondary genetic noise	No or few constant drivers exist	Drivers and markers exist, but vulnerable to reshuffling; the WE, however, never rescrambled
Purpose	None	None	None	Strongly purposed—"for survival"; explains phenotype
Solution	Inhibition of a few common drivers-naive	Individualized driver inhibition; either naïve or nihilistic	None suggested yet; nihilistic?	Overwhelming firepower via "absence" targeted therapies; combined with inhibition of WE, or of mutator phenotype

The standard model, however sophisticated, represents a "train wreck" perspective: everything is due to random accidental damage, nothing has purpose, every cancer is different, in the sense that individual DNA lesions are inconsistent, and the only known commonality is that at some level, at least, the cell cycle is stuck in the "go" position; more shadowy derangements are inferred to be present to account for invasion; and within the multiplicity of mutational abnormalities, the drivers are to be found, comprising an unknown (and perhaps unknowable) ensemble of (by definition) activated oncogenes and inactivated tumor suppressor genes. Except perhaps in childhood cancers, the path to carcinogenesis is gradualistic, and proceeds over decades. The minimum essential genetic lesions required to be targeted for therapy are unknown but discoverable in principle. Conscious of the need to explain ongoing post-transformational evolution, the model involves random damage to DNA homeostatic mechanism themselves, to explain clonal evolutionary exigencies like acquired drug resistance, as well as the emergence of much nonconsequential and noise-like variation; however, this DNA instability is not felt to be central to the malignant phenotype *per se*.

The key aspect of this model is that phenotypic variants can only, and do, appear entirely at random; if they are advantageous, they are selected and enrich the tumor thereafter. Those that are advantageous survive; those that survive must have been advantageous. In this view, the fact that all cancer cells, in all people, at all times, come to exhibit this nearly identical set of core traits has to be seen as a fantastically coincident type of convergent evolution, although this is not often acknowledged; rather, emphasis is placed on the proximate drivers that are known or believed to exist, in different embodiments, in different cancers, and different patients.

The standard model does not generally mention or analyze the fact that cancer cell proliferation occurs in an asexual fashion; that somehow the cells escape Muller's Ratchet (by which the disadvantageous mutations that occur in asexual organisms cannot be purged from the genome); that the autonomy observed implies that the cells have clearly ceased to be part of the multicellular animal, or that a highly characteristic, inevitable, and almost invariant metabolic program, the WE , suddenly appears as if by coincidence, in most and usually every aggressive cancer cell; and that very gross genomic changes occurring from cancer cell to cancer cell, and from one generation to the next, constitute a major departure from gradualist evolution, and threaten cozy assumptions about cause and effect. Nor does it note that occasionally—very occasionally, but categorically—descendant cancer cells can escape the dying host and be transmitted to a new host, as a whole cell, where they may successfully set up shop, and repeat the process over and over again, like any competent parasite (Murchison, 2008; Welsh, 2011).

As will be discussed by another contributor (Heng, 2009; Heng *et al.*, 2006, 2010), a major failing of the standard model, even the more sophisticated version which accommodates hypermutation, relates to its inability to account for the sheer amount of mutational heterogeneity between patients, within the same patient, and even from one generation to the next in a particular tumor cell lineage. Heng's viewpoint is that observations from the cancer genome sequencing project have failed to confirm that there is a universal pattern of genetic alterations shared among most cancers, revealing rather the opposite: genetic heterogeneity is much greater than expected, and the commonality much less.

This is regarded (by some, at least) as a threat to the gene-centric concept of cancer evolution, as contained in the standard model; and which should lead to alternative conceptualizations that account for the major consistent genetic finding, that is, macro-evolutionary instability, clearly evident at the genomic, and karyotypic level. In this latter viewpoint, cancer results as a combination of the more important macro-evolutionary rearrangements, evident karyotypically, and less important micro-evolutionary level (genetic and epigenetic) alterations that fine-tune adaptation. This viewpoint is unquestionably not only chaotic (sensu stricto), but especially saltationist, incompatible with organismal integrity, and a challenge to the standard model; primarily because the latter in no way envisages this large-scale whole genomic instability, cannot explain how this level of instability is compatible with ongoing cellular life, and requires that certain types of gene be consistently mutated in every cancer, by chance, but which is (apparently) not actually the case to the extent it should be.

It might, however, be argued that at least some sort of disablement of senescent, quiescent, apoptotic, or differentiation pathways is consistently found, and probably is necessary, and since massive genomic instability is most unlikely to reconstitute these normal exit strategies, that the standard model is not wholly incompatible with the "genome theory" of cancer elaborated by Heng *et al.* (ibid 2010). Also, it is wrong to suggest that at least some common drives (e.g., *K-RAS* in pancreatic cancer, and some lung and colorectal cancers) do not occasionally occur.

Nonetheless, what Heng *et al.* raise is that the precise mechanism of disablement is not necessarily constant from case to case or, more important, generation to generation within the same cancer and this hypervariability was never envisaged by the standard model, in which the malignant phenotype is seen as mediated by the continuing presence of particular genes, albeit with some potential for additional mutations to drive a more aggressive evolution. The reason that this is so serious is that drug treatments are going to fail inevitably sooner or later to the extent that driver functionality of any particular target becomes no longer necessary, or even disappears entirely, to be replaced by something else, and not just because the target molecule itself might mutate further. What this teaches is that targeting extant drivers is unlikely to have more than a temporary benefit (and actually, this is almost always confirmed in the clinic, see Table I); I will suggest later that, as an alternative strategy, what might be better targeted, perhaps, is the *absence* of something.

One philosophical concern with the standard model is that it does not attribute biological purpose to the cancer cell. This does not mean that it would have been designed with the future "in mind"; I am categorically not making an argument for intelligent design, just that the machinery of the cancer cell, and the phenotypic traits that result, could actually have a biological function. This would be just like any normal cell, constructed to fulfill a biological role, based not necessarily on the idea that the future is known but that it is at least overwhelmingly likely at any given time, to be very similar to the past.

In this respect, we should note the less famous quote by Dhobzhansky (in the same essay in which he more famously said that nothing makes sense except in the light of evolution). He said: "The enterprise of biology rests chiefly on two patterns of explanation. One is the organism-the-machine theory. The other is the theory of evolution . . . these explanations are complementary, without being either deducible from, or reducible to each other" (Dhobzhansky, 1964). In other words, there are two ways to understand anything in biology (or indeed anything that is naturally or artificially constructed): firstly, to strip it down into its component parts, itemize them, see how they fit together, etc.; and secondly, to understand what the whole structure is *for*. What it does, how it does it, and above all, why it does it; in other words, what is it *for*.

So dare we ask, *what is the cancer cell for?* Many would object to this question as invalid. Must it necessarily be for something? For some purpose? Why can't it just be a train wreck? After all, train wrecks exist, and they are not "for" anything but can only be understood as a failure to achieve the original objective. Engineering is not perfect, and so accidents happen; why should biological engineering be any better? The obvious production of occasionally nonviable offspring is a clear proof, if it were needed that biological engineering (i.e., "natural" history) is far from perfect and, for reasons of parsimony and efficiency, needs only to be "good enough." Surely, it could be argued that cancer must be like this, that is, a systems failure occasioned by the unattainably high price of, and lack of, necessity, for perfection. This view is not unreasonable, and a very much part of the standard model.

It may also be right. But we are entitled to at least ask the Dhobzhansky question "could the cancer cell be for something"? "And if so, what"? The standard model (at least the recent version) maintains that the target cell of transformation could be the normal stem cell; in a variety of tissues (see the paper by A. Pietras in this volume) normal stem cells possess a program of self-renewal. This program, in multicellular organisms, is normally constrained by a complex hierarchy of controls. Furthermore, this program is retained from the premetazoan era, that is, from the unicellular organisms that evolved into metazoa, and is clearly evident in unicellular organisms today. Mutations causing cancer disable one or more of the control systems regulating the cell cycle. As a result, the unrestrained cell cycle is unleashed. The transformed cell comes to behave as if it were a unicellular organism, since these same controls (now disabled) not only restrain the cycle, but tether the cell to the requirements of the organism as a whole. So, although the resulting cancer cell does not have a purpose, or function, *per se*, it manifests a program unbridled (self-renewal), which *at one point in evolutionary history,* certainly did have a purpose, and in the intact metazoan host, has a purpose but only to the extent it is harnessed to the needs of the entire organism.

Certainly, this particular interpretation of the standard model, while uncommon, is reasonable. But I shall try to push beyond this to see if other explanations are tenable. In particular, I will first look at an interesting paradigm, which I will call "Heng–Duesberg causality" in honor of its proponents; and then also my own perspective that carcinogenesis, and subsequent material events of clonal evolution, constitutes bona fide episodes of speciation, and even beyond that, a novel form of life itself. I shall now explain these theories, at the same time attempting to explore implications for therapy.

IV. ALTERNATIVE INTERPRETATIONS OF CANCER

A. Source Documentation

These deliberations are based on essays written by this author, one published (Vincent, 2010) and another in press (Vincent, Bio Essays in press 2011). I will also draw on a third essay (Vincent, in preparation). These essays deal with unorthodox conceptualizations of the malignant condition, and the therapeutic implications of these views. They are not offered as revealed facts, and even less as conventional wisdom, or indeed any kind of wisdom; but as alternative conjectures that are consistent with the known facts, with the general thrust of evolutionary theory, and which, in turn, have implications for those interested in therapy. However, I will first review some important precedent views, by two other sets of authors, which are somewhat consistent with, or bear upon, my own thoughts. Because these authors' views on the maintenance of the malignant phenotype substantially overlap, I will group them together.

B. Heng–Duesberg Causality

In fact, most of the conjectures that I entertain are not new. The concepts of cancer as a novel species has been raised by several authors as an analogy (Pearson, 1981; Vineis, 2003) and even as a matter of fact (Duesberg and Rasnick, 2000; Huxley, 1956). Duesberg and Rasnick make the argument (for speciation) based on karyotypic alterations, which are often gross and in fact (within parameters) continuously unstable. Recently published (Klein *et al.*, 2010) is a compelling set of data illustrating that initiating oncogenes may even, in certain cases, not be necessary for the subsequent maintenance of the malignant phenotype, and these authors further provide a review of the industrial-scale karyotypic diversity and phenotypic heterogeneity found even in tumors arising in the same target cell and initiated by the same oncogene. This is very similar content to the previously mentioned writings of Heng *et al.* (ibid 2010), for whom the malignant phenotype is maintained by whole-genome instability and not specific oncogenes. Both sets of authors contrast their positions with the "mutation theory" (i.e., standard model) of cancer, which predicts a *continuing* requirement for the initiating oncogene, in order for the cancer to be maintained, and which together with a relatively stable set of cooperating genetic alterations is meant to produce, in each cancer, a relatively constant phenotype (Lewin, 1997). Somewhat similar to the "genome theory" of Heng *et al.*, Duesberg's group proposes the "karyotypic theory," which can perhaps be summarized, to quote and paraphrase (Duesberg *et al.*, 2003; Fabarius, ibid 2003):

- The degree of malignancy is proportional to the degree of aneuploidy.
- Copy numbers of a given chromosome are often not the same in all cells of a given cancer as a result of the inherent instability of aneuploidy.
- Acquisition of drug resistance correlates with a specific karyotype.
- Oncogene-independent transformation occurs; that is, the initiating transgenic oncogene may later be dispensable, functioning as a carcinogen rather than a cancer gene.
- Carcinogens (including artificially activated oncogenes) initiate carcinogenesis by inducing random aneuploidy.
- Individual karyotypes maintain cancers, much as they maintain species.
- Aneuploidy destabilizes the karyotype by unbalancing teams of balance-sensitive mitosis genes; aneuploidy is inherently unstable.
- The resulting instability catalyzes the evolution of new cancer-specific karyotypes, *much like species-specific karyotypes* (emphasis added); and these cancer-specific karyotypes can function as "additional oncogenes."
- Clonal karyotypes (as opposed to the initiating oncogene) determine the phenotype of cancers, that is, maintain the (existence of) the cancers, as well as their specific characteristics.

- Transgenes function as carcinogens, catalyzing independent evolution of new cancers.

Duesberg and coworkers thus believe that "carcinogesis is a form of speciation," and that the appearance of new cancer-specific karyotypes is "much like species-specific karyotypes." Furthermore, and tellingly, "carcinogenesis is proposed to be much like phylogenesis," a theme to picked up below in connection with Huxley.

C. Speciation Criteria: Gene Flow versus Trait

As to the mechanisms underlying this genomic reshuffling and unstable aneuploidy, solutions have been suggested by Gisselsson and others (Gisselsson *et al.*, 2001; Gisselsson, 2011a,b), involving telomere dysfunction, defects in the spindle assembly checkpoint, merotelic attachments, segregation errors, and perhaps a combination of mechanisms such as multiplier mitosis and/or incomplete cytokinesis, with some sister chromatid nondisjunction. The issue here is not so much the mechanism(s) by which continuous genomic instability is produced, but the consequences; however, if it is to be part of some program, then it must be explained how the mechanism is consistent with this, and whether it offers any prospects for genomic restabilization.

It may appear to a "nonspecialist" that speciation is always associated with karyotypic differences, but this is not quite true. Rather, what is key is the cessation of gene flow, which may be brought about opportunistically (e.g., by isolation of subpopulations, for various reasons). Gene flow is, of course, a concept that applies best to sexual species, or those asexual ones that employ horizontal gene transfer (HGT); cancer cells are asexual organisms, and HGT is not well documented among them. The conventional model of speciation indicates that it is caused by incompatible alleles, which have become alternatively fixed, in two previously isolated populations; this is known as the BDM model (after Bateson, Dhobzhansky, and Muller) (Gavrilets, 2003). Chromosomal rearrangements are a relative and sometimes absolute barrier to successful recombination, or contribute to gamete nonviability by imbalance, and/or segregation problems in gametogenesis.

However, cancer cells are asexual and appear not to share genes by HGT. Therefore, each one is a genetic island, and to extend the metaphor, each lineage, an archipelago. Therefore since gene flow does not occur, except via lineage descent, karyotypic differences, which strongly imply speciation in sexual species, may matter less in the speciation discussion in respect of cancer cells versus their original host, or versus each other, in the same cancer lineage.

Consequently, while I do believe that karyotypic variation supports the idea of cancer as a different species from the originating host, and additional, clonally ordered karyotypic distinction post carcinogenesis supports the existence of subsequent rounds of speciation, there is an element of analogy in this, and so I chose to pursue the idea of cancer as species from a trait perspective, taking a cue from Julian Huxley, who focused on autonomy as a central phenotypic characteristic. "Once the neoplastic process has crossed the threshold of autonomy" he said, "the resultant tumor can be logically regarded as a new biologic species" (Huxley, ibid 1956), and even"... as constituting a special organic phylum or major taxonomic group ...." I shall return to this concept of "phylogenation," while noting here that Huxley was using it, I think, in a metaphorical sense to indicate the existence of a vast chasm between the originating host and its descendent cancer cell population, in that they are no longer a constituent part of the same individual; at the very least, it is legitimate to describe carcinogenesis as an act of "individuation." While this is not necessarily pathological (childbirth, gamete release, and budding might all be normal examples), in the case of cancer the descendent cells are not only autonomous but also destructive of the original host, and their cells are now in unequivocal competition. While intraspecific competition is not uncommon, for good reasons it tends to be limited, whereas interspecific competition (at least in the same niche, as is the case here) is much more likely to be destructive, and fully transformed cancer cells are indeed always destructive of their host. As to whether they constitute "... a special organic phylum ..." is arguably true since, although cancers in different victims are more closely related to their originating hosts than to each other, they are better conceived as unicellular organisms, or at most a loose and facultative colony, and most categorically not any more a multicellular animal, or part of a multicellular animal. This might, because of a striking resemblance among all cancer cells, constitute a separate "special organic phylum," in some literal sense. Or, going beyond even this, as I will later elaborate, cancer cells may constitute an entirely different form of life.

D. The Animal Within: Trait-Inferred Speciation

"The animal within" (Vincent, ibid 2010) illustrated various definitions of species (a strangely inflammatory topic) and concluded that although speciation in asexual organisms constituted a particular problem in definition, certain of the definitions did embrace the concept satisfactorily. In particular, the "cohesion species concept" (CSC) of Templeton (Templeton, 1989), together with the "evolutionary species concept (ESC) of Simpson (Simpson, 1961), proved most useful. The latter posits a lineage-relatedness series, evolving separately, with "its own unitary evolutionary role and tendencies." The CSC describes forces of genetic cohesion, which in asexual species would tend to be genetic drift with natural selection, to ensure genetic similarity. One can readily see how a population of asexual organisms, for example, bacteria, occupying a uniform ecological niche, and relatively isolated, could retain a large measure of genetic identity simply because of the convergent forces of natural selection. This does not, of course, necessarily imply genetic or genomic stagnation, as long as the entire population evolves more-or-less in concert, and the uniform application of the same forces of selection to the whole population might keep interindividual variation generally below a certain critical threshold, even in the absence of any HGT.

Nonetheless, it is probably worth re-examining the true extent of cohesion that actually exists in a population of cancer cells, either in culture, *in vivo*, or in the original patient. What appears to be the case is that although large scale differences can and often do occur between cancer cells in the same cancer, and a cancer cell and its descendants, there are broad karyotypic parameters that tend to bound this variation within certain limits. On the other hand, when, for example, a particular drug is introduced, the karyotype adjusts with the evolution of a drug-resistant subline; and the same may be true of a cancer cell as it comes to occupy and adapt to a different niche such as a different metastatic site. Trait differences of sufficient materiality might well constitute further rounds of speciation (or subspeciation, depending on the degree of difference and its clinicopathologic significance). However, it is reasonable to ask whether the obvious genetic plasticity of cancer cells to some extent undermines the notion that they should have attained the status of a novel species, if "species" is to be understood as a "type," and a "type" is to be invested with constancy, or consistency.

It does not, however, undermine the notion that cancer cells, whatever they are, are categorically different from the species of the original host animal (or human), and that at least a separational speciation event has taken place, and, with the emergence of unicellularity, possibly something more than "mere" speciation: the emergence of a holozoan opisthokont (animal-like) protist, with facultative colonial attributes, and which might be labeled with the title *dyskaryota*.

V. CANCER AS SPECIATION: IMPLICATIONS FOR BIOLOGY AND THERAPY

A. Words and Definitions Need to Matter

What, it might be asked, does it matter if we choose to label carcinogenesis as an act of speciation? After all, definitions can be arbitrarily changed; but this seems an empty exercise unless there are tangible gains in knowledge, efficiency, or cancer management. Let us now consider the implications for biology, and then therapy.

B. Biology

Does the issue of "cancer as species" tell us anything about speciation in general? It seems unlikely that speciation in sexual organisms, in which one expects a lineage to separate into two closely similar but stably distinct types following some geographic isolation, has much in common with carcinogenetic speciation, which requires not only a sudden, vast taxonomic leap but also genetic instability.

What about speciation in asexual organisms? It seems also unlikely that we have much to learn here either. Asexual speciation probably proceeds under similar conditions, that is, geographic isolation, or at least alternative niche occupancy, which, combined with mutation or at least genetic drift, and influenced by differential forces of natural selection, drives novel alleles to fixation, resulting in the eventual emergence of two different but genetically stable genomes. This does not appear to be a good description of carcinogenesis either, although it might be informative for subsequent rounds of clonal evolution in an already extant cancer.

What is distinctive about carcinogenetic speciation is:

1. the emergence of a unicellular, perhaps colonial organism from a multicellular organism
2. the emergence of an asexual organism from a sexual one
3. the emergence of a genomically (and genetically) highly unstable organism from a genetically highly stable one
4. the emergence of a novel type that, while *de facto* closely related by descent from its originating host, is grossly dissimilar to this host, and therefore unrelated by resemblance
5. the emergence of a novel type whose existence is not only competitive with its original host of origin, but is, under natural conditions, eventually incompatible with it
6. a process that occurs rapidly, much more rapidly than the other types of "usual" speciation
7. the process of carcinogenetic speciation is both temporally and morphologically nongradualistic, that is, is saltationist

This latter point, however, should be qualified: it is acknowledged that some cancers go through a premalignant phase, during which time they are partially transformed; the adenoma–carcinoma sequence in the colon is just one of several good examples. Yet the adenomas, already autonomous, unicellular, and superfluous with respect to growth, have already crossed the Rubicon; they are bona fide neoplasms, and, in my opinion, have already speciated. The bigger point, however, is that there are no intermediate body-forms—no "missing links"—between the metazoan host and the unicellular neoplastic colony; the process is essentially saltationist. The further progressions to frank malignancy are but additional rounds of speciation, and may or may not themselves be more gradualistic. But the biggest jump is saltationist especially when viewed against the more usual evolutionary time frames.

What I feel is most distinctive about carcinogenetic speciation is the lack of capacity to permanently restabilize the genome in an alternative configuration, all the while (even perhaps by this means) avoiding Muller's Ratchet (the invariant collection of adverse mutations, normally purged by sexual recombination, and which ultimately dooms most asexual organisms).

At the very least, however, the obvious taxonomic divergence of the cancer from its host of origin, whatever one actually calls it, stands as a stark lesson for those who continue to deny the natural evolvability of life.

C. Therapy

There are two major sources of therapeutic nihilism, when it comes to cancer:

1. The belief that cancer cells are too similar to normal cells for there ever to exist an exploitable therapeutic index (type I nihilism)
2. The belief that whatever molecular targets we attack in the cancer cell, resistant variants will usually evolve before eradication is possible (type II nihilism)

The juxtaposition of these two sentiments opens up something of a paradox. On the one hand, the taxonomic gulf between cancer and its host must be caused by something, and which (unless we are very much mistaken) must be reducible to a set of molecular alterations; even if we allow that the initiating molecular alterations may be different from those that ultimately maintain the malignant phenotype on an ongoing basis, there always have to be appreciable biomolecular differences between cancer and normal cells, and which should, therefore, be also regarded as potential targets.

On the other hand, the genomic/genetic instability, which somehow goes along with this, seems guaranteed to lead quite rapidly to the evolution of drug resistance. But if the material drivers of the cancer cell's existence, as targets, are themselves also alterable, perhaps this outcome may be incompatible with the continuing existence of the cancer cell, and if it happened a lot, the tumor as a whole might "go out of business." That this appears not to be the usual case supports the Heng–Duesberg model, and is a strike against the conventional mutation theory; at least it suggests that there are mechanisms to counter this, either compensatory hyperproliferation or driver invulnerability/redundancy. Almost certainly however, and which is of immense interest therapeutically, many cancer cells do become nonviable, slip silently out of existence, and are recycled by the tumor.

Insofar as a large taxonomic chasm opens up between cancer cells and their hosts of origin, then this has to be mediated by "a core set of molecular alterations that being both causal and distinctive, are potential targets" (Vincent, ibid 2010). Furthermore, "These fundamental changes have to exist and *ipso facto* must represent the ultimate targets for rational drug design"; that is, to the extent that "rational" means "based on causality reversal."

As to the nature of these molecular alterations, they are hypothesized to at least include the same genes involved in the evolution of multicellularity, since the development of cancer and the evolution of metazoa might be intimately related, as "bidirectional traffic on the same two-way street" (Vincent, ibid 2010). It turns out that this prediction is to a substantial extent backed up by the involvement of the cadherins, Hedgehog, Wnt, Notch, and others representing genes common to both processes, either through their constructive exploitation (evolution of multicellularity from unicellular organisms) or their deregulation (development of cancer in a multicellular host) (Farnie and Clarke, 2007; Lai, 2004; Rokas, 2008; Rubin and de Sauvage, 2006; Stemmler, 2008).

This notion that the genes involved in either multicellular evolution or cancer should at least to some extent be the same, not just because of the empirical evidence but because these macro-evolutionary processes represent opposite ends of the same processional pathway, is suggested as a potential "big idea"; one implication is that any new gene found to be involved in either one of these processes should be closely scrutinized for also being involved in the other.

I have suggested here that not only can the certain existence of categoric differences between normal cells and cancer be inferred, but their nature can be profitably suspected by extrapolation from a process that could be the reverse of carcinogenesis, that is, the genesis of the metazoan. This is not least because the properties of cell adhesion, positionality, cell population control, and intercellular communication are central to both metazoan life, when they are functional, and when they are dysfunctional, to cancer.

Lurking behind this is the issue of causality. The logic seems unassailable: since the cancer cells, by definition, behave differently from their cells of origin, there must be some causative molecular lesion(s), which maintain this malignant phenotype. This is true even if we are talking about the alterations that maintain the abnormal behavior, rather than these that first initiated it, and which may or may not be different, or even no longer operative.

This concept of causality, so entrenched into occidental thought, inevitably draws one to the concept of the "driver" therapeutic target, as distinct from the "marker" target concept. Essentially, the former ("driver" concept) implies an anticancer effect from successfully addressing and preventing the operation of the proximate cause ("causality-reversal"); the latter, the "marker" concept, does not depend on "causality-reversal" but derives efficacy from using some distinguishing feature, for example, as a simple "delivery box" or address, to direct and deposit an otherwise indiscriminately cytotoxic warhead to selectively kill the tumor cells; or to otherwise use this mark of distinction to enable selective cytotoxicity in a more indirect and abstract manner, as long as it does not involve causality-reversal.

For both driver- and marker-type targets, the molecular features may be exploitable either as a presence or as an absence. In the case of the driver, a "driver"-type target, as a "presence," would be exemplified by oncogenes such as the mutant EGFR, or the *BCR-ABL* fusion. Both of these are clearly responsible for an important causal role in, respectively, a subset of pulmonary adenocarcinoma, and CML; both are carcinogenic as early events, but more importantly seem to be necessarily required to maintain the malignant phenotype; and consequently, being druggable, engender extra-ordinary therapeutic effects when inhibited. In this, they exemplify the "driver-inhibitor" model, in which bona fide drivers (even when not as powerfully causal than these two classic "oncogene addiction" type situations) are therapeutic when inhibited. Likewise, for marker-type targets exploitable through their presence, they are features on or in the cancer cells (preferably all and only the cancer cells) to which a warhead (e.g., a toxin or radioisotope) can be directed by a homing device (e.g., a monoclonal antibody). In this situation, either the marker does not contribute in any way to causality of the malignant phenotype, or if it does, that is not the aspect of it that is exploited by a marker-directed targeted therapy, only its distinctiveness. In certain cases, for example, EGFR-wt amplified cancers treated with, say cetuximab, the driver and marker functionalities may both be exploited, or it may not be clear which is being exploited.

The "absence" situations, some of which may arise as an important prediction of the Heng–Duesberg model, constitute a somewhat more difficult opportunity. However, they are vital to confront because they almost certainly are numerically more common than the "presence" opportunities, especially in the case of the drivers. Driver-absences (which does not mean the absence of any drivers, but rather refers specifically to driver mutations characterized by loss-of-function, that is, associated with tumor suppressor gene abnormalities, rather than gain-of-function) are characterized by either the physical absence or the functional absence of a key cellular control system, for example, p53 itself or p53 functionality or some other defective brake on the cell cycle, which therefore fails in its normal function to halt the cell cycle. Marker-absences are characterized by features, in which the mark of distinction is the lack of a particular structure or function on the cancer cells, but which is present in or on normal cells. The absence may or may not drive the cancer in a causal

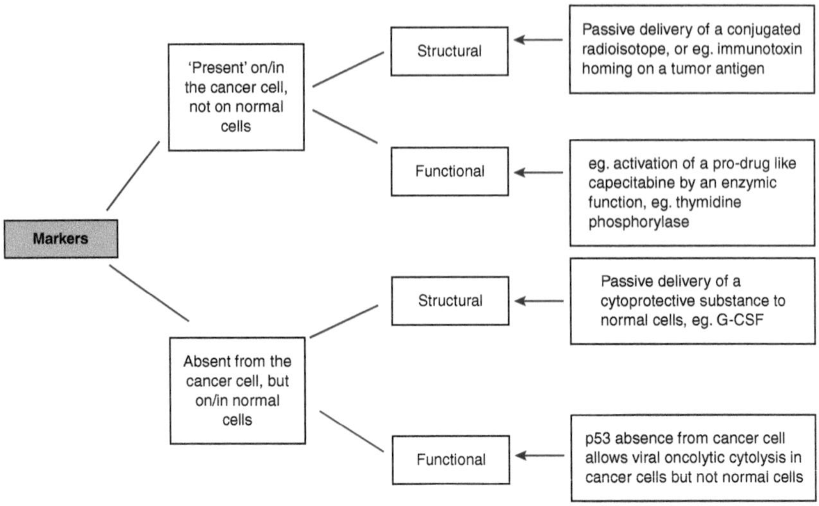

Fig. 1 Marker-Therapy Concept Map.

sense, but exploitation by a marker-type therapy would not attempt to rectify the defect, merely to exploit it for a kind of "negative targeting" (vide infra). In the case of the driver absence, by contrast, the therapeutic effect would be achieved by some strategy designed to restore functionality so that the malignant phenotype could no longer be enabled; for example, restoration of apoptosis, if some lack had disabled it. This is difficult, and clinically successful examples of this sort of strategy are hard to find. Fig. 1.

In the case of marker absences, the strategy would consist of targeting a protective agent to normal cells and then exposing all the cells to an otherwise nonspecifically cytotoxic agent; amifostine plus concomitant radiotherapy would be a good example, the amifostine being activated by the milieu of normal cells but not the milieu of cancer cells. Upon metabolism to the active derivative, free radicals generated by the radiotherapy are quenched but only in the normal cells, which are, therefore, protected (Citrin *et al.*, 2010). Amifostine is an approved drug and may also protect against radiomimetic and other chemotherapy, like platinum-based, as well as radiation itself.

These various mechanisms, incidentally, illustrate what it is about therapies that qualify them as "targeted"; not, as is commonly thought, that they are directed to interact with known specific molecules, rather that they are capable of inducing a *selectively* cytotoxic effect. It is possible for tumor selectivity to exist, even though the mechanisms for this might not actually be known; for example, that the addition of folate/B12 to a regimen of pemetrexed rescues the normal tissue but not the cancer cells is a fact without a readily available explanation. But the lack of a known mechanism has not prevented this strategy from incorporation into the gold standard chemotherapy regimen for nonsquamous non-small-cell lung cancer, and it remains an excellent example of targeting.

D. Speciation and the Target

While it seems inescapable that causative molecular lesions must exist as either or both initiating and/or maintenance factors for the properties that distinguish the malignant phenotype from its normal cell of origin, it is not to be concluded that these lesions are always known, druggable, or even knowable. In particular, it cannot currently be stated which lesion, or combination of lesions, must be "addressed" in order to effect a cure, although the standard model would seem to suggest otherwise. Furthermore, as mentioned above, it should be appreciated that druggable driver–inhibitor opportunities may be few and far between, as "low-hanging fruit," and should not constitute the sole aspirational model of drug designers; and while not unknown, it remains insufficiently appreciated that "absence" opportunities (both driver and marker), and marker-presence opportunities exist, but (the former especially) will require a particularly nuanced approach if they are to be exploited successfully. Most importantly, as per Heng–Duesberg, the proximate causality that maintains the malignant phenotype may involve, at least in part, a dynamically inconstant genotype (Heng) or massive genetic imbalance (Duesberg); in either case, the carcinogenic mechanisms are not readily obvious, and might even require excursions into unfamiliar domains of abstract thought to elucidate and exploit.

If the Heng–Duesberg model of proximate causality is correct, that is, that initiating oncogenes are later superfluous, that a constant set of common mutations does not exist, and that the neoplasm is maintained as such by massive genetic imbalance and/or the very lack of constancy in the genome, then I would submit that the traditionally staid driver–inhibitor model of targeted drug design, at least for the purposes of cure or markedly prolonged survival, is doomed to failure, and for reasons that are entirely distinct from the classical resistance mechanisms described. If the common advanced cancers are curable, it will probably require an unorthodox and probably nonobvious approach directed to this phenomenon of fluid causality.

Accepting a vast taxonomic gulf between normal cells and their malignant offspring exists, and that molecular changes must occur to cause this in an ongoing proximate fashion (and therefore banal "type I" nihilism is indeed inappropriate), I also believe that easy exploitation of this truth for curative purposes is not possible; and that even the ready-made opportunities for survival extension and other short-term palliative goals will be few, far between, and may already have been substantially discovered.

Type II nihilism, the inevitable emergence of resistant variants, will not be unfamiliar to anyone involved in the day-to-day systemic therapy of common cancers. With the possible exception of complete surgical extirpation, there is no drug or modality to which common types of cancer cells cannot or do not exhibit resistance, sooner or later. Although this may to some extent be due to lineage heterogeneity, and the existence of, and selection for, inherently more drug-resistant stem-like cells, most commentators believe that genetic plasticity is responsible for hard-core drug resistance. This is certainly consistent with the Heng–Duesberg model of gross genome-wide instability coupled with florid micro-evolutionary changeability. Without disagreeing, I will offer a certain perspective of the meaning and significance of this, which goes beyond not only speciation but the "hyperspeciation" implied by the acquisition of the unstable genome.

VI. BEYOND SPECIATION: THE NATURE OF CANCER

A. What Type of Life?

There is little doubt that carcinogenesis entails the opening up of some sort of taxonomic gap between the metazoan host and the cancer cell; indeed this is implicit in the well-named concept of "transformation." Furthermore, while it is true that not all speciation events involve karyotypic changes, the vast majority do (Heng, 2009), and the converse that karyotypic changes imply speciation is also probably true. It is also true that the autonomy, unicellularity, and asexuality, which characterize cancer cells, as well as certain other traits, mark them off as something very different from their host of origin.

As noted, the default taxonomic process ("speciation") is not without problems; chief among those is the failure of the cancer karyotype to stabilize, although to characterize it as 100% chaotic would also be an exaggeration (Lin *et al.*, 2009). Rather, cancer cells within a lineage appear to exhibit what has been described as "instability within stability" (Gusev *et al.*, 2001). Nonetheless, there is a growing consensus that cancers exhibit more genomic flexibility than any other type of eukaryotic normal cell, even though there are some recognizable parameters of stability.

Therefore, it is legitimate to ask what type of life is represented by a cancer? And is it valid to group them all together, given that they would almost always be more closely related to their originating hosts than to each other? Cancers here would be suggested to be a type, sharing certain descriptive and perhaps explanatory features, that is, similarity, but not the close lineage relatedness that is usually implied by a taxonomic subgrouping. Indeed, without this core group of shared traits and features, it would not be possible to talk about cancers as a "type" at all. It is precisely this shared suite of features that has to be explained and placed in evolutionary context. And while cancer cells within a particular patient are certainly in a close lineage relationship with one another, cancer cells in different individuals are not, despite their having more in common with each other than with the hosts of origin that actually derived them. This might be a unique problem in taxonomy, which adds gravitas to the question at the start of this paragraph—what type of life, indeed, is represented by cancer?

B. Speciation or Despeciation?

Part of the problem as usual is semantic. "Speciation" is a seductively simple-appearing word, but in fact it coheres two separate processes: firstly, the act or process of separating from an existing species; and secondly, the establishment of a novel species. There is an implicit assumption that the first automatically implies the second, even that it is not possible to have the first without the second. Yet, while the first process (which we can call de-speciation) always occurs in carcinogenesis, it may be that the second (respeciation) does not often occur because of persistent genomic instability.

In this representation, cancer cells exist in a sort of taxonomic limbo; having departed their original, stable karyotype, they engage in continuous, albeit partial, genomic reshuffling, without establishing the novel stability that would signal that they have achieved "type" status, that is, a new species. So what are they? Do we have the words or the concepts to describe them, let alone explain them?

C. An Alternative Form of Life

I have previously suggested (Vincent, ibid in press 2011) the idea that cancer cells are actually an alternative form of life, one that is latent, and encrypted in all eukaryotic cells, and which emerges as a solution to perceived existential threats, as a resilient survival machine. Rather than just a series of random mutations, the universal traits exhibited by cancer cells represent the re-emergence of an ancient survival program.

In this characterization, and because it promotes survival, ongoing genomic instability in fact is the whole point, part and parcel of the emergent program; and this is what poses the problem for the concept of speciation, at least as simplistically conceived.

Thus, returning to Dhobzansky, the cancer cell can be said to be "for something"—that is crude survival, as a better alternative to extinction. One can strip down an organism into its components and characterize them, but unless one knows the function of the whole, and what biological ends it promotes, the parts can only represent a very incomplete picture of reality.

D. Historical Perspective

The notion that the cancer cell is an atavistic throwback does not quite capture the viewpoint I am advancing here, in the sense that "throwbacks" are accidental rather than purposive; however, there is an element of this line of thinking that does inform what I am proposing, and it is reasonable to acknowledge it and record the contributors.

I am grateful here to Greaves (Greaves, 2000) who has traced this history. According to Greaves, both Sir Herbert Snow (Snow, 1893) and Morley Roberts (Roberts, 1926) "suggested that cancer cells were derived from cells that retain a memory for amoeba-like selfish behavior and that can still escape by losing allegiance to the community of cells within which they reside."

The following concepts are here enumerated:

A. Normal cells, at least some of them, are able to re-express archaic behaviors and characteristics, and which therefore have not been eradicated but just contained by subsequent evolutionary developments.
B. One of these characteristics consists of autonomy, and cells achieve this by escaping from the bonds that normally cause their activities to be subsumed in the interests of the whole organism.
C. The metazoan animal is in fact a "community of cells," which despite this fragmentation somehow is roped together by overarching coordinating forces, which can nonetheless occasionally be over-ridden.

Greaves then introduces Lucien Israël, a French oncologist, who, in a 1990s essay, seemed to extend or at least interpret this idea of atavism in the form of a re-emergent, ancient survival program (Israël, ibid 1996).

Israël begins by questioning the model in which "cancer is a blind process, because of random mutations appearing at any time in random order and selected for the unpredictable advantage that some offer . . .," that is, the standard model that he here credits to Nowell (Nowell, 1976) and Cairns (Cairns, 1975). Israël points to an alternative formulation that cancer's "formidable survival efficacy" can instead be considered a type of program (i.e., "combined and integrated genetic responses"), which results from environmental threats, whether these are mutagenic or not. The by-product of this program's expression is drug resistance that may go some way toward explaining the relatively poor results achieved by oncologists (still true in 2011 as it was in 1995).

Israël believes that this survival program is inherited from the archaic precursors of metazoa, that is, the unicellular eukaryotes, and before them, the prokaryotes. One prokaryote survival program he sees as a possible type of precursor is the SOS regulon (Witkin and Kogoma, 1984), a coordinated, multifaceted stress response, which includes, most crucially for our purposes, the induction of "error prone" DNA replication, and which has apparently permitted not only survival in the face of recurrent environmental catastrophes but also the fact of evolution. In Israël's view, cancer cells are the result of a similar set of "responses to danger," in which the ongoing display of a defense and survival program is a predominant motif.

He is, however, quick to acknowledge that no inducible system such as the SOS has been observed in animal cells to date. Nonetheless, mutational hotspots, that is, genetic loci which being more susceptible to mutation, might act as an "early warning" system, especially if contained in genes normally responsible for genome integrity like p53, and whose disablement would result in a pre-emptive or at least an early response to existential threats. Hints of this idea can be found in a 1990 publication by Israël in which broad-spectrum resistance, and other malignant traits such as autonomy, seem to appear at an accelerated rate not explicable in terms of random mutations (Israël, 1990).

Israël concludes with the opinion that the drug resistance so commonly featured by cancer cells is in fact similar to the survival attributes of unicellular cells, and which enabled their survival under stress. The key aspect of this hypothesis is that neither the phenotypic attributes of the cancer cells nor the genetic instability are the result purely of stochastic mutations, but of a de-repressed, ancient program of survival. This having been said, there is a role for random mutations in evoking this program and for classical Darwinian selection to operate via the usual methods.

In a later publication (Israël, 1998), these ideas were again elaborated, with the refinement that it was the specific function of the "antioncogene system" to restrain and control this program in healthy cells, but in a manner that would, under conditions of existential threat, permit its release, not as an accident but, so to speak, "intentionally."

This notion of inducible error-prone DNA replication seems to provide a connection with the Heng–Duesberg vision of karyotypic and macrogenomic instability, as described above, not only as an explanation for the phenomenon but as a reason for it as well, that is, as a survival strategy via genomic reshuffling, as with prokaryotes and the SOS regulon. This seemed to neatly account for both the cause of the genetic instability and one of its major consequences, drug resistance.

VII. THE NATURE OF CANCER: A TRAIT-BASED INFERENTIAL APPROACH

A. Cancer—A Consistent Suite of Traits

Cancer consistently manifests a highly characteristic suite of traits, ranging from ceaseless proliferation, through insult resilience, to aerobic glycolysis and aneuploidy; the usual explanation for this, that is, serial, blind, and random mutations followed by selection, might more parsimoniously be replaced by, or at least supplemented with, a programmatic explanation. This program would prioritize the continuation of cellular life over the preservation of the individual, the original species (or any species), and even, heretically, over the gene.

These various traits can be grouped under three broad categories: phylogenation, reprimitivization, and adaptive resilience. In turn, these three form the pillars of an overarching goal, which is essentially cellular survival at any cost (Vincent BioEssays, in press 2011). This viewpoint sees the cancer cell as a re-emergent protozoan-type organism, in whom specialized functions have been jettisoned as a means to sheer survival. Furthermore, this protozoan has a potential existence in every eukaryotic cell, able to be called into actuality by some trigger such as a perceived threat to the life of the cell. Clinical cancer is thus seen as a competitive struggle between this reawakened Precambrian organism, and the residual normal metazoan body; and one in which the metazoan usually loses, for a very coherent and compelling set of reasons.

B. Phylogenation

The concept of *phylogenation* is traceable to Julian Huxley, who opined that autonomy, *per se*, conferred upon the cancer cell a considerable taxonomic distance from the originating host. Now, it is true that collectively, cancer cells in different patients are more closely related to their originating hosts than to each other, and this constitutes a barrier to the literal interpretation of Huxley's words. However, it is at the same time true that a cancer is no longer the same individual as the originating animal, nor even a member of the same species, either phenotypically or genetically, irrespective of the lineage descent; and considering the gross phenotypic differences, especially the aspects of metazoan deconstruction (unicellularity and host destruction), autonomy (growth factor sufficiency and growth inhibitor insensitivity), niche finding (invasion and metastasis), speciation (chromosomal changes and distinct evolutionary destiny), and parasitism (nonsymbiosis, and occasional transmissibility + absence of host anticipatory alterations), it is arguable that each cancer in each patient represents a newly emergent phylum, in which the constituent cancer cells sit in an entirely different locale in the *tree of life* from the metazoan host.

The fact that different cancers in different individuals are less related to each other than to their respective hosts does not gainsay the taxonomic gulf between each cancer and its own host, which is the main point. The similarity of each cancer between individuals might then be misconstrued as a case of convergent evolution, but this similarity is more a consequence of the similarity of the universally inherited, underlying (and now de-repressed) program than pure randomness and selection.

C. Reprimivitization

Allied to this concept of phylogenation is the next collective category, *reprimitivization.* This category groups the traits that appear truly atavistic ("recurrence in an organism of a trait or character typical of an ancestral form ...") (Merriam Webster). These include re-emergent unicellularity (autonomy and metazoan deconstruction), immortality (telomerase de-repression and apoptotic evasion), fermentation (the WE), asexual reproduction/abandonment of the soma (maturation block, "stemness," and retreat into the germ line), abrogation of the self (gross genomic instability and debasement), and inhospitability tolerance (WE).

The WE is prominently displayed here, and it is Otto Warburg who must be credited for providing the ancient but empirical theory of atavism with a biochemical grounding. It was Warburg who noted that cancer cells preferentially use glycolysis to produce ATP (i.e., ferment) even when oxygen is present: the WE. Upon this most central and fundamental of insights, which describes a universal cancer attribute, is founded not only the entire PET scan industry but what some consider one of the best hopes for the development of truly targeted anticancer drugs.

What I am focused on here is not these practical benefits for human health, actual or potential, but the philosophical significance of the WE. Warburg, apart from noting the retreat into a primitive type of metabolism more appropriate for the early and preoxygenated earth's atmosphere, believed that sophistication and complexity, in animals, absolutely required oxidative phosphorylation. Absent the latter, therefore, a retreat into primivitization was inevitable, although he admitted ignorance as to how and why exactly this should be the case.

Warburg went on to proclaim that this metabolic "defect" was fundamental to cancer, in a causative sense: "... the prime cause of cancer is the replacement of the respiration of oxygen in normal body cells by a fermentation of sugar" (Warburg, 1989). This position became known as the Warburg Hypothesis. With the discovery of oncogenes and the rise of the molecular biology of the gene, and the gene-centric emphasis in evolution, Warburg's Hypothesis was believed discredited and was neglected.

However, Warburg never discounted the role of genetic initiators, and nuanced statements about the precise mode of causality and aerobic glycolysis probably suffered in translation from the original German. The misjudgment of the Warburg Hypothesis, dating back to the middle decades of the twentieth century, was probably based on a linguistic misinterpretation of the distinction between originating and proximate causality, and may well have cost us five or six decades in the war on cancer. Warburg, however, probably did err in his opinion that respiration in cancer cells was necessarily defective (Koppenol *et al.*, 2011); it does appear that some degree of intact respiration can and does coexist with the accelerated and anaerobic glycolysis, which is typical of cancer cells.

My sense, reviewing this material (Warburg, 1928, 1956, 1956), is that Warburg was simply pointing out what is patently true, which is the indispensability of his effect in the process of carcinogenesis and the maintenance of the malignant phenotype: "By a prime cause of the disease I mean one that is found in *every* case of the disease" (Warburg, 1989 ibid). Indeed, a large quantum of venture capital is betting he was right, and, as a target, that one or more key components in this type of metabolism holds the potential, not only to cripple the malignant phenotype, but to do this in a highly specific and nontoxic way—an opinion long shared by this author. I shall also discuss (vide infra) the emerging work that indicates that the WE is deeply implicated in resistance to systemic therapies, and radiation.

For our immediate purposes, however, we are interested in the pragmatic aspects of Warburg metabolism and its ancient origins on the hypoxic earth. The modern understanding of the WE is still evolving and, along with its key role in drug resistance, will be more fully described below; but it is enough to note here the near-universality of this phenomenon in cancer, and its relative consistency across all (especially aggressive) cancer cells as a "package," argue strongly against its purely stochastic rediscovery in every new case of malignancy, as almost fantastically unlikely.

Unicellularity is another defining trait of cancer and has been one of the signatures that have evoked analogies with ameba, for instance. However, it might be questioned whether cancers, which usually grow as collections of cells, are truly unicellular. Certainly, they are no longer team members of the originating metazoan host, and unquestionably represent a breakdown in the original version of multicellularity—a repudiation, perhaps, and which may have been occasioned by a perception that membership in the metazoan team may be associated with too high a chance of death and extinction. In this respect, the increased incidence of cancer with age makes a certain sense if the cancer cell is seen as trying to escape the sinking ship.

But are cancers unicellular? If by that is meant are they necessarily obligated to adopt a solitary existence, the answer is no; they usually grow in colonies of their siblings and close relatives, the dynamics of which are characterized by a mixture of cooperation and competition (Axelrod *et al.*, 2006).

Furthermore, cancers exhibit organoid lineage hierarchies that resemble those found in normal tissues, especially in relation to stem cells, which suggest that the forces underlying these organizational paradigms in normality might continue to function in cancer. On the other hand, cancer cells seem fully viable on their own, and some of them at least are competent to repopulate an entire tumor from a single cell. If this is what unicellularity means, then cancer cells, especially cancer "stem" cells, certainly embody that within their repertoire.

As perceptively, and provocatively, noted by E.O. Wilson, it is sometimes difficult to know what, in biology, constitutes an "individual" (Wilson, 1975). Is the cancer an "individual"? If yes, it must be a multicellular animal, albeit of a different kind. But I am rather inclined to think that the defining autonomy is not only with respect to the originating metazoan (now transformed into a cross between a food source, habitat, waste-removal service, and killing field), but also with respect to each other cancer cell. But as Wilson notes, "There may be more than one way for nature to constrict a multicellular organism."

An example of cooperation, for instance, is the secretion of VEGF, which serves to stimulate angiogenesis, without which tumors would remain relatively fixed in position and small in size (approximately <2 mm). A single cancer cell would not likely be able to secrete enough VEGF to attract capillary ingrowth; the same principle applies to the cooperation required of tumor cells in the production of a sufficient supply of diffusible products to overcome host defenses (Axelrod, 2006 ibid). So some degree of cooperation may well be essential to their growth and spread. But at the same time, the genetic heterogeneity in cancer cells seems poorly coordinated, and they almost certainly have the capacity to evolve independently, and outcompete each other. Yet it is a matter of no small interest that the members of the expanding cancer mass seem to be themselves resistant to the external killing effect mediated by milieu acidification and which "dissolves away" the normal tissue to make room for the expanding cancer; on the interface with normal cells, this may perhaps be due in part to lactate/proton export by tumor cells (Mathupala *et al.*, 2010). Probably the correct description is of basically unicellular organisms with facultative colonial attributes, and which allow them to most effectively "hunt in packs" as they set about dismantling the host.

A "colony," according to Wilson, has to either be physically united, or differentiated "into reproductive or sterile casts," or both. At what point a well-functioning colony becomes indistinguishable from a "super-organism" is perhaps arbitrary. Suffice to say I can see advantages for cancer cells, as units of survival, to be capable of cooperation with each other, but ultimately capable of an independent existence, and, under duress, just as capable of ruthless competition with sibling cancer cells. And as the lineage hierarchy within cancers comprises self-renewing "clonogenic" stem-like cells, as well as nonclonogenic end-stage members, which can be crudely seen as "reproductive and sterile casts," respectively, and also because metastatic foci are composed of cells in a sort of united physical continuity, one should not underestimate the colony as a form of social organization still intact in cancer cells. Nonetheless, this is a very different form of multicellularity than the one that characterized the now-transformed cell's previous life in the metazoan team, and the colonial attributes of cancer cells in no way undermine the concept of reprimitivization, and in fact reinforce it.

The reprimitivization argument is further buttressed by the reappearance of asexual reproduction. Of course, the somatic cells in metazoan tissue also require replenishment from stem cells, which occurs by asexual clonal expansion. But the difference here is that this process is perforce interrupted intergenerationally by an episode of sexual reproduction, and which may serve, while purging the genome of extreme variants, to permit a cautious amount of genetic diversity, followed by rigidly conservative rounds of somatic cell renewal. In the case of cancer cells, there is no longer any cleansing sexual recombination, and in each cancer cell the forces of genetic conservation are substantially disabled. Hence the cancer cells, freed from the obligation, or the ability, to conserve their genomes, essentially must abrogate their own identity.

Likewise immortality, a feature that must characterize at least some unicellular organisms, is a typical and striking feature of cancer and easily construed as an emergent primitive trait.

Thus, from a mortal, complex, sexual, and oxygen-respiring multicellular organism is born this immortal, fermenting, colonial, asexual, and stripped-down organism, and it is difficult to conceive of this as anything other than a primitive type of animal, in search of the late Precambrian. Whether genetic instability was a characteristic of the preoxygenated world might be unanswerable; but the absence of an ozone layer might at least imply that the earth's surface was a much more mutagenic environment than it is today, as might the first appearance of oxygen-free radicals.

D. Adaptive Resilience

The final organizing principle is *adaptive resilience*. Adaptive resilience results from the combined effects of the sheer number of cancer cells (the "demographic cushion"), coupled with genetic hypervariation ("mutator phenotype"). These attributes allow the tumor to overcome obstacles to its survival, including (but not limited to) therapeutic challenges like chemotherapy or radiation. The genetic variability performs the role of novelty-generation, which is then heuristically tested against the environmental vicissitudes. Of necessity, many of the genomic variants will be nonviable, a fact that is compensated by the unlimited cellular proliferation.

The population explosion is enabled not only by the ceaseless proliferation but by biomass expansion as well. Ceaseless proliferation arises from oncogene activation and tumor suppressor gene inactivation; whether these might sometimes be only initiators, or always required for the maintenance of the proliferation, is an open question (Klein *et al.*, 2010). Biomass expansion is enabled by the WE (Koppenol *et al.*, 2011; Shlomi *et al.*, 2011; Vander Heiden *et al.*, 2009) (comprising glucose siphoning, accelerated glycolysis, subversion of oxidative phosphorylation, provisioning of substrates like acetyl CoA and ribose, and apoptosis evasion). Also contributing to biomass expansion is host destruction (nutrient diversion, milieu acidification, metastasis, invasion, enzymatic digestion, and biomass interconversion) together with host enslavement (as evidenced by angiogenesis, induced autophagy, and immunosuppression).

The source of genetic hypervariation is believed by many to be an attribute known as the "mutator phenotype." This attribute is not without controversy (Loeb, 2010), but most commentators seem now to accept the empirical reality of high and *persistent* levels of point mutation, altered gene expression and aneuploidy, which together provide the basis for both micro- and macro-evolution (Campbell *et al.*, 2010; Duesberg and Li, 2003; Fabarius *et al.*, 2003; Heng, 2006, ibid 2009; Klein *et al.*, 2010 ibid; Negrini *et al.*, 2010; Salk *et al.*, 2010;). I propose the mutator phenotype to comprise a "generator" function (consisting of defective DNA repair, and a tunable endogenous mutator in the form of reactive oxygen species, or ROS) but also a "tolerator" function, consisting of apoptosis evasion (probably due to binding of hexokinase II to the mitochondrial VDAC), which is a feature of the WE. The "tolerator" can be inferred because normal cells would either apoptose or repair the gross genome alterations, and cancer cells seem to be relatively delinquent in the performance of both of these tasks. Likely the very common p53 disablement contributes to both defective DNA repair and apoptosis evasion.

Thus is constructed a machine in which the associated DNA instability is not just a pointless collection of errors, but in fact the principal means to the goal of cellular survival. The cancer cell is not primarily a "growth machine," as Warburg had thought (Warburg ibid, 1989), but rather a *survival machine*. Why it should be constructed like this probably relates to the extreme measures required to endure an environment much more challenging than our own, that is, the Precambrian. The list of perils would have included nutrient deprivation, food-chain collapse, acidification, climate extremes, volcanism, radiation, and, as the ambient oxygen levels began to rise, reactive oxygen and other chemical species.

The latter phenomenon is believed to be the reason for the successful endosymbiosis with the alpha-proteobacterial mitochondria, acting as an oxygen detoxification system and only secondarily as a source of ATP. The availability of these "tame mitochondria," in addition to the archaic glycolytic pathway, provided these early eukaryotes with the means to "tune" metabolism by toggling between oxidative phosphorylation and glycolysis. The relative utilization of each pathway would influence the cellular economy in terms of glucose consumption, pH, lipid and nucleic acid precursor generation, ATP production, ROS generation, and, closely related to mitochondrial function, apoptotic threshold; all factors relating to the WE.

ROS, it turns out, are quite mutagenic for both mitochondrial and nuclear DNA. By altering ROS production (and quenching), the cell would have a way to endogenously alter its rate of mutation. The WE provides a pragmatic package for the cell to switch to a high ROS, high substrate provision economy, resulting in massive biomass production (to sustain the population explosion), and genetic chaos required to "test drive" new gene–environment pairings. At the same time, the elevated glucose consumption ensures just enough ATP to sustain this, while the lactic acid by-product is exported into the milieu to facilitate the digestion of surrounding tissues in the project of lebensraum creation.

Thus the WE, a term created to describe the narrow (but fundamental) observation of aerobic glycolysis, is actually a whole package of far-reaching changes, all of which contribute to the malignant phenotype in a variety of ways: invasion, biomass interconversion and expansion; persistent DNA reshuffling, with the attendant possibilities for evolution; the ability to tolerate inhospitable environments characterized by hypoxia and acidosis, and most important, from our perspective, insult resilience (including drug and radiation resistance), especially by means of apoptosis evasion, and the aforementioned DNA heuristics.

At the same time, however, something has to be given up: multicellularity. Multicellularity requires genetic continuity, and above all, predictability. Otherwise the cooperative entrainment of cellular fate, and investment in the organism as a whole (as a propagator of genetic similarity) is utterly undermined. This is why metazoan deconstruction is the inevitable accompaniment of malignancy.

Recently, investigators have begun to elucidate a direct relationship between aspects of the WE and therapeutic resistance. For example, lactate dehydrogenase (LDH), an important glycolytic enzyme, is vital in the WE as it is responsible for the interconversion of pyruvate and lactate as well as generating NAD+, an essential cofactor in upstream glycolysis. Breast cancer cells resistant to paclitaxel were shown to express elevated levels of LDH-A, and downregulation of LDH-A by siRNA increased sensitivity to paclitaxel (Zhou *et al.*, 2010). The combination of oxamate (an LDH inhibitor) with LDH-A downregulation substantially potentiated paclitaxel in these cells, involving increased apoptosis. Paclitaxel therapy, in fact, actually induces LDH mRNA, which in turn promotes resistance. The actual mechanism may involve activation of the pentose phosphate shunt pathway, greatly enabled by the WE, and which in turn generates reduced glutathione, which then prevents the oxidative activation of proapoptotic cytochrome C (Vaughn and Deshmukh, 2008). Zhao *et al.* have also shown that HER-2 overexpressing cells are associated with increased glycolytic activity, and that HER-2 downregulation likewise downregulates glycolysis (Zhao *et al.*, 2009).

Another approach to selectivity, which exploits the WE at the same time as overcoming cisplatin resistance, is the synthesis of a molecule that results from the fusion of cisplatin and dichloroacetate (DCA). DCA inhibits pyruvate dehydrogenase kinase (PDK), a key WE enzyme. The release of DCA from the fusion drug induces mitochondrial release of cytochrome C and apoptosis-inducing-factor (AIF), promoting both DNA damage and selective restoration of apoptosis in the cancer cells, but not the normal cells. DCA, by inhibiting PDK, disinhibits pyruvate dehydrogenase (PDH), which is then free to shuttle the pyruvate-derived metabolite acetyl CoA into the mitochondria, thus forcing the previously glycolytic cell back toward mitochondrial oxidative phosphorylation. Since this process is functioning normally in nonmalignant cells, this is not toxic to them. In cancer cells, however, the mitochondrial membrane potential is hereby reduced, which then opens the mitochondrial transition pores, releasing cytochrome C and AIF, and lowering the barrier to apoptosis (Dhar and Lippard, 2009).

The relationship of the WE to resistance has also been reviewed by Sattler *et al.* (2010). They also highlight the elevated levels of free radical scavenging metabolites, induced by the WE, and which tend to frustrate the effects of therapies that work by the generation of free radicals. These authors therefore advance the notion that specific inhibitors of the WE should be explored in combination with "established cancer treatments," a sentiment that I strongly believe should be extended to any kind of cancer treatment, established or not.

Viewing the WE as a program, however multifaceted, opens the door to the highly attractive prospect that, if it can be activated as a whole in the process of carcinogenesis, it might *ipso facto* be able to be switched off as a whole, in the process of therapy, instead of piecemeal. The recent discovery that STAT1 may be responsible for the regulation of glycolysis, the Krebs cycle and oxidative phosphorylation, in such a way that its knockdown potentiates radiation, perhaps by energy deprivation (Pitroda *et al.*, 2009), is indicative of what might be possible.

The author has also advanced the WE as an exploitable target in the selective potentiation of natural product cytotoxics, extruded by resistance pumps like p-gp and MRP-1, by an experimental drug, tesmilifene; this drug only potentiates taxanes, vinca alkaloids, and anthracyclines in pump-positive (resistant) cells and not their pump-negative, sensitive parent cell lines. The mechanism is likely to involve the substantial ATP consumption by these pumps, when activated by the cytotoxics, a condition aggravated to the point of lethality in cancer cells, but not normally respiring noncancer cells, by tesmilifene (Reyno *et al.*, 2004; Vincent, 2006;).

These various examples are intended to provide a flavor of what might be possible with resistance reversal by inhibition, or exploitation, of the WE, and the favorable therapeutic ratio (i.e., selectivity) that might accompany it.

VIII. RESISTANCE TO THERAPY: THE LONG ARM OF THE PROTEROZOIC

A. Resistance Reversal—A Sorry Tale

A multitude of drug and radiation resistance mechanisms have been and continue to be described. Given this wealth of knowledge, it is surprising how rarely this has been translated successfully into clinical application. As an explanation, it is tempting to speculate that whenever a mechanism has been successfully addressed, another one, hydra-like, will take its place. Insofar as the cancer cell is seen as embodying the above model of pragmatic hypermutation, as something that is either its very *raison d'etre*, or close to it, it is unlikely that a piecemeal approach to resistance-reversal will result in anything other than temporary victories, and more often than not abject failure to solve the problem. Therapeutic resistance is indeed a hard problem—the hard problem of oncology—in the sense that apparent greater understanding seems not to lead to any solution.

B. What the Cancer Cell Is For

We should take this opportunity to emphasize here a paradox. I have, along with others (Bergstein, 2003; Campbell *et al.*, ibid 2010; Heng, 2010; Israël, ibid 1996; Nicholson and Duesberg, 2009), taken the view that oncogenesis is not only initiated by random mutations but that oncogenesis *results* in massive genetic instability, including random mutations, while preserving, in a peculiar sort of privileged way, the core de-repressed program that is actually causative, in the proximate, mediating sense. Understanding that the cancer cell is actually *for* something allows one to put this in perspective: the "purpose" of the genetic instability is heuristic experimentation with novelty, probably in response to some dissatisfaction with the status quo ante. Without this perspective, the cancer cell would simply be seen as a collection of blind, stochastic errors, hopefully with a small core of constant but strangely elusive changes that could act as targets, buried in a cacophony of genetic noise, whose only importance is to act as a source of epiphenomenal drug resistance, that is, the "standard model."

C. Why Targeted Drugs Usually Fail

There are several reasons why targeted approaches usually fail, sooner or later (more likely the former). Firstly, the targets may not actually exist, or not exist any longer (Klein, ibid 2010). Alternatively, the cell may lose dependency on a particular target because of redundancies. Or the target itself may mutate, so it can no longer bind the drug, but can, and does still, drive the cancer cell. Finally, some of the driver targets that do actually exist, like *K-RAS*, or defective tumor suppressor genes, are really hard to "drug."

Reasons for the elusive nature of these much sought after "constant targets" might include the following:

1. The real cause is a preordained program, including Warburg metabolism, controlled by a physiological switch.
2. Several different alterations could result in the same (type of) defect, that is, the details differ but the downstream pathway perturbations are the same, as a "final common pathway," but perhaps less easy to "drug."
3. The chromosomal instability is what really "causes" the malignant phenotype, at least in the sense of maintaining it, even if not initiating it. In this model, the malignant phenotype is not maintained by a small coterie of aberrant genes; rather, the cancer is actually maintained by the continuous, genome-level disablement of control pathways, and, which therefore, never re-establish meaningful restraints on the cell cycle (even assuming that a novel equilibrium, allowing the emergence of stabilizing effectors anew, could actually be created if only there was enough time) (Ptashne, 2009).

These explanations are not mutually exclusive.

D. More Difficult to Kill

In this view of the cancer cell, drug resistance is rendered inevitable in respect not only of single-targeted drugs but also multifaceted agents like classical chemotherapy/radiation, because the fidelity of DNA repair/

DNA constancy is no longer required by the simplified cancer organism (having sacrificed differentiation, identity and metazoan loyalties), and which, not coincidentally, reduces the vulnerability of the organism to genetic insult. For example, a much larger proportion of the DNA, no longer required, could then act as a giant decoy. Furthermore, the prolific genetic novelty provides excellent prospects at finding multiple ways to circumvent the effects of any drug or drug combination. And for as long as the cancer cell is a reduced, simplified object, it will be difficult to kill, not because it is sophisticated but because it is not.

E. Precambrian Training

Although the stem-cell nature of cancer contributes to resistance, this alone is not sufficient. Normal stem cells strive to protect their genomic integrity; but life's challenges were variable, unpredictable and highly lethal. Hence the need to have evolved a broad spectrum, rapid response defense mechanism, of which stochastic genomic reshuffling was a major feature, and to which everything else might, at least temporarily, be sacrificed. I am not saying that all unicellular organisms are necessarily tolerant of egregious DNA damage, but I am saying the only cells that can tolerate, and propagate, egregious DNA damage are unicellular organisms, asexual as well, and in cancer, the DNA instability is the whole point. In this respect, cancer stem cells are very different from normal stem cells.

Furthermore, since, as this theory maintains, the resilient phenotype evolved in the Precambrian to deal with a certain suite of perils, we should not be surprised to discover that therapies exploiting similar mechanisms of action as these perils might not be very successful (Table IV). For instance, radiotherapy mimics extraterrestrial radiation absent the ozone layer; alkylating agents and some other forms of chemotherapy are radiomimetic; both radiation and certain chemotherapies induce ROS; antimetabolites mimic nutrient deprivation; thermal treatments (cryosurgery, etc.)

Table IV Correspondence of Modern Therapies to Ancient Threats

Ancient threat	Modern therapeutic equivalent
Extraterrestrial radiation unshielded by an ozone layer	Radiotherapy and radiomimetic chemotherapy, e.g., alkylator, platinums
Reactive oxygen species and other reactive species	Radiotherapy, radiomimetic chemotherapy including anthracyclines
Food chain collapse, nutrient deprivation	Antimetabolites
Climate changes	Cryotherapy, radiofrequency ablation
Interspecies competition via chemical warfare	Natural product" cytotoxics

mimic climate extremes; natural product chemotherapies recapitulate interspecies chemical warfare; and vascular disruption mimics food-chain collapse and anoxia. Therefore resistance to these agents is probably wired in to the repertoire of all cells that survived this period in the earth's history.

F. Lessons from Mass Extinctions

Are cancer cells killable? One approach to this all-important question is to look back to the mass extinctions on earth of which there have been five, along with many minor ones. The one unifying theme seems to have been the combination of a sudden shock, superimposed upon a chronic stress—so-called press-pulse (Arens and West, 2008; Yedid *et al.*, 2009). I have previously suggested (Vincent, BioEssays in press 2011) that eradication of a cancer should in some ways be akin to a mass extinction event, except that the requirement for selectivity is also imposed. In any event, we may have chanced upon this empirically, for example, with 5FU + radiation, representing, respectively, chronic nutrient deprivation and "sudden shock." These methods do seem to be at least relatively effective in killing large numbers of cancer cells, and any selectivity there might arise from the same defects in DNA repair, which paradoxically sooner or later enable resistance to emerge.

IX. IS THERE AN ANSWER?

A. The Problem Restated

Despite sporadic interest in induced differentiation as a treatment strategy, most would still back eradication of the cancer as the surest route to cure. This would generally entail the very ambitious engineering challenge of designing a therapy that would seek, find, and specifically eradicate all potentially clonogenic transformed cells, while at the same time keeping damage to normal cells, not necessarily at zero, but at least below the threshold of intolerability. This is not to suggest that cytostasis, or lesser degrees of debulking, are not helpful, just that if cure is the goal, there is likely only one final way to guarantee it; anything else is inevitably a temporary palliative, or at best results in a "chronic disease" outcome.

It is safe to say that the first major chemotherapeutic successes were constructed mostly on an empirical basis. The theory seemed to consist mainly of a sense that it would be rational to treat a proliferative disease with antiproliferative drugs, coupled with a realization that it was possible to combine agents with nonoverlapping toxicities, and in the hope that available resistance mechanisms could not deal simultaneously with different classes of drug administered in combination. However, the reason these drugs should be selectively toxic for malignant but not normal cells was not only poorly understood (Harrap, 1978), but even poorly misunderstood, as being based on the incorrect notion of rapidity of cell division (i.e., cell cycle time), rather than the more important growth fraction. Since most traditional anticancer drugs are more toxic to proliferating cells, "rapid proliferation" might be a reasonable hypothesis, but it is the latter (growth fraction) rather than the former (cycle time) which actually determines "rapidity of proliferation" for the tumor as a whole. In fact, some normal cells have a much shorter cell-cycle transit time (e.g., 12–24 h for bone marrow precursor cells, vs. about 2–3 days for many cancer types) (Donovan et al., 2005).

It is not actually possible to make sweeping generalizations about the growth fractions of tumors compared to normal tissues; some tumors have relatively low growth fractions and some normal tissues quite high growth fractions (Elion, 1985), or vice versa. In fact, it is quite clear that these purely kinetic considerations are not enough to explain selectivity, and while they most certainly speak to part of it, there is much they do not account for. Cisplatin, for example, is not affected by the proliferative status of the cells, and the same is true of the nitrosureas and bleomycin (Boyer and Tannock, 2005), yet these drugs are much more active against selected types of cancer than normal cells.

Some additional factor is obviously required to explain such selectivity as occurs, and that factor may well be differential DNA repair (Evers *et al.*, 2010), although supplemental hypotheses exist in respect of apoptotic threshold (Wilson *et al.*, 2009). The three most curable solid cancers (germ cells tumors, lymphomas, and gestational choriocarcinomas) apparently share highly intact proapoptotic machinery, retained from their normal tissues of origin, and which was necessary there because genetic rearrangements occur physiologically in each of these tissues, and result in errors that regularly require removal. This was hypothesized to be the reason that chemotherapy is so effective in curing this limited range of malignancies (Savage *et al.*, 2009). Treatment of these cancers, for example, advanced testicular cancer or Hodgkin's disease, is characterized by massive debulking, and usually permanent eradication of the cancers; yet only transient (albeit significant) toxicity occurs in normal tissues that recover fully, or almost fully in many patients (Muggia, 2009). It is unlikely that a lower apoptotic threshold in the cancers versus normal tissues is solely responsible for this, and possible that DNA repair is more intact in the normal tissues than the tumors; however much of this remarkable phenomenon remains to be explained (di Pietro *et al.*, 2005).

Although the idea of differential DNA repair capacity has proven durable in general, the inadequacies of the whole conceptual package became clear in proportion to the inadequacies of these initial drug combinations when applied to the common cancers like non-small-cell lung, stomach cancer, and colorectal cancer. Although transient regressions were sometimes apparent, especially in breast and small-cell lung cancer, the overwhelming majority of all these cancer cases was either utterly drug resistant *de novo*, or rapidly became highly resistant.

In particular, it was realized that, while the initiating "causes of cancer" were readily apparent, they were generally of the "hit and run" variety and did not help to understand the day-to-day inner workings of the cancer cell. Nonetheless, the blossoming discipline of molecular biology seemed to offer a pathway to a deeper understanding, fuelling the optimism that once the proximate causation and mediation of the malignant phenotype was revealed, druggable and specific targets would rapidly emerge, and cures would automatically follow.

Although an immense amount has been learned as a consequence of this "war on cancer," in only a handful of cases has this led to really effective targeted treatments, based on fundamental causality, and then only in rather unusual types of cancer such as CML, and less successfully, EGFR mutated or ALK-activated non-small-cell lung cancer subgroups, in which the benefits appear temporary, as with hormonal therapies in some breast and nearly all prostate cancers. Other targeted therapies (especially rituximab, trastuzumab, cetuximab, and bevacizumab) have been moderately successful, but seem to function at the level of relatively specific tumor markers, or contributory causality, rather than fundamental causality.

The philosophy underlying these and other sorts of targeted drugs is often based on a number of premises, some of which may, to a greater or lesser extent, be incorrect:

1. All cancers have dominantly acting driver abnormalities, which can be identified, called oncogenes.
2. These oncogenes are "necessary" causes of the cancer (whether or not they are sufficient).
3. These oncogenes have to be present in a continuous fashion to drive the cancer.
4. Inhibiting these driver oncogenes is sufficient to regress the cancer and may lead to a cure.
5. Oncogene products are druggable.
6. Because they drive the cancer, in a causative sense, and because cancer is abnormal, these drivers themselves must be qualitatively abnormal, and therefore side effects will generally be minimized.
7. If we cannot find the targets, we are just not looking hard enough.

Permeating this set of propositions is the notion of causality; if only the proximate cause can be identified, it can be inhibited, and the tumor cells will respond in some desirable way. In future, it was thought that even the critical mutational aberrations in the nuclear DNA itself will be identifiable, and amenable to manipulation.

These ideas reflect much of Aristotelian causality: the "material" cause is the tumor cells, the cause of the illness, and which must therefore be made to disappear; the "formal" cause is represented by the entrenched abnormal blueprint in the tumor DNA, and which is felt to represent the ultimate target for corrective manipulation; and the "efficient" cause is the chain of events that mediate the malignant phenotype, among which specific, inhibitable, and necessary drivers are supposed to reliably exist. Only the final, Aristotelian "purposive" causal aspect is, apparently, irrelevant.

Furthermore, in this paradigm, drug resistance is conjectured as based on specific mechanisms that can be diagnosed and addressed; for example, for *BCR-ABL* mutations leading to imatinib resistance, there can always be a dasatinib, or nilotinib; and for EGFR-TKI resistance, for example, based on the T790M mutation, there will always be drugs like afatinib; each of these successor drugs being allegedly efficacious against the newly mutated forms of the driver oncogene products.

B. What's So Wrong with This?

Unfortunately, serious questions can be raised about all of the above premises.

1. ALL CANCER HAVE . . . ONCOGENES

Just because some cancers have identifiable, mutated oncogenes, which do exist as a "presence" in a dominantly acting fashion, it may not be true that all cancers must necessarily have them, or necessarily retain them (Klein *et al.*, ibid 2010; Perez-Caro *et al.*, 2009) even if they needed initiating oncogenes. What we need here is much more factual information, including information about "gene sets" or pathways, as well as individual genes, before we can properly evaluate the contributions of individual genes versus "whole genome" causality (Boca *et al.*, 2010).

2. THESE ONCOGENES ARE "NECESSARY" . . .

The implication is that in order to halt or regress the cancer, it is sufficient merely to inhibit the oncogene(s) in question; that is, it presupposes that if oncogenes exist, there must be a state of absolute dependence on one or more of them; a condition known as "oncogene addiction." In the case of *BCR-ABL*-associated CML, this situation may apply at least in many if not all patients. In EGFR mt+ cancers, massive tumor reductions certainly occur with small molecule EGFR-TKI blockade, but those patients are never cured, either because the target EGFR mutates to resistance via nonbinding (the T790M mutation) or alternative pathway availability (c-Met amplification) or for other reasons not yet understood (Choi *et al.*, 2010; Sequist *et al.*, 2011; Suda *et al.*, 2010). A similar situation is unfolding in ALK+ lung cancers for drugs like crizotinib.

It is not the case that unambiguous oncogene addiction-type situations exist for most cancers. Although *K-RAS* mutations seem to drive about 60% of colorectal cancers, and most cancers of the pancreas, the impossibility of successfully drugging *ras* has precluded the acid test in human patients. And while oncogenes have been known about for decades, very few have been successfully drugged, and it is possible, even likely, that pathway redundancies and other problems will continue to frustrate these efforts.

Therefore, although exploitable oncogene addiction situations do exist, so far they have been few in number, and it remains at best an open question on the generalizability of this approach. For instance, it might be possible to at least initiate transformation of cells by the inactivation of tumor suppressor genes, but whether full transformation can proceed without any help from or reliance on oncogenes is an open question; but it might be the case, for example, in retinoblastoma of childhood (Little *et al.*, 2011).

3. THESE ONCOGENES ARE PRESENT IN A CONTINUOUS MANNER, TO DRIVE THE CANCER

As noted, Klein *et al.* document experimental situations in which the oncogene is required but only for initiation of the cancer, and thereafter is not only redundant but is actually lost altogether. This supports at least the possibility of the Heng–Duesberg concept of causality in which it is the instability of, or imbalance in, the whole genome that continues to drive the cancer, and not the presence of any particular gene. And since most human cancers present well into their tumor life spans (i.e., at about 20–25 tumor doublings, out of the maximum of 40 beyond which the host capitulates), the opportunity to direct a targeted agent against a mutated oncogene might be long past in real-life situations, if oncogene transience is indeed a real phenomenon.

While it is only fair to point out that how often this initiation-loss sequence occurs in actual human cancer, and indeed the whole Heng–Duesberg model of causality, has not even begun to be critically assessed, it is not a surprise that given the known level of genomic plasticity, cause-and-effect relations can alter with ongoing tumor evolution. Therefore we can anticipate not only loss of absolute dependency on a particular oncoprotein but also even, sometimes, actual loss of that protein. Either way, a targeted drug directed against this particular oncogene product would soon cease to be efficacious, and may even, via selection, accelerate its own failure.

Therefore, although the notion of causality as a therapeutic lead is seductive, it is certainly true that causal relations in a complex system are themselves complex, not easily disentangled, and will, unavoidably, be naive in their initial formulations. Perhaps the generally rather modest benefits to date of "targeted therapies" (Table II) reflect this.

4. INHIBITING THESE ONCOGENES WILL REGRESS THE CANCER, AND LEAD TO A CURE

This follows logically from the second premise above, and therefore suffers from the same weaknesses. It of course also depends on the feasibility of successful inhibition, below some ill-defined threshold of activity. If oncogenes are more loosely defined, as contributing to the malignant phenotype, rather than being absolutely necessary on an ongoing basis, one can allow that temporary but worthwhile benefits might materialize. But to imagine that benefit is guaranteed presupposes that an oncogene addiction exists in at least most of the tumor cells; and to suppose that oncogene addiction will enable a cure (via inhibition) is to suppose that cancers have no strategies to circumvent the obstacle, a supposition that appears demonstrably false. Even with CML, once the tumor has evolved to the accelerated or blast phase, imatinib and other *BCR-ABL* tyrosine kinase inhibitors are less and less likely to be efficacious (Skorski, 2011).

5. ONCOGENE PRODUCTS ARE DRUGGABLE

Of course, they may be, and some have been. But as noted, mutated *K-RAS* remains undrugged, despite decades of effort (McCormick, 2010). It might be necessary to resort to antisense technology, an avenue that is being actively explored, but with no guarantees of success, either. Furthermore, there are several well-described oncogenes, for example, Myc, for which specific inhibitors do not appear to be even near the clinic.

6. BECAUSE THEY DRIVE . . . THE DRIVERS THEMSELVES MUST BE QUALITATIVELY ABNORMAL . . .

This may sometimes be true, but often it is not. *BCR-ABL* is mutationally abnormal, as are the activating mutated forms of EGFR (e.g., del 19 and exon 21 point mutations). However, neither gefitinib nor erlotinib are specific for the mutated form, and also inhibit the normal protein, which in fact may actually be also required for efficacy, in addition to inhibition of the mutant protein. Not surprisingly, on-target side effects occur, as a result of EGFR-wt inhibition in normal tissues (rash and diarrhea), but interestingly, these are generally mild-moderate.

Of more concern are situations in which increased production of an unmutated product contributes to cancer growth, as with EGFR wt amplification in colorectal and many cases of non-small-cell and other cancers. In these cases, there is only a vague hope that the cancers are "more reliant" on these gene products than normal cells, so that their inhibition will result in a proportionately greater effect on the cancer, thus enabling what is known as a "therapeutic index." Whether or not this situation exists can only be discovered empirically, in the intact human being. The level of expression is not necessarily a guide here; for example, for ER expression in breast cancers, the higher the level, the greater the sensitivity to inhibition (Bonomi *et al.*, ibid 1988), whereas for thymidylate synthase, the opposite is true (Inoue *et al.*, 2005; Ozasa *et al.*, 2010; Salonga *et al.*, 2000).

Thus although it might be argued that at some level, qualitative differences might exist, or even might have to exist between tumor and normal cells, this does not always translate into qualitative differences at the level of the protein actors that are the immediate and concrete targets; rather, the qualitative differences may only be apparent at "formal cause" level, that is, the DNA, which for the time being remains undruggable.

Furthermore, if it is true that transformation may be brought about by gene dose effects only, that is, without actual mutation at the DNA level, then there may never be any qualitative differences to exploit at all, and the only opportunities for selectivity might exist, for example, because of an unusual degree of "reliance" of the cancer cell on the gene in question, or that the really vital normal tissues do not depend on these genes, for example, the ER or androgen receptor, and expression and dependency are confined to nonvital organs, such as the remaining breast tissue; or that cancers may depend on oncofetal proteins normally repressed in the adult (Mizejewski *et al.*, 2010).

7. IF WE CANNOT FIND THE TARGETS, WE ARE JUST NOT LOOKING HARD ENOUGH

This might be true, since an average driver might only afford a tiny growth advantage to a cancer cell (Bozic *et al.*, 2010), and it might in fact be almost impossible to discern this. This type of driver, however, is unlikely to be "necessary" in the sense that as a single target, it could ever be addressed therapeutically in such a way that the cancer cell would then cease to exist.

Furthermore, it has been recently estimated that among the many mutations in a cancer cell, only a small fraction are actually driver mutations (Bozic *et al.*, ibid 2010; Carter *et al.*, 2009). And, I would agree, at least to some extent with Heng *et al.* (ibid 2010) that the data suggest with respect to most detected mutations in cancers in which this has been looked at (e.g., cancer of the pancreas (Blackford *et al.*, 2009)), very few of the mutations occur in more than one cancer sample. In this pancreas survey ~89% of mutations did not occur in >1 patient, and a further 10% were mutated in only "2,3 or 4 tumor samples" out of 114. On the other hand, *K-RAS* was mutated in 99%, *TP53* in 82%, *TTN* in about 7%, *SMAD4* in about 12.8%, and *CDKN2A* in about 20%.

From this, one might conclude that *K-RAS* mutation is probably a necessary, ongoing dominant driver (and p53 also required in most patients, but as a function-absence abnormality); and *SMAD4,* and CDKN2A (i.e., *p16*) would be of interest. Nonetheless, the vast majority of mutations appear to be unique and as such, it will prove to be very difficult to know if any particular one is or is not an important driver. And, needless to say, only the important drivers will be worth pursuing for drug design purposes, and with current technology and preferences, only dominantly acting "presence" abnormalities would be considered suitable for drug design efforts, which would rule out *TP53, CDKN2A*, and perhaps even *K-RAS* that is, although oncogenic as a single, dominant allele, actually locked in an enzymatically inactive configuration. Finally, gene activity can be affected by epigenetic mechanisms, as might also be the case with *CDKN2A* in pancreatic cancer (Lomberk, 2011), and which might well be missed in a sweep of the DNA; therefore purely mutational analysis might underestimate the common involvement of certain genes influenced by epigenetic mechanisms (e.g., DNA methylation/demethylation aberrations, altered micro RNA, etc.)

Therefore, while I agree that undiscovered drivers might exist, it is not easy to discover them, nor easy to drug them; all these factors explain why so few drugs emerge from so simple and attractive a concept as the "driver-inhibitor" model.

C. Why Is Traditional Chemotherapy Generally Not Curative?

Combination chemotherapy came of age in the 1970s, and its successes and failures are by now empirically very well documented. As noted above, much remains to be learned about why it is highly active in certain situations and highly inactive in others; differential DNA repair and variable apoptotic threshold are candidate explanations. It is perhaps easier to comprehend why tumor cells in general might be resistant: apoptotic pathways are partially disabled, DNA repair and/or tolerance of DNA damage might be "good enough," stem cell properties confer specific resistance mechanisms like extrusion pumps, etc., parts of the genome might function as a decoy and, above all, the unstable genome is more likely to generate resistant variants, which are then selected to survive.

It is obviously true that the curable cancers lack the slew of general and specific resistance mechanisms, but why this should be the case is unknown. It is equally obvious that the less sensitive cancers are equipped with resistance mechanisms of various sorts, some of which are specific for certain drugs. However, the central question as to *why* these sensitive cancers lack resistance mechanisms, and the resistant ones do not, is not often asked or answered; the standard model would suggest "random mutations" as the explanation, but the well-known general similarity of drug sensitivity in most members of a particular tumor type compared to members of a different type indicates otherwise; for example, small cell versus non-small-cell lung cancer.

As suggested (Vincent, ibid BioEssays, in press 2011; Israël, 1996 ibid), to the extent that modern therapies resemble ancient existential threats, the cancer cells, manifesting an equally ancient survival program, may be well trained to circumvent them. In fact, these similarities might be quite striking (Table IV).

The Heng–Duesberg model, or any other model identifying accelerated macroevolution as the central fact of cancer, would also account for the rapid emergence of resistant variants, as would "adaptive resilience" as an ancient trait.

There may be an interesting reciprocal relationship between prognosis and drug sensitivity, which is pertinent here. In a recent trial of adjuvant chemotherapy in resected non-small-cell lung cancer, comparing a cisplatin-based regimen to observation, patients with a high level of ERCC1 (and hence, presumably, more proficient DNA repair) had an inherently better prognosis (i.e., absent chemotherapy) but no added benefit from the addition of chemotherapy. Conversely, those with low levels of ERCC1, and therefore less proficient DNA repair, had an inherently worse prognosis (absent treatment), but the biggest benefit from the chemotherapy; in fact, the chemotherapy could turn the worst group into the best (Olaussen *et al.*, 2006).

What this appears to be telling us is that (subject to confirmation) cancers with poor DNA repair are more likely to evolve aggressive subclones, likely via mutation; whereas those that are less mutagenically prone will pursue a somewhat more indolent or at least less aggressive course absent therapy, that is, a *prognostic* determinant. On the other hand, the former, with less efficient DNA repair, may be more easily killed by DNA-damaging agents, whereas the latter, with more efficient repair, will tend to be more refractory to the effects of the chemotherapy—a *predictive* determinant.

That less efficient DNA repair should lead to the evolution of more diverse subclones makes sense, and it is also to be expected that upon exposure to chemotherapy, more damage would remain unrepaired and therefore more cells would be "pushed over the edge" and succumb (Helleday, 2010). On the other hand, with more DNA instability and hence diversity, there might also be more chance for resistant forms to evolve and therefore this situation may represent a trade-off, with the potential to generate conflicting results.

Therefore, and surprisingly, to the extent that traditional chemotherapy does exercise selective cytotoxicity to cancer cells, and is therefore targeted, at least part of the selectivity involves exploiting a "marker absence" type of mechanism: the (relative) absence, or at least diminished capacity for DNA repair, compared to normal cells. The final determinant of cellular fate after exposure to chemotherapy or radiation, however, involves a poorly measurable and opaque relationship between DNA damage and apoptotic threshold; intensive efforts to manipulate the latter have not, to date, proven successful (Gimenéz-Bonafé *et al.*, 2009). There may be more to gain by enhancing DNA damage via manipulation of DNA repair mechanisms (O'Connor *et al.*, 2007).

As regards traditional chemotherapy, however, it is worth emphasizing that differential sensitivity remains underexplored and underexplained, and the potential for trade-offs may cloud interpretation of the result, especially with trial designs that prevent the clear delineation of prognostic effects from predictive ones.

D. A Shift in Emphasis?

As noted above, there is little to suggest that more hope should be placed in the emergence of novel driver-inhibitor type drugs, especially if the ambition is to cure rather than to temporize. (One exception to this might be novel drugs to disable Warburg-type metabolism, since it is apparently, a near-universal accompaniment and mediator of malignancy, and also fairly specific to cancer as well.) On the other hand, repair of the cell transformed by a molecular-absence type of driver situation (which are more common) may be beyond the reach of current technology. So where do the answers lie? I believe a case can be made for more intensive efforts to develop marker (as distinct from driver) opportunities, especially marker-absence opportunities that are more consistent with reality under conditions of a rapidly evolving cancer genome (i.e., Heng–Duesberg causality, or adaptive resilience).

For greater clarity, I am suggesting that while it might be possible to define at least some of the causative molecular lesions, either as aberrant presences or aberrant absences, successful therapeutic exploitation of these "driver" discoveries will either prove much more difficult than expected, or yield in general much less enduring remissions than we might wish. An alternative approach consists in exploiting distinguishing "marker" features, irrespective of causality. Analogous to driver opportunities that are due either to a presence or an absence of some molecular functionality, cancer marker opportunities stem either from the presence of a distinguishing feature, on or in the cancer cells, or an absence of one compared to normal cells. Furthermore, such marker features (whether present or absent) can be exploited in a variety of ways. One method involves using a simple structural type of approach, for example, the marker as an address box to which to deliver a warhead (address present only on the cancer cell) or a cytoprotective device (address present on normal cells but absent on the cancer cells). Alternatively, these marker features can be exploited in a functional way, as long as the function is not expected to act via a causality mechanism. For example, a marker "presence" opportunity could be used to catalyze the activation of a cytotoxic drug, just like the relatively higher levels of thymidine phosphorylase in the tumor cell are used to activate the prodrug capecitabine (Van Cutsem *et al.*, 2001) or the use of SPARC and the gp60 receptor to actively deliver nab-paclitaxel to the tumor milieu (Desai *et al.*, 2009; Gradishar, 2006;). In a marker-absence example, the concept (if not the actual reality) of an oncolytic virus that is kept inactivated by intact p53 in normal cells but the absence of p53 allows the progression of the viral life cycle to the cytolytic stage only in the cancer cells (O'Shea *et al.*, 2004). Another marker absence function example previously quoted is the activation of amifostine in a normal cellular milieu but not the cancer cell milieu. This is a "functional" example because the amifostine is not simply deposited in the normal milieu, but has to be actively changed from its prodrug form to the entity actually capable of quenching radiation effects, that is, WR-1065 (vander Vigh and Peters, 1994).

It is important to appreciate the nuance that molecular particularities can be exploited as markers, even if they are causal drivers, if the result is an anticancer effect that is not mediated by a causality-reversal mechanism; if the latter, then it should be categorized as a "driver-type" therapy.

Obviously, marker-presence mechanisms are much easier to envisage: simply identify a distinguishing feature on or in the cancer cell, and use it to deposit or activate a cytotoxic mechanism, which will then selectively kill the tumor cells. This is a perfectly rational strategy, and several successful drugs are either approved, based on this mechanism (e.g., capecitabine), or in development (e.g., maytansine linked to a trastuzumab-like antibody) (Isakoff and Baselga, 2011). It is also true that most immunotherapy approaches involve a "marker-presence" type of strategy whether humoral or cellular, and whether the cytolytic capability is artificial or some endogenous branch of the immune system (like ADCC).

The problem with marker-presence opportunities is that they appear in fact to be infrequent. Although tumor-specific antigens may exist, it is hard to find a universal one, one that does not differ from patient to patient. It is very difficult to redesign a drug for each patient, and this in fact imposes a limit on "personalized" medicine. Candidate target markers in the tumor cells include any enzyme expressed at levels higher than the surrounding tissue; oncofetal antigens like αFP, repressed in normal cells; components of markedly upregulated signal transduction pathways, like EGFR, but used for delivery purpose or immune attack; or mutant macromolecules that can be used for delivery or catalysis purposes. Note that the difference between normal and cancer cells need not be absolute but merely relative, provided careful attention is paid to lethal threshold effects.

It is, however, my belief that the really major type of missed opportunity is provided by the marker-absence category. This is because of the nature of cancer genomic instability; while it is true that all sorts of markers might be gained as a "presence" in the cancer cell, whether as oncogenic drivers or simply bystander phenomena, they are rarely, given the random nature of this process, going to be universal cancer markers. Loss of genetic information, however, is likely to be at least as frequent if not more so, and will include consistent areas of the genome across different patients, especially if the chromosomal loss enables a "driver" function. While it is not yet possible to think about "restoring" the missing chromosomal function to each tumor cell, it may well be possible to exploit consistent absences of genetic information in every cancer cell, at least of that type of cancer.

For this, Heng–Duesberg causality and the adaptive resilience/mutator phenotype both imply constant, roiling genetic reshuffling in which part or whole chromosome losses are frequent. And it matters little if one does not accept the Heng–Duesberg model of causality (i.e., it is the karyo-genotype that maintains the cancer via a mixture of gene imbalance, instability, mutational fine-tuning, and "context"); all one has to do is accept the empiric facts of massive genetic loss, irrespective of whether one regards it as a cause, or a consequence of the cancer, or both. But it is readily apparent that random macro- and micro-evolutionary change, accelerated as it is, will produce more distinctive constancy in the form of loss than of gain; and that the losses are extremely unlikely to be reconstituted, as a resistance mechanism and this is what needs to be exploited.

E. Are "Marker-Absence" Strategies Feasible?

The answer depends on two important factors: the constancy of identifiable loss of genetic information in cancer and the ability to design efficacious and selectively cytotoxic drugs based on this principle.

Genetic information can be embodied, or expressed, at the DNA, RNA, or protein level. If homozygously absent at the DNA level, then automatically it will also be absent at the RNA and protein levels. More commonly, however, is a hemizygous loss often accompanied by a second hit (the "two hit" hypothesis) in the remaining allele, which prevents expression (e.g., promoter methylation), or which results in the expression of a loss-of-function, mutant mRNA (e.g., point mutated, frame shifted, or truncated), so that no protein, or no functional protein, is expressed. The fact that these homo- or hemi- zygously deleted sequences often involve tumor suppression genes, that is, are "absence-driver" situations is not particularly germane to the general marker approach, except to highlight areas of relatively predictable genetic loss.

The ideal situation is where a homozygous DNA deletion accompanies a known type of cancer in a relatively high proportion of cases. In lung cancer, for example, about 30% of cases exhibit an homozygous deletion in chromosome arm 9p (Sato *et al.*, 2005; Zhao *et al.*, 2005). Cox *et al.* surveyed 636 cancer cell lines (Cox *et al.*, 2005), finding 281 homozygous deletions. They also note that homozygous deletions may arise in or near fragile sites, compounded perhaps by DNA repair defects, or simply be random events. They also note, incidentally, that hemizygous loss was often accompanied by duplication of the remaining allele, finding "several thousand." For homozygous deletions, glioma and pancreatic cancer cell lines seemed particularly prone to develop them, whereas cervical cancer and hepatocellular cancer cell lines were less likely. Furthermore, homozygous deletions were particularly likely to cluster around known tumor suppressor genes, especially *CDKN2A*, *PTEN*, *RB1*, and *LMAD4*, and/or bordered on seven of nine fragile sites known and sequenced at the time (e.g., FRA3B). However, 178 deletions were not "directed" in these ways. In the minority of these cell lines where this could be determined, it was clearly established that these deletions were somatically acquired (i.e., in the cancer) and spared normal cells, which is of course crucial to the prospects for selectivity.

Short of the "homozygous loss" ideal, however, loss of heterozygosity (LOH) with associated failure to express the cognate mRNA, and protein, seem much more common, and may be suitable. For example, 31 primary ovarian cancers were surveyed, revealing >380 small regions of "copy gain or loss" (and incidentally, 27 homozygous deletions) (Gorringe *et al.*, 2007).

Some of the hemizygous loss situations will represent tumor suppressor genes that require full inactivation in order to drive the cancer, and which might reasonably be expected to lack any gene expression. However, others represent haplo-insufficient situations (e.g., *TP53* or *PTEN*), in which LOH might still be a driver, even if some reduced level of normal expression off the remaining normal allele occurs, since protein levels supported by both alleles may be required for normal functioning (Berger and Pandolfi, 2011).

A variety of methodologies to identify genetic losses have been described, the more sophisticated and recent of which seem capable of finer and finer resolution (Caren *et al.*, 2008; Gorringe *et al.*, ibid 2007; Rothenberg and Settleman, 2010). Furthermore, it is also clear that microRNA might be selectively reduced or deleted in cancer cells (Davidson *et al.*, 2010).

As to how drugs could be designed to exploit cancer-specific loss of information, a variety of strategies were noted above, some of which rely, for activation, on a specific function in normal tissues, lacking in cancer, as with activation of the cytoprotectant amifostine that is said to occur selectively in normal cells because of the presence of an alkaline phosphatase and a permissive pH, neither of which occur in the tumor cell milieu; and others on lack of function in the cancer cells, present in normal cells (e.g., the concept, if not the reality, of oncolytic virus cytotoxicity, which "activates" the cytolytic virus life cycle only in the cancer, in the permissive milieu of absent p53); yet others might rely on the marker's mere presence in normal cells for the simple delivery of a protective agents (e.g., G-CSF binding to its receptor in the normal myeloid line, as partial protection from otherwise myelosuppressive chemotherapy). It is easy to see (perhaps easier to see than to do) the possibility of designing a shield, enabled in some way by the genetic information present in normal cells, but absent in the cancer cells, which would then lack the shield; and then to apply an otherwise unselective cytotoxic, against which only the normal cells would be protected.

In 2007 was described a method (Varshavsky, 2007), envisaging a "deletion-specific vector," consisting of a ring of DNA encoding a cytotoxic protein, together with an inactivating enzyme package triggered by the presence of specific sequences present in normal cells, but not cancer cells. Described as "a brilliant idea" by Bert Vogelstein, of John Hopkins, "because it exploits the Achilles heel of cancer. Deletions are likely to be present in every cancer" (Goymer, 2008 ibid). Varshansky was awarded the $1 million Gotham prize for this achievement. I have good reason to know that other variations are possible, at least conceptually.

The key concept to appreciate, however, is that gain of function mutations might be variable, and unstable; but loss of genetic information is unlikely ever to be reconstituted and repaired by the cancer cell, no matter how unstable its genome. Therefore, the target will not disappear, removing one important source of resistance. Not to underestimate the cancer cell, however, I would also advocate the accompanying cytotoxic insult be massive, overwhelming, and sudden, in the spirit of "shock and awe." This, of course, puts a high demand on the mechanism of selectivity, whether the cytoprotective capabilities of the constructed shield or however embodied, and which will not be so easy to engineer in practice. But I think this is the best and perhaps only route to cure of the common cancers, especially if it is combined, as I think it should be, with emerging approaches aimed at disabling the WE (Vander Heiden *et al.*, 2010).

X. WHAT FORM OF LIFE?

This question is rarely asked and never answered, at least explicitly. It is, however, implicit in the various evolutionary theories of cancer. The basic evolutionary view (standard model) is that cancer is simply the result of a series of random mutations, and that selection for fitness enables the exhibition of a remarkable convergence of traits; a refinement of this acknowledges that an endogenous program of stem-like self-renewal is unleashed in this process, as exit strategies are blocked (i.e., by random, disabling mutations).

A somewhat different perspective, but still compatible with Darwinian selection, is provided by the atavism theories, common to which are that some kind of a survival "memory" (i.e., a de-repressed, ancient program) is reactivated under duress, leading to what is quite literally a re-emergent organism, albeit genetically unstable; and that this is a more parsimonious explanation than convergent evolution, for the many consistent aspects of the malignant phenotype.

I have here provided a viewpoint very much within this tradition, but characterizing the re-emergent program as none other than the WE, whose multiple facets explain much of the character of the cancer cell, not the least of which is the genetic instability referred to in the Heng–Duesberg model of oncogenic causality.

Furthermore, important therapeutic consequences flow from this, namely that missing genetic information specific to the cancer cell can be exploited to create a shield in normal cells to an otherwise indiscriminate cytotoxic, that this method of selective toxicity will be difficult to avoid, thus lowering the probability of resistance, provided the cytotoxic assault is abrupt and overwhelming. The WE is itself also an important driver target, whose inhibition might, if successful, lead to conversion to a chronic disease, or even, with concomitant chemotherapy, such as the "marker-absence" concept quoted above, a cure.

The concept of speciation is not enough to describe the cancer cell, conveying perhaps only half the truth, and the easy half of that: the despeciation from the host. What makes cancers difficult to classify taxonomically, is their continually unstable genetic makeup, which violates their depiction as a concrete type; unless, risking casuistry, it is their very instability that is their type-characteristic. The *tree of life*, as we perceive it, is an ensemble of generally stabilized genomic vehicles, and which allows the conclusion that purpose is served, or even defined by the continuation of genetic, or genomic identity. Cancer cells are unlike any other living creature, which raises the issue of where they reside on the *tree of life*, and which brings me back to Dhobzhansky's question: what are they for?

The only conclusion that can be drawn is that, as engines of biomass interconversion and purposed genomic innovation, they are "for" survival; of the cell, or even the cytoplasm; and exist in this fashion because they are, or were, for whatever reason, ill-served by the genetic stability with which they were previously associated. In effect, the cytoplasm has fired the old genome and is actively looking for a new one.

There is no obvious reason why this approach to life should not have proven a superior alternative, under conditions of constant environmental hostility and change, than the genetic stability that ultimately produced our lineage; except perhaps its very success seems to depend on the availability of large, multicellular organisms that, as resource outlets and residencies, can be parasitically exploited. Absent the host, where would the cancer cell be? Are they well enough adapted for the scarcity they would encounter as free-living organisms? Could there ever be enough biomass creation to support the demographic cushion necessary for this sort of genomic roulette, absent a host? Will the genomic rescrambling slow down in the face of resource scarcity? Should we be looking for free-ranging, unicellular "dyskaryotes" in nature? We should at least retain an open mind on this, notwithstanding how far-fetched it appears.

Cancer cells probably need us, certainly more than we need them; and unless a method exists to restabilize their genome, once a good "fit" has been discovered, it is hard to see how this form of life could continue to exist independently. Perhaps there really is an "off switch," perhaps this is the only conclusion that makes any sense, and perhaps we should be devoting more resources to looking for it.

REFERENCES

Arens, N. C., and West, I. D. (2008). Press-pulse: a general theory of mass extinction? *Paleontol. Soc.* **34**, 456–471.

Arranz, J. A., Climent, M. A., Gonsalez-Larriba, J. L., Leon, L., and Maroto, J. P. (2011). Sorafenib in renal cell carcinoma. *Crit. Rev. Oncol /Hematol.* [Epub ahead of print]

Axelrod, R., Axelrod, D. E., and Pienta, K. J. (2006). Evolution of cooperation among tumor cells. *Proc. Nat. Acad. Sci. USA* **103**, 13474–13479.

Berger, A. H., and Pandolfi, P. P. (2011). Haplo-insufficiency: a driving force in cancer. *J. Pathol.* **223**, 137–146.

Bergstein, I. (2003). A non-Darwinian role for mutagenesis in stem cell-derived cancers. *Mol. Carcinog.* **36**, 1–5.

Blackford, A., Parmigiani, G., Kensler, T. W., Wolfgang, C., Jones, S., Zhang, X., Parsons, D. W., Lin, J. C., Leary, R. J., Eshleman, J. R., Goggins, M., and Jaffee, E. M., et al., (2009). Genetic mutations associated with cigarette smoking in pancreatic cancer. *Cancer Res.* **69**, 3681–3688.

Blanke, C. D., Rankin, C., Demetri, G. D., Ryan, C. W., von Mehren, M., Benjamin, R. S., Raymond, A. K., Bramwell, V. H., Baker, L. H., Maki, R. G., Tanaka, M., and Hecht, J. R., et al., (2008). Phase III randomized, intergroup trial assessing imatinib mesylate at two dose levels in patients with unresectable or metastatic gastrointestinal stromal tumors expressing the kit receptor tyrosine kinase: S0033. *J. Clin. Oncol.* **26**, 626–632.

Boca, S. M., Kinzler, K. W., Velculescu, V. E., Vogelstein, B., and Parmigiani, G. (2010). Patient-oriented gene set analysis for cancer mutation data. *Genome Biol.o* **11**, R112.

Bonomi, P., Gale, M., Von Roenn, J., Anderson, K., Johnson, P., Wolter, J., and Economou, S. (1988). Quantitative estrogen and progesterone receptor levels related to progression-free interval in advanced breast cancer patients treated with megestrol acetate or tamoxifen. *Seminars Oncol.* **15**, 26–33.

Bonneterre, J., Buzdar, A., Nabholtz, J. M., Robertson, J. F., Thurlimann, B., von Euler, M., Sahmoud, T., Webster, A., and Steinberg, M. (2001). Anastrozole is superior to tamoxifen as first-line therapy in hormone receptor positive advanced breast carcinoma. *Cancer* **92**, 2247–2258.

Boyer, M. J., and Tannock, I. F. (2005). Chapter 17. *In* "The Basic Science of Oncology" (J. Tannock., et al., , 4th Edn. pp. 349–375. McGraw-Hill, New York.

Bozic, I., Antal, T., Ohtsuki, H., Carter, H., Kim, D., Chen, S., Karchin, R., Kinzler, K. W., Vogelstein, B., and Nowak, M. A. (2010). Accumulation of driver and passenger mutations during tumor progression. *Proc. Nat. Acad. Sci. U.S.A.* **107**, 18545–18550.

Butts, C. A., Ding, K., Seymour, L., Twumasi-Ankrah, P., Graham, B., Gandara, D., Johnson, D. H., Kesler, K. A., Green, M., Vicnent, M., Cormier, Y., and Goss, G., et al., (2010). Randomized phase III trial of vinorelbine plus cisplatin compared with observation in completely resected stage IB and II non-small-cell lung cancer: updated survival analysis of JBR-10. *J. Clin. Oncol.* **28**, 29–34.

Cairns, J. (1975). Mutation selection and the natural history of cancer. *Nature* **255**, 197–200.

Campbell, P. J., Yachida, S., Mudie, L. J., Stephens, P. J., Pleasance, E. D., Stebbings, L. A., Morsberger, L. A., Latimer, C., McLaren, S., Lin, M. L., McBride, D. J., and Varela, I., et al., (2010). *Nature* **467**, 1109–1113.

Cao, Y. (2010). Off-tumor target–beneficial site for antiangiogenic cancer therapy? *Nat. Rev.: Clin. Oncol.* **7**, 604–608.

Caren, H., Erichsen, J., Ollsson, L., Enerback, C., Sjoberg, R. M., Abrahamsson, J., Kogner, P., and Martinsson, T. (2008). High-resolution array copy number analyses for detection of deletion, gain, amplification and copy-neutral LOH in primary neuroblastoma tumors: four cases of homozygous deletions of the CDKN2A gene. *BMC Genomics* **9**, 353.

Carter, H., Chen, S., Isik, L., Tyekucheva, S., Velculescu, V. E., Kinzler, K. W., Vogelstein, B., and Karchin, R. (2009). Cancer-specific high-throughput annotation of somatic mutations: computational prediction of driver missense mutations. *Cancer Res.* **69**, 6660–6667.

Choi, Y. L., Soda, M., Yamachita, Y., Ueno, T., Takashima, J., Nakajima, T., Yatabe, Y., Takeuchi, K., Hamada, T., Haruta, H., Ishikawa, Y., and Kimura, H., et al., (2010). EML4-ALK mutations in lung cancer that confer resistance to ALK inhibitors. *New Engl. J. Med.* **363**, 1734–1739.

Citrin, D., Cotrim, A. P., Hyodo, F., Baum, B. J., Krishna, M. C., and Mitchell, J. B. (2010). Radioprotectors and mitigators of radiation-induced normal tissue injury. *The Oncologist* **15**, 360–371.

Cox, C., Bignell, G., Greenman, C., Stabenau, A., Warren, W., Stephens, P., Davies, H., Watt, S., Teague, J., Edkins, S., Birney, E., and Easton, D. F., et al., (2005). A survey of homozygous deletions in human cancer genomes. *Proc. Nat. Acad. Sci. U.S.A.* **102**, 4542–4547.

Davidson, M. R., Larsen, J. E., Yang, I. A., Hayward, N. K., Clarke, B. E., Duhig, E. E., Passmore, L. H., Bowman, R. V., and Fong, K. M. (2010). MicroRNA-218 is deleted and downregulated in lung squamous cell carcinoma. *PLoS One* **5**, e12560.

Dawkins, R. (1976). The Selfish Gene. Oxford University Press, NY City ISBN 0-19-286092-5.

Desai, N., Trieu, V., Damascelli, B., and Soon-Shiong, P. (2009). SPARC expression correlates with tumor response to albumin-bound paclitaxel in head and neck cancer patients. *Trans. Oncol.* **2**, 59–64.

Dhar, S., and Lippard, S. J. (2009). Mitaplatin, a potent fusion of cisplatin and the orphan drug dichloroacetate. *Proc. Nat. Acad. Sci. U.S.A.* **106**, 22199–22204.

Dhobzansky, T. (1937). Genetics and the origin of species. Columbia University Press, New York, NY.

Dhobzhansky, T. (1964). Biology, molecular and organismic. *Amer. Zool.* **4**, 443–452.

di Pietro, A., Vries, E. G., Gietema, J. A., Spierings, D. C., and de Jong, S. (2005). Testicular germ cell tumours: the paradigm of chemo-sensitive solid tumours. *Int. J. Biochem. Cell Biol.* **37**, 2437–2456.

Donovan, J. C. H., Slingerland, J., and Tannock, I. F. (2005). Cell Proliferation and Tumor Growth. *In* "The Basic Science of Oncology" (L. Harrington., R. G. Bristow., R. P. Hill, and I. F. l. Tannock, eds.),, 4th Edn. pp. 1–3. McGraw-Hill, New York, NY.

Druker, B. J., Guilhot, F., O'Brien, S. G., Gathmann, I., Kantarjian, H., Gattermann, N., Deininger, M. W., Silver, R. T., Goldman, J. M., Stone, R. M., Cervantes, F., and Hochhaus, A., et al., (2006). Five-year follow-up of patients receiving imatinib for chronic myeloid leukemia. *N. Engl. J. Med.* **355**, 2408–2417.

Duesberg, P., and Li, R. (2003). Multistep carcinogenesis: a chain reaction of aneuploidizations. *Cell Cycle* **2**, 202–210.

Duesberg, P., and Rasnick, D. (2000). Aneuploidy, the somatic mutation that makes cancer a species of its own. *Cell Motility Cytoskeleton* **47**, 81–107.

Elion, G. B. (1985). Selectivity—key to chemotherapy: presidential address. *Cancer Res.* **45**, 2943–2950.

Evers, B., Helleday, T., and Jonkers, J. (2010). Targeting homologous recombination repair defects in cancer. *Trends Pharmacol. Sci.* **31**, 372–380.

Fabarius, A., Hehlmann, R., and Duesberg, P. H. (2003). Instability of chromosome structure in cancer cells increases exponentially with degrees of aneuploidy. *Cancer Genet. Cytogenet.* **143**, 59–72.

Falkson, C. I., and Falkson, H. C. (1996). A randomised study of CGS 16949A (fadrozole) versus tamoxifen in previously untreated postmenopausal patients with metastatic breast cancer. *Annals Oncol.* **7**, 465–469.

Farnie, G., and Clarke, R. B. (2007). Mammary stem cells and breast cancer—role of Notch signalling. *Stem Cell Rev.* **3**, 169–175.

Fisher, R. (1930). The Genetical Theory of Natural Selection. Oxford University Press, London.

Gavrilets, S. (2003). Perspective: models of speciation: what have we learned in 40 years? *Evolution* **57**, 2097–2215.

Gimenéz-Bonafé, P., Tortosa, A., and Pérez-Tomás, R. (2009). Overcoming drug resistance by enhancing apoptosis of tumor cells. *Current Cancer Drug Targets* **9**, 320–340.

Gisselsson, D., Jonson, T., Peterson, A., Strombeck, B., Dal Cin, P., Hoglund, M., Mitelman, F., Mertens, F., and Mandahl, N. (2001). Telomere dysfunction triggers extensive DNA fragmentation and evolution of complex chromosome abnormalities in human malignant tumors. *Proc. Natl. Acad. Sci. U.S.A.* **98**, 12683–12688.

Gisselsson, D. (2011a). Aneuploidy in cancer: sudden or sequential? *Cell Cycle* **10**, 359–361.

Gisselsson, D. (2011b). Mechanisms of whole chromosome gains in tumors—many answers to a simple question. *Cytogenet. Gen. Res* **133**, 190–201.

Goldschmidt, R. (1940). The Material Basis for Evolution. Yale University Press, New Haven, CT pp205–6, ISBN 0300028237.

Gorringe, K. L., Jacobs, S., Thompson, E. R., Sridhar, A., Qiu, W., Choong, D. Y., and Campbell, I. G. (2007). High-resolution single nucleotide polymorphism array analysis of epithelial ovarian cancer reveals numerous microdeletions and amplifications. *Clinical Cancer Res.* **13**, 4731–4739.

Goymer, P. (2008). Natural selection: the evolution of cancer. *Nature* **454**, 1046–1048.

Gradishar, W. J. (2006). Albumin-bound paclitaxel: a next-generation taxane. *Expert Opin. Pharmacother.* 7, 1041–1053.

Greaves, M. (2007). A Darwinian eye view of cancer. *In* "Evolution and Medicine: How New Applications Advance Research and Practice" (R. Nesse ed.), The Biomedical & Life Sciences Collection, Henry Stewart Talks Ltd., London.

Greaves, M. (2000). Cancer: The Evolutionary Legacy. Oxford University Press, London.

Gusev, Y., Kagansky, V., and Dooley, W. C. (2001). Long-term dynamics of chromosomal instability in cancer: A transition probability model. *Math. Comput. Model.* **33**, 1253–1273.

Haldane, J. B. S. (1924–1934). A mathematical theory of natural and artificial selection. *Trans. Proc. Cambridge Philos. Soc.; Genet.* **19**, 412–429.

Hall, M. D., Handley, M. D., and Gottesman, M. M. (2009). Is resistance useless? Multidrug resistance and collateral sensitivity. *Trends Pharmacol. Sci.* **30**, 546–556.

Hanahan, D., and Weinberg, R. A. (2000). The hallmarks of cancer. *Cell* **100**, 57–70.

Hanahan, D., and Weinberg, R. A. (2011). Hallmarks of cancer: the next generation. *Cell* **144**, 646–674.

Harrap, K. R. (1978). Towards selectivity in cancer chemotherapy: a biochemical overview. *Adv. Enzyme Regul.* **17**, 457–478.

Helleday, T. (2010). Homologous recombination in cancer development, treatment and development of drug resistance. *Carcinogenesis* **31**, 955–960.

Heng, H. H. (2009). The genome-centric concept: resynthesis of evolutionary theory. *BioEssays* **31**, 512–525.

Heng, H. H., Bremer, S. W., Stevens, J., Ye, K. J., Miller, F., Liu, G., and Ye, C. J. (2006). Cancer progression by non-clonal chromosome aberrations. *J. Cell. Biochem.* **98**, 1424–1435.

Heng, H. H., Stevens, J. B., Bremer, S. W., Ye, K. J., Liu, G., and Ye, C. J. (2010). The evolutionary mechanism of cancer. *J. Cell. Biochem.* **109**, 1072–1084.

Hurwitz, H., Fehrenbacher, L., Novotny, W., Cartwright, T., Hainsworth, J., Heim, W., Berlin, J., Baron, A., Griffing, S., Holmgren, E., Ferrara, N., and Fyfe, G., et al., (2004). Bevacizumab plus irinotecan, fluorouracil, and leucovorin for metastatic colorectal cancer. *N. Engl. J. Med.* **350**, 2335–2342.

Huxley, J. (1942). Evolution: The Modern Synthesis. Allen & Unwin, London.

Huxley, J. (1956). Cancer biology: comparative and genetic. *Biol. Rev.* **31**, 474–514.

Inoue, T., Hibi, K., Nakayama, G., Komatsu, Y., Fukuoka, T., Kodera, Y., Ito, K., Akiyama, S., and Nakao, A. (2005). Expression level of thymidylate synthase is a good predictor of chemosensitivity to 5-fluorouracil in colorectal cancer. *J. Gastroen.* **40**, 143–147.

Isakoff, S. J., and Baselga, J. (2011). Trastuzumab-DM1: building a chemotherapy-free road in the treatment of human epidermal growth factor receptor 2-positive breast cancer. *J. Clin. Oncol.* **29**, 351–354.

Israël, L. (1990). Accelerated genetic destabilization and dormancy: two distinct causes of resistance in metastatic cells; clinical magnitude, therapeutic approaches. *Clin. Exp. Metastasis* **8**, 1–11.

Israël, L. (1996). Tumour progression: random mutations or an integrated survival response to cellular stress conserved from unicellular organisms? *J. Theor. Biol.* **178**, 375–380.

Israël, L. (1998). Malignant progression and resistance of cancer cells: an inducible survival program similar to the SOS system of unicellular organisms induced by environmental assaults. *Bulletin de l'Academie Nationale de medicine* **182**, 49–57.

Jain, R. K. (2005). Normalization of tumor vasculature: an emerging concept in antiangiogenic therapy. *Science* **307**, 58–62.

Kim, J. E., Ryoo, B. Y., Ryu, M. H., Chang, H. M., Suh, D. J., Lee, H. C., Lim, Y. S., Kim, K. M., and Kang, Y. K. (2011). Sorafenib for hepatocellular carcinoma according to Child-Pugh class of liver function. *Cancer Chemother. Pharmacol.* [Epub ahead of print]

Klein, A., Li, N., Nicholson, J. M., McCormack, A. A., Graessmann, A., and Duesberg, P. (2010). Transgenic oncogenes induce oncogene-independent cancers with individual karyotypes and phenotypes. *Cancer Gen. Cytogen.* **200**, 79–99.

Koppenol, W. H., Bounds, P. L., and Dang, C. V. (2011). Otto Warburg's contributions to current concepts of cancer metabolism. *Nat. Rev. Cancer* **11**, 325–337.

Kwak, E. L., Bang, Y. J., Camidge, D. R., Shaw, A. T., Solomon, B., Maki, R. G., Ou, S. H., Dezube, B. J., Janne, P. A., Costa, D. B., Varella-Garcia, M., and Kim, W. H., et al., (2010). Anaplastic lymphoma kinase inhibition in non-small-cell lung cancer. *New Engl. J. Med.* **363**, 1693–1703.

Lai, E. C. (2004). Notch signaling: control of cell communication and cell fate. *Development* **131**(5), 965–973.

Larson, E. J. (2004). Evolution: The Remarkable History of a Scientific Theory. The Modern Library, New York.

Lewin, B. (1997). Genes VI. Oxford University Press, London.

Lin, L., McCormack, A. A., Nicholson, J. M., Fabarius, A., Ruediger, h., Sachs, R. K., and Duesberg, P. H. (2009). Cancer-causing karyotypes: chromosomal equilibria between destabilizing aneuploidy and stabilizing selection for oncogenic function. *Cancer Genet Cytogenet* **188**, 1–25.

Linardou, H., Dahabreh, I. J., Kanaloupiti, D., Siannis, F., Bafaloukos, D., Kosmidis, P., Papadimitriou, C. A., and Murray, S. (2008). Assessment of somatic k-RAS mutations as a mechanism associated with resistance to EGFR-targeted agents: a systematic review and meta-analysis of studies in advanced non-small-cell lung cancer and metastatic colorectal cancer. *Lancet Oncology* **9**, 962–972.

Little, M. P., Kleinerman, R. A., Stiller, C. A., Li, G., Kroll, M. E., and Murphy, M. F. (2011). Analysis of retinoblastoma age incidence data using a fully stochastic cancer model. *Int. J. Cancer* doi: 10. 1002/ijc. 26039. [Epub ahead of print].

Loeb, L. A. (2010). Mutator phenotype in cancer: origin and consequences. *Semin. Cancer Biol.* **20**, 279–280.

Lomberk, G. A. (2011). Epigenetic Silencing of Tumor Suppressor Genes in Pancreatic Cancer. *J. Gastrointestinal Cancer* **42**(2) 93–99.

Malthus, T. (2000). An Essay on the Principle of Population: Library of Economics. *Liberty Fund, Inc.*

Mao, C., Qiu, L. X., Liao, R. Y., Du, F. B., Ding, H., Yang, W. C., Li, J., and Chen, Q. (2010). KRAS mutations and resistance to EGFR-TKIs treatment in patients with non-small cell lung cancer: a meta-analysis of 22 studies. *Lung Cancer* **69**, 272–278.

Mathupala, S. P., Ko, Y. H., and Pedersen, P. L. (2010). The pivotal roles of mitochondria in cancer: Warburg and beyond and encouraging prospects for effective therapies. *Biochimica et Biophysica Acta* **1797**, 1224–1230.

Mayr, E. (1942). Systematics and the Origin of Species. Columbia University Press.

McCormick, F. (2010). Science of Oncology Award Lecture. American Society of Clinical Oncology.

McHugh, K., and Callaghan, R. (2008). Clinical trials on Mdr reversal agents. *In* "Multidrug Resistance: Biological and Pharmaceutical Advance in Antitumour Treatment" (N. A. Colabufo ed.), pp. 321–353. Research Signpost, Kerala, India.

Miller, V.A., Hirsch, V., Cadranel, J., Chen, Y.-M., Park, K., Kim, S.-W., Caicun, Z., Oberdick, M., Shahidi, M., and Yang, C.-H. (2010). Afatinib benefits lung cancer patients whose cancer progressed after treatment with EGFR inhibitors. *35th European Society for Medical Oncology Congress*, Milan, Italy. Abstract LBA1.

Mizejewski, G. J., Mirowski, M., Garnuszek, P., Maurin, M., Cohen, B. D., Poiesz, B. J., Posypanova, G. A., Makarov, V. A., Severin, E. S., and Severin, S. E. (2010). Targeted delivery of anti-cancer growth inhibitory peptides derived from human alpha-fetoprotein: review of an International multi-center collaborative study. *J. Drug Targe* **18**, 575–588.

Mok, T., Wu, Y. L., Thongprasert, S., Yang, C. H., Chu, D. T., Saijo, N., Sunpaweravong, P., Han, B., Margono, B., Ichinose, Y., Nishiwaki, Y., and Ohe, Y., et al., (2009). *N. Engl. J. Med* **361**, 947–957.

Muggia, F. (2009). Platinum compounds 30 years after the introduction of cisplatin: implications for the treatment of ovarian cancer. *Gynecol. Oncol.* **112**, 275–281.

Murchison, E. P. (2008). Clonally transmissible cancers in dogs and Tasmanian devils. *Oncogene* **27**(Suppl 2), S19–S30.

Negrini, S., Gorgoulis, V. G., and Halazonetis, T. D. (2010). *Nat. Rev. Mol. Cell Biol* **11**, 220–228.

Nicholson, J. M., and Duesberg, P. (2009). On the karyotypic origin and evolution of cancer cells. *Cancer Gen. Cytogen.* **194**, 96–110.

Nowell, P. C. (1976). The clonal evolution of tumor cell populations. *Science* **194**, 23–28.

O'Connor, M. J., Martin, N. M., and Smith, G. C. (2007). Targeted cancer therapies based on the inhibition of DNA strand break repair. *Oncogene* **26**, 7816–7824.

Olaussen, K. A., Dunant, A., Fouret, P., Brambilla, E., Andre, F., Haddad, V., Taranchon, E., Filipits, M., Pirker, R., Popper, H. H., Stahel, R., and Sabatier, L., et al., (2006). DNA repair by ERCC1 in non-small-cell lung cancer and cisplatin-based adjuvant chemotherapy. *New Engl. J. Med.* **355**, 983–991.

O'Shea, C. C., Johnson, L., Bagus, B., Choi, S., Nicholas, C., Shen, A., Boyle, L., Pandey, K., Soria, C., Kunich, J., Shen, Y., and Habets, G., et al., (2004). Late viral RNA export, rather than p53 inactivation, determines ONYX-015 tumor selectivity. *Cancer Cell* **6**, 611–623.

Ozasa, H., Oguri, T., Uemura, T., Miyazaki, M., Maeno, K., Sato, S., and Ueda, R. (2010). Significance of thymidylate synthase for resistance to pemetrexed in lung cancer. *Cancer Sci.* **101**, 161–166.

Parkin, D. M., Bray, F., Ferlay, J., and Pisani, P. (2005). Global cancer statistics, 2002. *CA: A Cancer J. Clin.* **55**, 74–108.

Paz-Ares, L., Soulieres, D., Melezinek, I., Moecks, J., Keil, L., Mok, T., Rossell, R., and Klughammer, B. (2010). Clinical outcomes in non-small-cell lung cancer patients with EGFR mutations: pooled analysis. *J. Cell Mol. Med.* **14**, 51–69.

Pearson, R. D. (1981). Tumourigenesis: the subterfuge of selection. *Acta Biotheoretica* **30**, 171–176.

Perez-Caro, M., Cobaleda, C., Gonsalez-Herrero, I., Vicente-Duenas, C., Bermejo-Rodriguez, C., Sanchez-Beato, M., Orfao, A., Pintado, B., Flores, T., Sanchez-Martin, M., Jimenez, R., and Piris, M. A., et al., (2009). Cancer induction by restriction of oncogene expression to the stem cell compartment. *EMBO J.* **28**, 8–20.

Pitroda, S. P., Wakim, B. T., Sood, R. F., Beveridge, M. G., Beckett, M. A., MacDermed, D. M., Weichselbaum, R. R., and Khodarev, N. N. (2009). STAT1-dependent expression of energy metabolic pathways links tumour growth and radioresistance to the Warburg effect. *BMC Medicine* **7**, 68.

Ptashne, M. (2009). Binding reactions: epigenetic switches, signal transduction and cancer. *Current Biol.* **19**, R234–R241.

Reyno, L., Seymour, L., Tu, D., Dent, S., Gelmon, K., Walley, B., Pluzanska, A., Gorbunova, V., Garin, A., Jassem, J., Pienkowski, T., and Dancey, J., et al., (2004). Phase III study of *N*, *N*-diethyl-2-[4-(phenylmethyl) phenoxy]ethanamine (BMS-217380-01) combined with doxorubicin versus doxorubicin alone in metastatic/recurrent breast cancer: National Cancer Institute of Canada Clinical Trials Group Study MA.19. *J. Clin. Oncol* **22**, 269–276.

Roberts, M. (1926). Malignancy and Evolution. Grayson and Grayson Publishers, London.

Rokas, A. (2008). The origins of multicellularity and the early history of the gentic toolkit for animal development. *Ann. Rev. Gen.* **42**, 235–251.

Rosell, R., Moran, T., Queralt, C., Porta, R., Cardenal, F., Camps, C., Majem, M., Lopez-Vivanco, G., Isla, D., Provencio, M., Insa, A., and Massuti, B., et al., (2009). Screening for epidermal growth factor receptor mutations in lung cancer. *New Engl. J. Med.* **361**, 958–967.

Rothenberg, S. M., and Settleman, J. (2010). Discovering tumor suppressor genes through genome-wide copy number analysis. *Current Genomics* **11**, 297–310.

Rubin, L. L., and de Sauvage, F. J. (2006). Targeting the Hedgehog pathway in cancer. *Nature Rev. Drug Discovery* **5**, 1026–1033.

Salk, J. J., Fox, E. J., and Loeb, L. A. (2010). Mutational heterogeneity in human cancers: origin and consequences. *Ann. Rev. Pathol.* **5**, 51–75.

Salonga, D., Danenberg, K. D., Johnson, M., Metzger, R., Groshen, S., Tsao-Wei, D. D., Lenz, H. J., Leichman, C. G., Leichman, L., Diasio, R. B., and Danenberg, P. V. (2000). Colorectal tumors responding to 5-fluorouracil have low gene expression levels of dihydropyrimidine dehydrogenase, thymidylate synthase, and thymidine phosphorylase. *Clinical Cancer Res.* **6**, 1322–1327.

Sanofi Aventis Press Release (2011). Sanofi-aventis reports top-line results from phas III study with BSI-201 in metastatic triple-negative breast cancer. www.sanofi-aventis.com.

Sato, M., Takahashi, K., Nagayama, K., Arai, Y., Ito, N., Okada, M., Minna, J. D., Yokota, J., and Kohno, T. (2005). Identification of chromosome arm 9p as the most frequent target of homozygous deletions in lung cancer. *Genes, Chromosomes, Cancer* **44**, 405–414.

Sattler, U. G., Hirschhaeuser, F., and Mueller-Klieser, W. F. (2010). Manipulation of glycolysis in malignant tumors: fantasy or therapy? *Current Med. Chem.* **17**, 96–108.

Savage, P., Stebbing, J., Bower, M., and Crook, T. (2009). Why does cytotoxic chemotherapy cure only some cancers? *Nature Clin. Practice. Oncol.* **6**, 43–52.

Scagliotti, G. V., Parikh, P., von Pawel, J., Biesma, B., Vansteenkiste, J., Manegold, C., Serwatowski, P., Gatzemeier, U., Diqumarti, R., Zukin, M., Lee, J. S., and Mellemgaard, A., et al., (2008). Phase III study comparing cisplatin plus gemcitabine with cisplatin plus pemetrexed in chemotherapy-naïve patients with advanced-stage non-small-cell lung cancer. *J. Clin. Oncol.* **26**, 3543–3551.

Schiller, J. H., Akerley, W. L., Brugger, W., Ferrari, D., Garmey, E. G., Gerber, D. E., Orlov, S. V., Ramlau, R., Von Pawel, J., and Sequist, L. V. (2010). Results from ARQ 197-209: A global randomized placebo-controlled phase II clinical trial of erlotinib plus ARQ 197 versus erlotinib plus placebo in previously treated EGFR inhibitor-naïve patients with locally advanced or metastatic non-small cell lung cancer (NSCLC). *J. Clin. Oncol.* **28**, 18s Abstract LBA7502.

Sequist, L. V., Waltman, B. A., Dias-Santagata, D., Digumarthy, S., Turke, A. B., Fidias, P., Bergethon, K., Shaw, A. T., Gettinger, S., Cosper, A. K., Akhavanfard, S., and Heist, R. S., et al., (2011). Genotypic and histological evolution of lung cancers acquiring resistance to EGFR inhibitors. *Sci. Transl. Med.* **3**, 75ra26.

Shlomi, T., Benyamini, T., Gottlieb, E., Sharan, R., and Ruppin, E. (2011). Genome-scale metabolic modeling elucidates the role of proliferative adaptation in causing the Warburg effect. *PLoS Comput. Biol.* **7**, e1002018.

Simpson, G. G. (1961). In, Principles of Animal Taxonomy. Columbia University Press, New York 153.

Skorski, T. (2011). Chronic myeloid leukemia cells refractory/resistant to tyrosine kinase inhibitors are genetically unstable and may cause relapse and malignant progression to the terminal disease state. *Leukemia & Lymphoma* **52**(Suppl 1), 23–29.

Snow, H. (1893). Cancers and the Cancer Process. J. & A. Churchill Publishers, London.

Stemmler, M. P. (2008). Cadherins in development and cancer. *Mol. Biosyst.* **4**, 835–850.

Suda, K., Murakami, I., Katayama, T., Tomizawa, K., Osada, H., Sekido, Y., Maehara, Y., Yatabe, Y., and Mitsudomi, T. (2010). Reciprocal and complementary role of MET amplification and EGFR T790M mutation in acquired resistance to kinase inhibitors in lung cancer. *Clinical Cancer Res.* **16**, 5489–5498.

Sullivan, R. J., and Atkins, M. B. (2010). Molecular targeted therapy for patients with melanoma: the promise of MAPK pathway inhibition and beyond. *Expert Opin. Invest. Drugs* **19**, 1205–1216.

Templeton, A. R. (1989). The meaning of species and speciation: a genetic perspective. *In* "Speciation and its consequences" (D. Otte, and J. A. Endler, eds.), pp. 3–27. Sinauer Associates, Inc, Sunderland, Massachusetts.

Van Cutsem, E., Cunningham, D., Hoff, P. M., and Maroun, J. (2001). Thymidine phosphorylase (TP) activation: convenience through innovation. *Oncologist* **4**, 1–2.

Vander Heiden, M. G., Cantley, L. C., and Thompson, C. B. (2009). Understanding the Warburg effect: the metabolic requirements of cell proliferation. *Science* **324**, 1029–1033.

Vander Heiden, M. G., Christofk, H. R., Schuman, E., Subtelny, A. O., Sharfi, H., Harlow, E. E., Xian, J., and Cantley, L. C. (2010). Identification of small molecule inhibitors of pyruvate kinase M2. *Biochem. Pharmacol.* **79**, 1118–1124.

van der Vigh, W. J., and Peters, G. J. (1994). Protection of normal tissues from the cytotoxic effects of chemotherapy and radiation by amifostine (Ethyol): preclinical aspects. *Seminars Oncol.* **5**(Suppl 11), 2–7.

Varshavsky, A. (2007). Targeting the absence: homozygous DNA deletions as immutable signposts for cancer therapy. *Proc. Nat. Acad. Sci. U.S.A.* **104**, 14935-1440.

Vaughn, A. E., and Deshmukh, M. (2008). Glucose metabolism inhibits apoptosis in neurons and cancer cells by redox inactivation of cytochrome c. *Nature. Cell Biol.* **10**, 1477–1483.

Vincent, M. D. (1987). The definition of a neoplasm. *Medical Hypotheses* **24**, 151–160.

Vincent, M. (2006). Tesmilifene may enhance breast cancer chemotherapy by killing a clone of aggressive, multi-drug resistant cells through its action on the p-glycoprotein pump. *Medical Hypotheses* **66**, 715–731.

Vincent, M. (2010). The animal within: Carcinogenesis, and the clonal evolution of cancer cells are speciation events sensu stricto. *Evolution* **64-4**, 1173–1183.

Vincent, M. (2011). Cancer: A de-repression of a default survival program common to all cells? *BioEssays (In press).*

Vincent, M. (in preparation). Cancer: Towards a general theory of the target.

Vineis, P. (2003). Cancer as an evolutionary process at the cell level: an epidemiological perspective. *Carcinogenesis* **24**, 1–6.

Vogel, C. L., Cobleigh, M. A., Tripathy, D., Guthell, J. C., Harris, L. N., Fehrenbacher, L., Slamon, D. J., Murphy, M., Novotny, W. F., Burchmore, M., Shak, S., and Stewart, S. J., et al., (2002). Efficacy and safety of trastuzumab as a single agent in first-line treatment of HER2-overexpressing metastatic breast cancer. *J. Clin. Oncol.* **20**, 719–726.

Warburg, O. (1928). The chemical constitution of respiration ferment. *Science* **68**, 437–443.

Warburg, O. (1956 a). On the origin of cancer cells. *Science* **123**, 309–314.

Warburg, O. (1956 b). On respiratory impairment in cancer cells. *Science* **124**, 269–270.

Warburg, O. (1989). The Prime Cause and Prevention of Cancer—Part 1. Revised lecture at the meeting of the Novel laureates on June 30, 1966 at Lindau, Lake Constance, Germany, 2nd revised Edition. Pub Konrad Triltsch, Wurzburg, Germany.

Welsh, J. S. (2011). Contagious cancer. *Oncologist* **16**, 1–4.

Willett, C. G., Duda, D. G., Czito, B. G., Bendell, J. C., Clark, J. W., and Jain, R. K. (2007). Targeted therapy in rectal cancer. *Oncology* **21**, 1055–1065.

Williams, G. C. (1966). Adaptation and Natural Selection: A Critique of Some Current Evolutionary Thought. Princeton University Press, Princeton, NJ.

Wilson, E. O. (1975). Sociobiology: The New Synthesis. The Belknap Press of Harvard University Press, Cambridge, Massachusetts and London, England.

Wilson, T. R., Johnston, P. G., and Longley, D. B. (2009). Anti-apoptotic mechanisms of drug resistance in cancer. *Curr. Cancer Drug Targets* **9**, 307–319.

Witkin, E. M., and Kogoma, T. (1984). Involvement of the activated form of RecA protein in SOS mutagenesis and stable DNA replication in Escherichia coli. *Proc. Nat. Acad. Sci. U. S.A.* **81**, 7539–7543.

Yedid, G., Ofria, C. A., and Lenski, R. E. (2009). Selective press extinctions, but not random pulse extinctions, cause delayed ecological recovery in communities of digital organisms. *The American Naturalist* **173**, E139–E154.

Young, A., Lyons, J., Miller, A. L., Phan, V. T., Alarcon, I. R., and McCormick, F. (2009). Ras signaling and therapies. *Adv. Cancer Res.* **102**, 1–17.

Zhao, X., Weir, B. A., LaFramboise, T., Lin, M., Beroukhim, R., Garraway, L., Beheshti, J., Lee, J. C., Naiki, K., Richards, W. G., Sugarbaker, D., and Chen, F., et al., (2005). Homozygous deletions and chromosome amplifications in human lung carcinomas revealed by single nucleotide polymorphism array analysis. *Cancer Res* **65**, 5561–5570.

Zhao, Y. H., Zhou, M., Liu, H., Ding, Y., Khong, H. T., Yu, D., Fodstad, O., and Tan, M. (2009). Upregulation of lactate dehydrogenase A by ErbB2 through heat shock factor 1 promotes breast cancer cell glycolysis and growth. *Oncogene* **28**, 3689–3701.

Zhou, M., Zhao, Y., Ding, Y., Liu, H., Liu, Z., Fodstad, O., Riker, A. I., Kamarajugadda, S., Owen, L. B., Ledoux, S. P., and Tan, M. (2010a). Warburg effect in chemosensitivity: targeting lactate dehydrogenase-A re-sensitizes taxol-resistant cancer cells to taxol. *Mol. Cancer* **9**, 33.

Zhou, C., Wu, Y., Chen, G., Feng, J., Liu, X., Wang, C., Zhang, S., Wang, J., Zhou, S., and Ren, S. (2010b). Efficacy results from the randomized phase III OPTIMAL (CTONG 0802) study comparing first-line erlotinib versus carboplatin (CBDCA) plus gemcitabine (GEM), in Chinese advanced non-small-cell lung cancer (NSCLC) patients (pts) with EGFR activating mutations. *European Society for Medical Oncology* Abstract LBA13.

Index

A

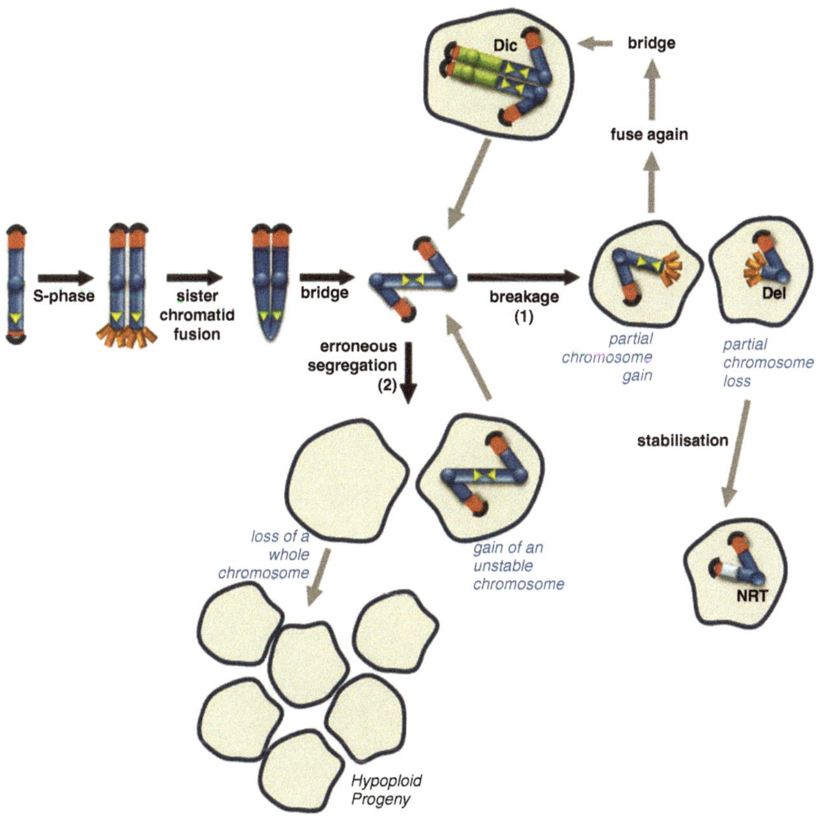

Fig. 3, Anna Genescà et al. (See Pages 24–26 of this volume.)

B

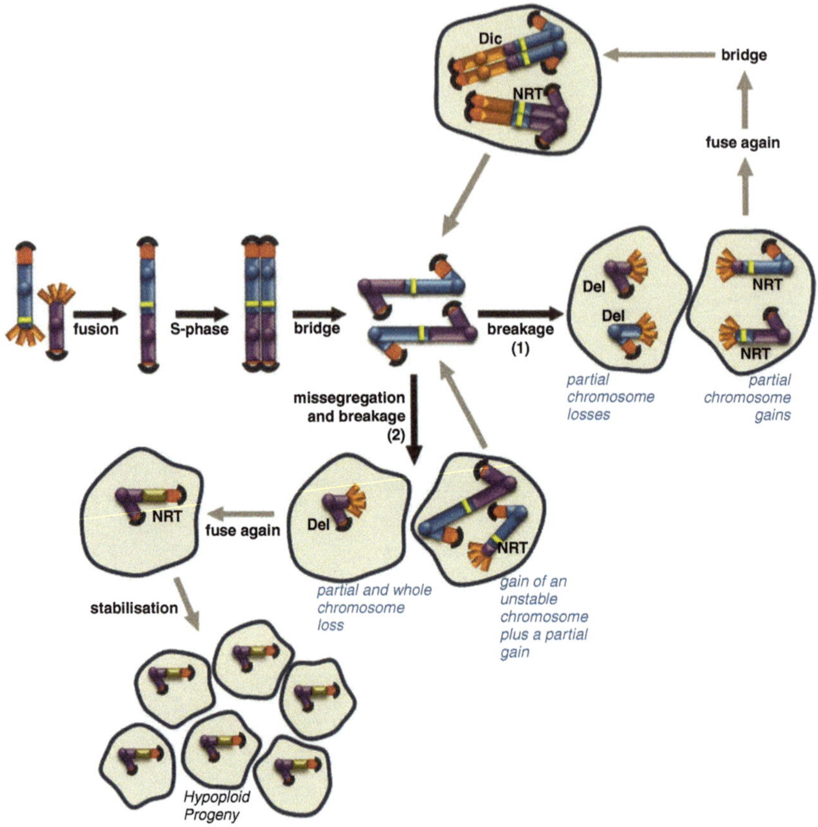

C

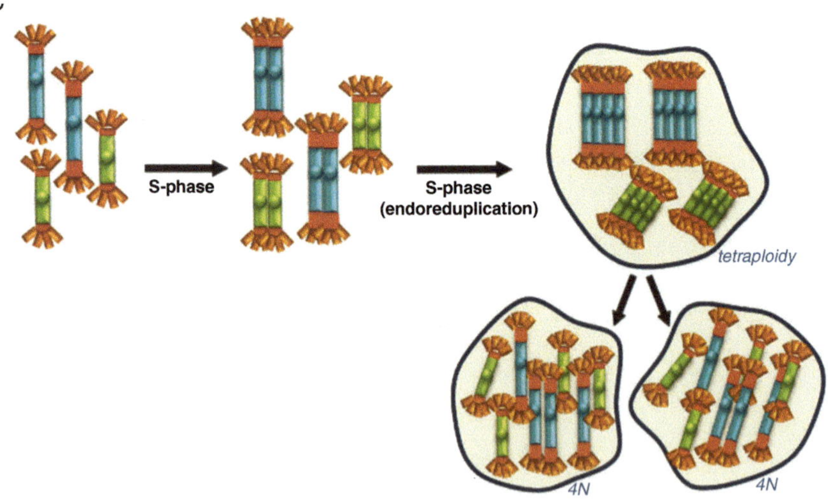

Fig. 3, Anna Genescà et al. (Continued).

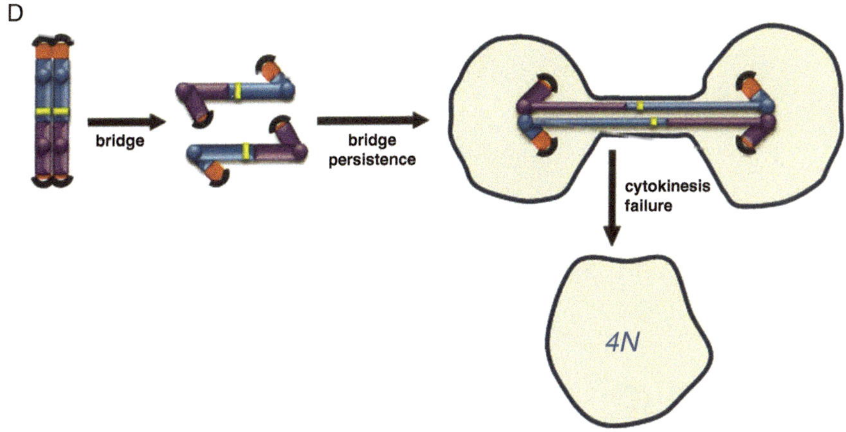

Fig. 3, Anna Genescà et al. (Continued).

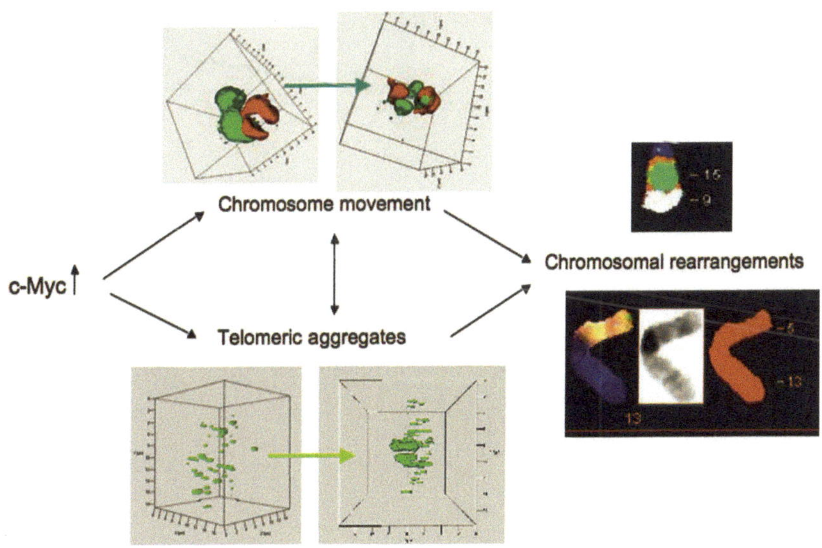

Fig. 2, Macoura Gadji et al. (See Page 85 of this volume.)

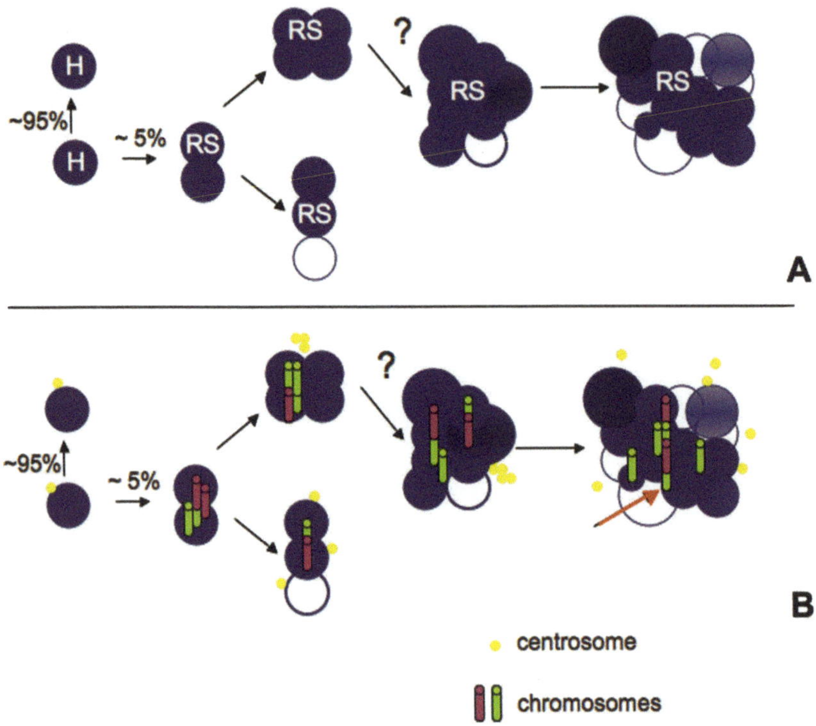

Fig. 3, Macoura Gadji et al. (See Page 91 of this volume.)

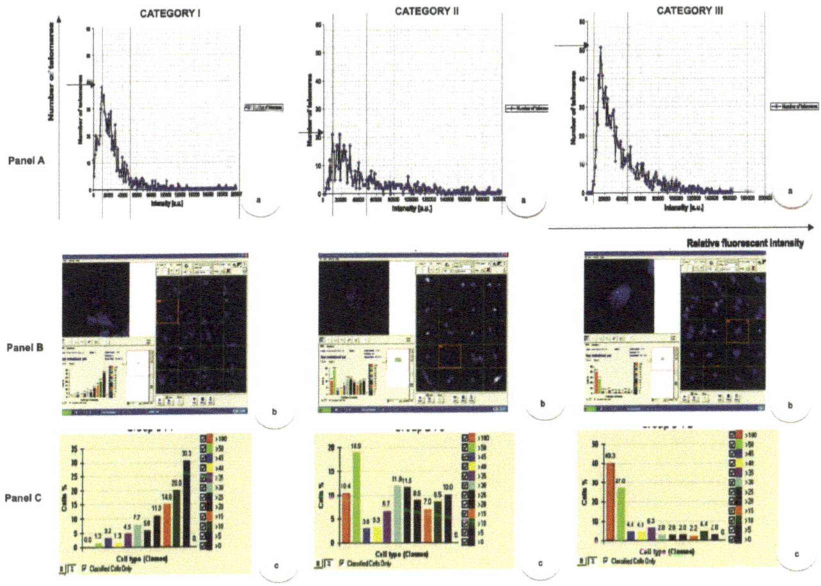

Fig. 4, Macoura Gadji et al. (See Page 94 of this volume.)

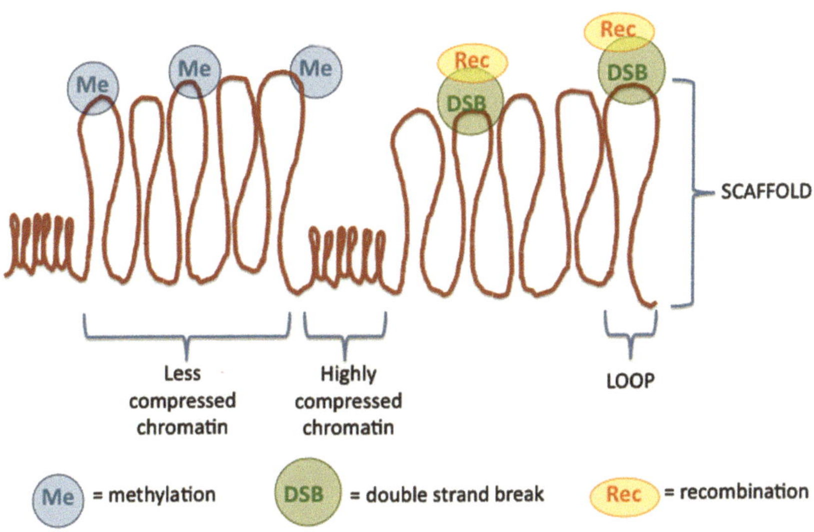

Fig. 7, Macoura Gadji et al. (See Page 102 of this volume.)

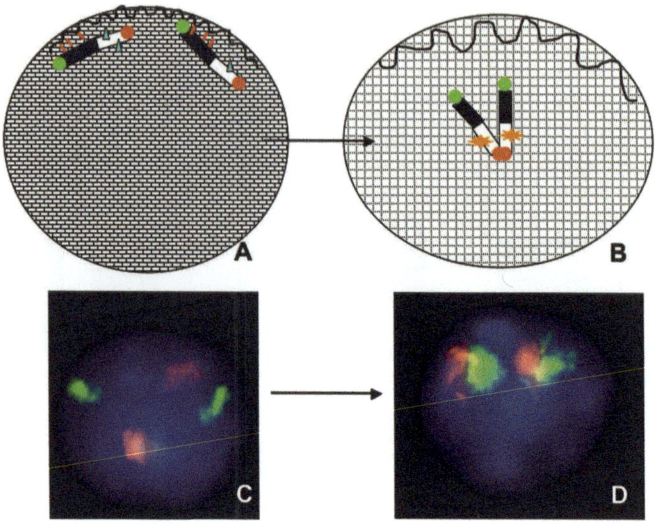

Fig. 8, Macoura Gadji et al. (See Page 107 of this volume.)

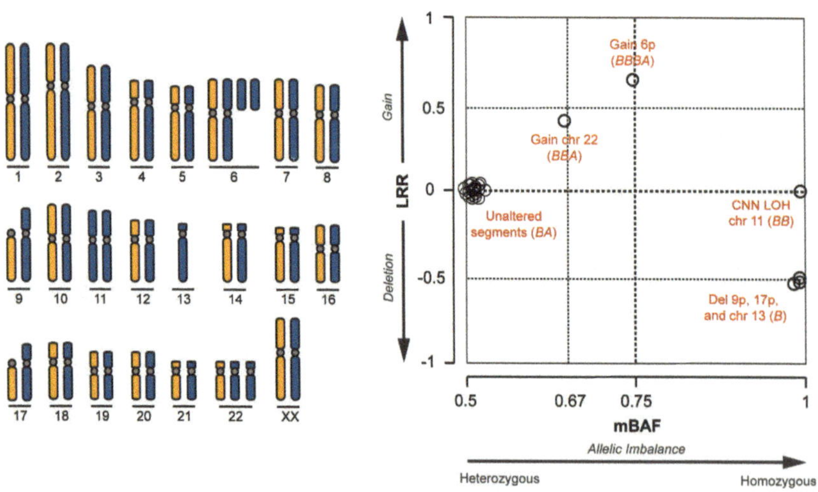

Fig. 5, David Lindgren et al. (See Page 167 of this volume.)

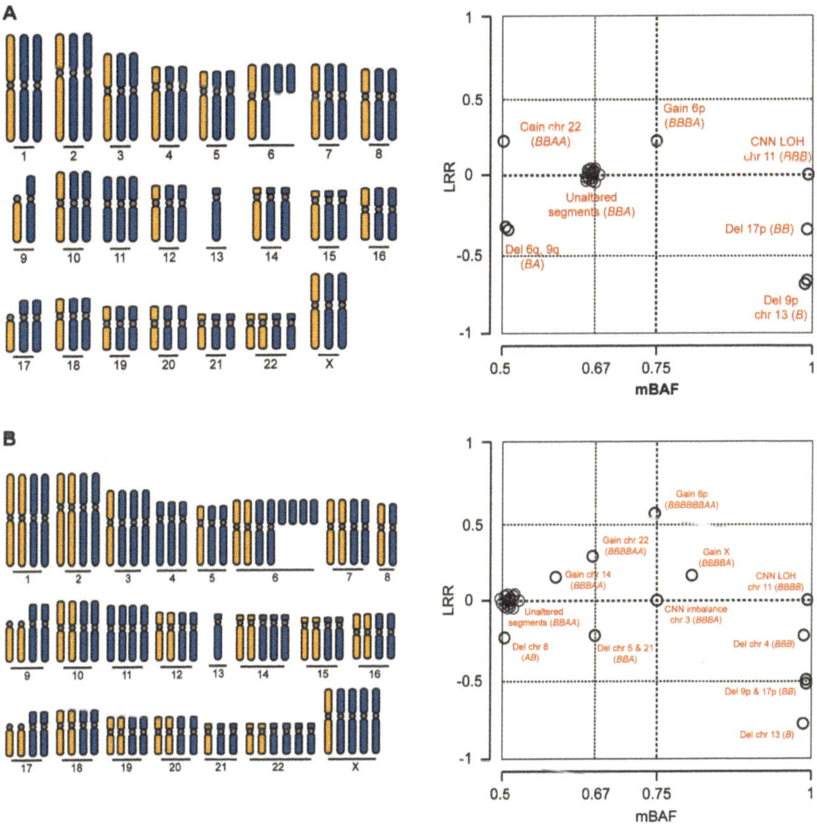

Fig. 7, David Lindgren et al. (See Page 170 of this volume.)

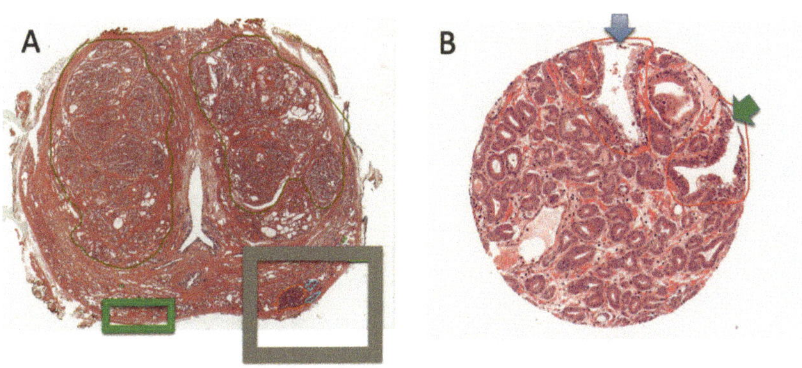

Fig. 1, Jeremy A. Squire et al. (See Page 186 of this volume.)

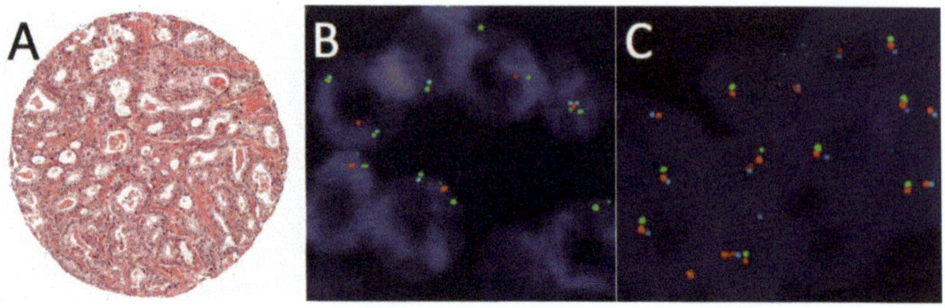

Fig. 2, Jeremy A. Squire et al. (See Page 193 of this volume.)

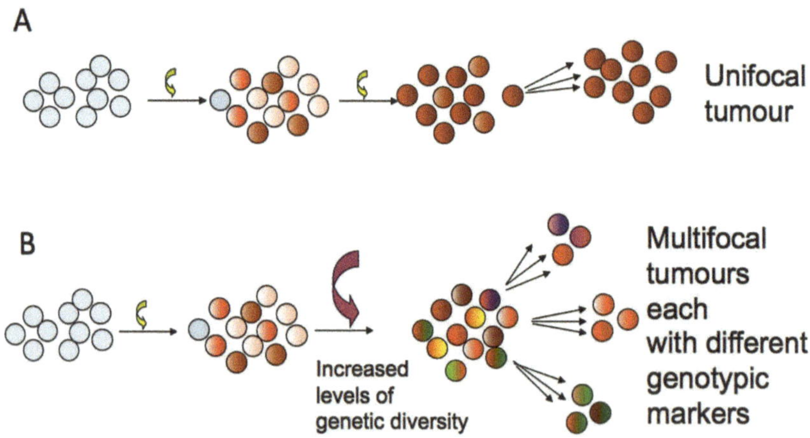

Fig. 4, Jeremy A. Squire et al. (See Page 200 of this volume.)

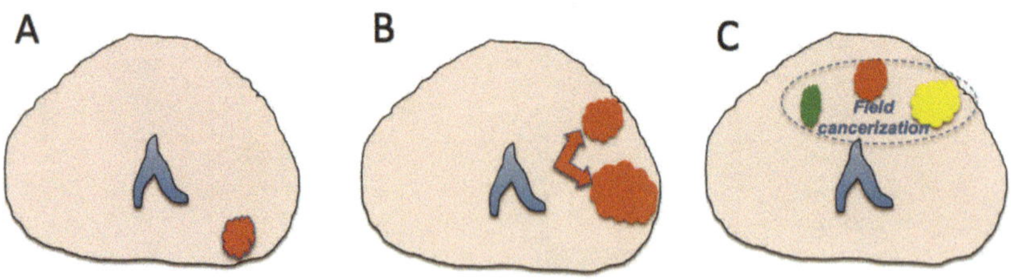

Fig. 5, Jeremy A. Squire et al. (See Page 201 of this volume.)